D0163903

OPEN SOURCE PHYSICS:
A USER'S GUIDE WITH EXAMPLES

OPEN SOURCE PHYSICS:
A USER'S GUIDE WITH EXAMPLES

Wolfgang Christian
Davidson College

Includes

Physics Curricular Material by Mario Belloni

Tracker Video Analysis and OSP XML by Doug Brown

BQ Database by Anne Cox and William Junkin

Easy Java Simulations by Francisco Esquembre

PEARSON

Addison
Wesley

San Francisco   Boston   New York
Cape Town   Hong Kong   London   Madrid   Mexico City
Montreal   Munich   Paris   Singapore   Sydney   Tokyo   Toronto

Editor-in-Chief: Adam Black
Assistant Editor: Deb Greco
Production Supervisor: Nancy Tabor
Managing Editor: Corinne Benson
Executive Marketing Manager: Christy Lawrence
Manufacturing Manager: Pam Augspurger
Project Management and Composition: Techsetters, Inc.
Cover Illustration: Blake Kim
Cover Designer: Lisa Devenish
Text Printer: R.R. Donnelley, Crawfordsville
Cover Printer: Phoenix Color

Library of Congress Cataloging-in-Publication Data

Christian, Wolfgang.
  Open source physics : a user's guide with examples / Wolfgang Christian.–1st ed.
     p. cm.
  ISBN 0-8053-7759-X (alk. paper)
  1. Physics–Data processing. 2. Physics–Simulation methods. 3. Physics–Computer simulation. 4.
Object-oriented methods (Computer science) 5. Open source software. I. Title.
  QC52.C57 2006
  530.01′13–dc22

                        2005030859

(ISBN) 0-8053-7759-X

Copyright © 2007 Pearson Education, Inc., publishing as Addison Wesley, 1301 Sansome St., San
Francisco, CA 94111. All rights reserved. Manufactured in the United States of America. This
publication is protected by Copyright and permission should be obtained from the publisher prior to
any prohibited reproduction, storage in a retrieval system, or transmission in any form or by any
means, electronic, mechanical, photocopying, recording, or likewise. To obtain permission(s) to use
material from this work, please submit a written request to Pearson Education, Inc., Permissions
Department, 1900 E. Lake Ave., Glenview, IL 60025. For information regarding permissions, call
(847) 486-2635.

Many of the designations used by manufacturers and sellers to distinguish their products are
claimed as trademarks. Where those designations appear in this book, and the publisher was aware
of a trademark claim, the designations have been printed in initial caps or all caps.

1 2 3 4 5 6 7 8 9 10—RRD—08 07 06 05
www.aw-bc.com/physics

OLSON LIBRARY
NORTHERN MICHIGAN UNIVERSITY
MARQUETTE, MI 49855

Preface

OPEN SOURCE PHYSICS

The Open Source Physics project is a synergy of curriculum development, computational physics, and physics education research. One goal of the project is to make a large number of Java simulations available for education using the GNU Open Source model. This Guide describes some of the classes and interfaces that are being used in this project.

Portions of Open Source Physics are being incorporated into the following projects:

- The third edition of *An Introduction to Computer Simulation Methods* by Harvey Gould, Jan Tobochnik, and Wolfgang Christian.

- The *Statistical and Thermal Physics* project by Harvey Gould and Jan Tobochnik.

- The *Easy Java Simulations* high-level modeling tool by Francisco Esquembre.

- The *Tracker* video analysis program by Doug Brown.

Programmers wishing to adopt Open Source Physics tools for their projects are encouraged to do so, provided that they release their source code under the GNU Open Source GPL license. Open Source Physics curricular material and developer resources are being distributed from the Open Source Physics server hosted at Davidson College and from other servers. Links to this material are available on the Open Source Physics server.

```
http://www.opensourcephysics.org
http://www.opensourcephysics.org/developer
```

LICENSE

Open Source Physics software is free software; you can redistribute it and/or modify it under the terms of the GNU General Public License (GPL) as published by the Free Software Foundation either version 2 of the License or (at your option) any later version.

Code that uses any portion of the code in the `org.opensourcephysics` package or any subpackage (subdirectory) of this package must also be released under the GNU GPL license.

A copy of the GNU General Public License in included on the CD accompanying this book. If this CD is not available, write to the Free Software Foundation, Inc., 59 Temple

Place, Suite 330, Boston, MA, 02111-1307 USA or view the license online at

http://www.gnu.org/copyleft/gpl.html

AUDIENCE

This Guide is intended as a guide to the tools and philosophy of the Open Source Physics project. It does not intend to teach computational physics, numerical analysis, or Java programming. There are many books for these purposes.

Prose descriptions of a class are inexact, and there is no substitute for a short demonstration program. This Guide provides many such programs. Another reason for the Guide is to document the test suite we use to maintain the integrity of the Open Source Physics library before distributing it on the Internet.

SUPPORT

To keep the objects simple, the Open Source Physics library does not include advanced features, such as serialization and synchronization. Although we have tried to make the Open Source Physics software as bug free as possible, we cannot guarantee that the library is free of bugs. If you find a bug and would like to contact the author(s), email is your best option. The following email address can be used:

wochristian@davidson.edu

All reasonable questions will be answered and bugs fixed as time allows. But please do not make incredible demands and expect them to be fulfilled.

ACKNOWLEDGEMENTS

There are very many people and institutions that have contributed to the Open Source Physics project, and I take great pleasure in acknowledging their support and their interest. The Open Source Physics project would not have been possible without the support and collaboration of my longtime friend and Davidson colleague Mario Belloni. Harvey Gould at Clark University and Jan Tobochnik at Kalamazoo College have pioneered the teaching of computational physics to both undergraduate and graduate students. This Guide would not have been possible without their early adoption of the Open Source Physics libraries in their teaching and curriculum development projects. Anne Cox has provided many insightful and thoughtful comments about the Guide and the physics examples and has hosted an Open Source Physics developers workshop at Eckerd College.

Douglas Brown at Cabrillo College and Francisco Esquembre at the University of Murcia, Spain have contributed code, fixed bugs, and made suggestions on the OSP architecture. Joshua Gould of the Broad Institute at MIT and Harvard and Kipton Barros of Boston University were invaluable in helping to design and test the Open Source Physics libraries. Aaron Titus, Matt DesVoigne, Bruce Mason, and Slavomir Tuleja tested the OSP libraries and suggested numerous improvements to this Guide. Glenn Ford developed the JOGL implementation of the OSP 3D API. Carlos Ortiz prepared the index and provided numerous other suggestions. We are especially thankful to students and faculty at Davidson College, Clark University, and Kalamazoo College who have generously commented on the Open Source Physics project as they suffered through early versions of the library.

The Open Source Physics project makes use of a number of software packages released under open source licenses. In particular, we acknowledge the following authors:

- Bruce Miller for the Fast Fourier Transformation (FFT) package `jnt.fft`.

- Nathan Frank for the Java expression parser package `org.nfunk.jep`.

- Andrew Gusev and Yuri B. Senichenkov at Saint Petersburg Polytechnic University, Russia for the high-order differential equation solvers in the ode package `org.opensourcephysics.ode`.

These packages are included on the Open Source Physics Software CD, but newer versions may be available on the Web or from their respective authors.

The Open Source Physics project is indebted to Leigh Brookshaw for publishing his graphics package under the GNU GPL license. Portions of this package were used in developing our own x-axis and y-axis components. Thanks to Paul Mutton for making the `EpsGraphics2D` package available under the GPL license. This package makes it easy to create high-quality EPS graphics for use in documents and papers. Thanks to Yanto Suryono for the source code to the Function Plotter program that is the basis of the `ContourPlot` and `SurfacePlot` components. Yanto Suryono also developed the expression parser that is included in the Open Source Physics numerics package.

Davidson College has supported the Open Source Physics in many ways, including the hosting of the Open Source Physics Web server and by providing an academic home for the author of this Guide. OSP would not be possible without this generous support.

The Open Source Physics project is supported in part by the National Science Foundation grants DUE-0126439 and DUE-0442481. These grants have helped us to write books, to provide workshops at professional meetings, and to develop Open Source Physics curricular materials for distribution on the Internet.

We thank Adam Black and Deb Greco at Addison–Wesley for their encouragement while writing the third edition, Nancy Tabor at Addison–Wesley and Carol Sawyer and the rest of the staff at Techsetters for their work in producing the book, and Robin Rider of The Perfect Proof for her insightful copyediting.

Finally, I thank my wife, Barbara, and my children, Beth, Charlie, and Konrad Rudolf, for their love and support. Although such acknowledgements may appear gratuitous, anyone who has ever written a book knows otherwise.

Contents

Contents

CHAPTER

1

Introduction to Open Source Physics

We describe the tools and philosophy of the Open Source Physics project.

1.1 ■ INTRODUCTION

The switch from procedural to object-oriented (OO) programming has produced dramatic changes in professional software design. However, object-oriented techniques have not been widely adopted by scientists teaching computational physics and computer simulation methods. Although most scientists are familiar with procedural languages such as Fortran, few have formal training in computer science and therefore few have made the switch to OO programming. The continued use of procedural languages in education is due, in part, to the lack of up-to-date curricular materials that combine science topics with an OO framework. Although there are many good books for teaching computational physics, most are not object oriented. The Open Source Physics project was initially intended to be a small object-oriented library that would provide input/output and graphing capabilities for the third edition of *An Introduction to Computer Simulation Methods* by Harvey Gould, Jan Tobochnik, and Wolfgang Christian. After showing this library to other educational software developers, it became apparent that it has a potentially wider audience than computational physicists. What is needed by the broader science education community is not a computational physics, numerical analysis, or Java programming book (although such books are essential for discipline-specific practitioners), but a synthesis of curriculum development, computational physics, computer science, and physics education that will be useful for scientists and students wishing to write their own simulations and develop their own curricular material. The Open Source Physics (OSP) project was established to meet this need.

Open Source Physics is a National Science Foundation-funded curriculum development project that seeks to develop and distribute a code library, programs, and examples of computer-based interactive curricular material. The goal of the project is to create and disseminate a large number of ready-to-use Java simulations for education using the GNU open source model for code distribution. The OSP project also maintains a website that serves as an in-depth guide to the tools, philosophy, and programs developed as part of this project.

http://www.opensourcephysics.org

The Open Source Physics project is based, in part, on a collection of Java applets written by Wolfgang Christian known as *Physlets*. Although Physlets are written in Java, they are not open source. We often receive requests for Physlet source code but have declined to

1

distribute this code. Physlets are compiled Java applets that are embedded into html pages and controlled using JavaScript. This paradigm works well for general-purpose programs such as a Newton's law simulation, but fails for more sophisticated one-of-a-kind simulations that require advanced discipline-specific expertise. Users and developers of these types of programs often have specialized curricular needs that can only be addressed by having access to the source code. However, anyone who has ever written a program in Java knows that writing code for opening windows and creating buttons, text fields, and graphs can be tedious and time consuming. Moreover, it is not in the spirit of object-oriented programming to rewrite these methods for each application or applet that is developed. The Open Source Physics project solves this problem by providing a consistent object-oriented library of Java components for anyone wishing to write their own simulation programs. In addition, the Open Source Physics library provides a framework for embedding these programs into an html page and for controlling these programs using JavaScript.

The basic Open Source Physics library includes the following packages:

- The *controls* package (`org.opensourcephysics.controls`) builds user interfaces and defines the OSP xml framework.

- The *display* package (`org.opensourcephysics.display`) draws 2D objects and plots graphs on the DrawingPanel class using the Drawable interface.

- The *2D display* package (`org.opensourcephysics.display2d`) plots 2D data using contour and surface plots.

- The *3D display* package (`org.opensourcephysics.display3d`) defines and implements an API for three-dimensional objects.

- The *Ejs* package (`org.opensourcephysics.ejs`) builds custom user interface components.

- The *numerics* package (`org.opensourcephysics.numerics`) contains numerical analysis tools, such as ordinary differential equation solvers.

- The *tools* package (`org.opensourcephysics.tools`) contains utility programs such as Launcher and LaunchBuilder that work with other OSP programs.

Open Source Physics libraries are based on Swing and Java 1.4. Although the focus of the basic libraries is traditional computational physics, they have already been extended to include topics not covered in computational physics texts, such as video analysis. These packages are available from other OSP developers and from the OSP website.

A number of curriculum authors have adopted OSP for their projects and have agreed to let the OSP project distribute their programs as examples. These projects include:

- *An Introduction to Computer Simulation Methods* (3rd ed.) by Harvey Gould, Jan Tobochnik, and Wolfgang Christian.

- *Statistical and Thermal Physics* by Harvey Gould and Jan Tobochnik.

- *Easy Java Simulations* high-level modeling tool by Francisco Esquembre.

- *Tracker* video analysis program by Doug Brown.

The development of these programs has enabled the Open Source Physics project to improve the core library in many ways. The API has been improved and made more consistent, bugs have been found and squashed, and we have learned what tools are useful to the developer community. But most importantly, the development of Tracker and Ejs have resulted in great tools for the physics education community.

1.2 ■ MODEL-VIEW-CONTROL DESIGN PATTERN

Experienced programmers approach a programming project not as a coding task but as a design process. They look for data structures and behaviors that are common to other problems that they have solved. Separating the physics from the user interface and the data visualization is one such approach. In keeping with computer science terminology, we refer to the user interface as the *control*. It is responsible for handling events and passing them on to other objects. Plots, tables, and other visual representations of data are examples of *views*. Finally, the physics can be thought of as a *model* which maintains data that defines a state and provides methods by which that data can change. There is usually only one model and often only one control. It is, however, common to have multiple views. We could, for example, show a plot and a table view of the same data. Separating programs into models, views, and controls is known as the Model-View-Control (MVC) design pattern. The various OSP frameworks are designed to support this pattern.

The MVC design pattern is one of the most successful software architectures ever devised. Java is an excellent language with which to implement this pattern because it allows us to isolate the model, the control, and the view in separate classes. Object-oriented programming makes it easy to reuse these classes and easy to add new features to a class without having to change the existing code. These features are known as encapsulation, code reuse, and polymorphism. They are the hallmarks of object-oriented programming.

1.3 ■ SIMULATIONS AND PEDAGOGY

Good programming practice is best taught by having students modify, compile, test, and debug their own code. A typical exercise in *An Introduction to Computer Simulation Methods* begins with a discussion of theory, the presentation of an algorithm, and the necessary Java syntax. We implement the algorithm in a sample application and then ask the reader to test it with various parameters. The user then modifies the model, adds visualizations such as graphs and tables, and performs further analysis.

Consider the classical three-body model of helium shown in Figure 1.1. The book narrative asks the reader to do the following:

- Modify the program `PlanetApp.java` to simulate the classical helium atom. Let the initial value of the time step be $\Delta t = 0.001$. Some of the possible orbits are similar to those we have already studied in our mini solar system.

- The initial condition $r_1 = (1.4, 0)$, $r_2 = (-1, 0)$, $v_1 = (0, 0.86)$, and $v_2 = (0, -1)$ gives "braiding" orbits. Most initial conditions result in unstable orbits in which one electron eventually leaves the atom (autoionization). Make small changes in this initial condition to observe autoionization.

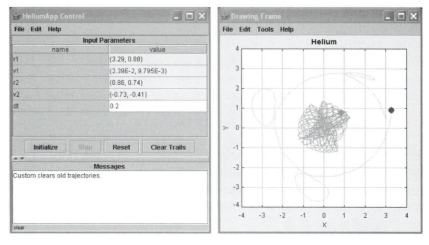

(a) A simple user interface.   (b) Helium electron trajectories.

Figure 1.1   A classical model of two electrons orbiting about a nucleus. From the third edition of *An Introduction to Computer Simulation Methods* by Gould, Tobochnik, and Christian.

- The classical helium atom is capable of very complex orbits (see Figure 1.1). Investigate the motion for the initial condition $r_1 = (3, 0)$, $r_2 = (1, 0)$, $v_1 = (0, 0.4)$, and $v_2 = (0, -1)$. Does the motion conserve the total angular momentum?

- Choose the initial condition $r_1 = (2, 0)$, $r_2 = (-1, 0)$, and $v_2 = (0, -1)$. Then vary the initial value of v_1 from $(0.6, 0)$ to $(1.3, 0)$ in steps of $v_{1x} = 0.02$. For each set of initial conditions, calculate the time it takes for autoionization. Assume that ionization occurs when either electron exceeds a distance of six from the nucleus. Run each simulation for a maximum time equal to 2000. Plot the ionization time versus v_{1x}. Repeat for a smaller interval of v_{1x} centered about one of the longer ionization times.

As the helium example shows, the focus is on both the physics and programming. The user interface need not be very sophisticated, and a simple control that includes a few buttons and a table for data entry is all that is needed for a computational physics textbook. Modifications to the program, such as a custom user interface or Web delivery, allow the program to be used with different pedagogies such as in-class demonstrations, Peer Instruction, traditional homework, and Just-in-Time Teaching. Because the code is released under a GNU GPL license, users can write their own narrative for other contexts, such as astronomy and classical mechanics.

Another goal of the OSP project is to make Java simulations widely available for physics education. We do this by converting the applications such as those in the Java edition of *An Introduction to Computer Simulation Methods* into Web-deliverable applets. Compiled applets, sample scripts, and source code are available on the OSP Web server.

1.4 ■ USING LAUNCHER

Java programs are usually distributed in archives that contain compiled code, resources such as images, security information, and a *manifest* that specifies the path to the class file containing the main method. These archives end with the extension `*.jar` and are referred to as *jar* files. If a Java VM is installed on a computer and if a Java application has been properly packaged in a jar file, then the program specified in the jar file's manifest can be executed from the command line by passing the name of the jar file to the Java VM. (See Section 1.5 for Java installation instructions.) For example, the `osp.jar` archive can be executed by typing the following commands in the console (terminal) after the jar file has been downloaded from the OSP website and copied into the `apps` directory. You will, of course, need to invoke the change directory `cd /apps` command (or `cd \apps` on Windows) to navigate to this directory.

```
cd /apps
java -jar osp_demo.jar
```

Operating systems with graphical user interfaces allow users to easily execute a jar file by double-clicking on the jar's icon displayed within a filesystem browser. As far as the user is concerned, an archived Java program behaves like any other installed application.

If the jar file contains more than one program, the user can override the manifest's target class using other command line parameters. For example, the following command adds the `osp_demo.jar` archive to the *classpath* and executes the `FirstPlotApp` program in the archive.

```
java -classpath osp_demo.jar demo.FirstPlotApp
```

To execute the helium example we copy the `osp_guide.jar` into the current directory and type

```
java -classpath osp_guide.jar org.opensourcephysics.manual.ch09.HeliumApp
```

Although it is possible to distribute programs in a single jar file or multiple jar files, this approach is not well suited for the distribution of curricular material. Few users want to deal with the complexities of command line syntax and resource management. Although jar files can contain resources such as html-based documentation, images, and sound, they are difficult to create and modify by casual users, particularly if the archive has been digitally signed. A large curriculum development project creates hundreds of programs and each program may be used in multiple contexts with different initial conditions. The *Launcher* and *Launch Builder* programs developed by Doug Brown at Cabrillo College solve many of these problems. *Launcher* (shown in Figure 1.2) is a Java application that can launch (execute) other Java programs. We use *Launcher* to organize and distribute collections of ready-to-use programs, documentation, and curricular material in a single easily modifiable package (see Table 1.1).

Executing the `osp_demo.jar` archive starts an instance of *Launcher* containing curricular material described in Chapter 15. Double-clicking on a leaf node in the Launcher's left-hand "tree" launches a Java application. The right-hand panel shows html-based documentation for the given program. Other Launcher packages are available on the CD and from the OSP website. The `osp_guide.jar` archive contains examples from this Guide and the `osp_csm.jar` file contains examples from *An Introduction to Computer Simulation Methods*.

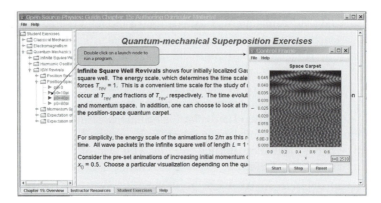

Figure 1.2   The *Launcher* program executes other Java programs.

Table 1.1   Java archive (jar) files available on the CD.

Ready-To-Use Open Source Physics Programs

osp.jar	Runs Launcher and provides access to LauchBuilder using the Edit menu item in Launcher. This archive contains the basic OSP library without any programs or curricular material. This jar is also used to show how to execute programs from the command line.
osp_csm.jar	Examples from the third edition of *An Introduction to Computer Simulation Methods* by Gould, Tobochnik, and Christian.
osp_demo.jar	A small collection of OSP-based physics curricular material developed by Mario Belloni and Wolfgang Christian at Davidson College for demonstrations and modification at faculty development workshops.
osp_guide.jar	Examples from this Guide.
osp_stp.jar	Statistical and thermal physics examples from Clark University and Kalamazoo College developed by Harvey Gould, Jan Tobochnik, and their students.
tracker.jar	Tracker video analysis program by Doug Brown.

Optional Libraries

osp_ode.jar	High-order differential equation solvers developed by Andrew Gusev and Yuri B. Senichenkov.
osp_jogl.jar	A Java for Open GL (JOGL) implementation of the OSP 3D API developed by Glenn Ford.
osp_media.jar	Video tools developed by Doug Brown.

The *Launcher*'s configuration, application data, and associated documentation are organized using an Extensible Markup Language (xml) text file named `launcher_default.xset`. Although this file is often packaged inside a Java archive, it can also be distributed as a stand-alone file.[1] *Launcher* uses the external configuration file if it is placed in the same directory as the jar file. Because this file can be modified without recompiling code or rebuilding the jar file, teachers and authors can easily adapt Open Source Physics curricular material for their own needs.

Although *Launcher* configuration xml files can be opened with any text editor, they are most easily created and edited using the *Launch Builder* program as discussed in Chapter 15. The impatient reader can select the *Edit* menu item under the *Launcher* frame's *File* menu to explore the capabilities of *LaunchBuilder*.

1.5 ■ INSTALLING JAVA

Java developers know how finicky Java can be. An incorrect installation, forgotten environment variables, or out-of-date libraries can cause pages of error messages. Many beginning programmers report that installing the Java Development Kit (JDK) and running their first program is the most difficult part of learning to use Java and Open Source Physics. Even if the the JDK is properly installed, finding a simple syntax error can be frustrating when using the Java compiler from the command line. For this reason, almost all Java developers use an integrated development environment such as *Eclipse* as described in Section 1.7. You can almost certainly begin Java programming by double-clicking on file icons to install the JDK and Eclipse. But it is always useful to know what is going on under the hood, and this section will help you debug your installation should something go wrong.

Java programs can be readily compiled and run on all standard operating systems. Although some operating system vendors, such as Apple, ship with Java preinstalled, users of most operating systems will need to install Java. Users wishing to only run programs should install the Java runtime environment, but developers should install the complete Java Development Kit. Up-to-date versions of the JDK are available from Sun Microsystems for Solaris, Linux, and Windows at the following website:

`http://java.sun.com/j2se/`

Look for version 5.0 or later and download the Java 2 Standard Edition (J2SE) in contrast to the Java 2 Enterprize Edition (J2EE) or the Java 2 Micro Edition (J2ME). Currently, you can also download an integrated development environment (IDE) named *NetBeans* which includes the JDK, but this is not necessary. This IDE can be downloaded and installed at a later time from `http://netbeans.org/`. The *Eclipse* IDE is an alternative to *NetBeans* and is described in Section 1.7.

After downloading, double-clicking on the downloaded archive (zip, gzip, or tar) in a filesystem browser is all that is required to install the JDK on most platforms. However, because Java has Unix origins, we recommend that users not install Java programs in directories containing spaces in the directory path. For example, Windows users may wish to install the JDK in `c:\jdk5.0` rather than the default `c:\program files\java\jdk5.0`.

[1]Files in Java archives can be extracted and examined using the *jar utility* program from Sun or by using a decompression program such as *WinZip* or *StuffIt Expander*.

Table 1.2   Typical subdirectories within a JDK installation.

JDK Directory Structure	
bin	Compiler and other developer tools.
demo	Java demonstration programs from Sun.
docs	Java documentation from Sun.
include	Tools for compiling native methods.
jre	Java runtime environment files.
lib	Compiled libraries.
sample	Examples of Java-related programs and tools.
src	Source code extracted from src.zip.

The JDK contains the source code for the core Java libraries in the src.zip archive. Because we will want to peak under the hood to see how Sun implements a particular class, it is a good idea to unpack the source code archive in the JDK directory. Create a directory named src and unpack the archive in this directory. You may also want to download and install Sun's Java documentation from http://java.sun.com/docs. A typical JDK directory structure is shown in Table 1.2.

In order to run and compile programs without typing the complete path to the programs in the JDK directory, it is convenient to add the jdk/bin directory to the search (execution) path by setting the *path* environment variable. This variable is set automatically by the JDK installer, but it can also be set from the console.

Setting environment variables is operating system dependent and you should consult your operating system manual to determine the proper syntax. On Unix/Linux environment variables are set by editing a user's environment file. On Windows XP environment variables are set using the advanced tab in the "System Properties" dialog box in the control panel. A good way to determine if the path contains the location of your JDK is to examine the current setting by entering the following in the command in the Windows console.

```
set path
```

You can modify the existing path by appending directories to the current path.

```
set PATH=\%PATH\%;c:\jdk1.5.0_04
```

If the Java VM is properly installed, you should now be able to open a console (terminal) window and type the java command to test the runtime installation.

```
java -version
```

The javac command tests the compiler in the JDK installation.

```
javac -version
```

In a Windows console the java -version responds with output similar to the following:

```
Z:\>java -version

java version "1.5.0_04" Java(TM) 2 Runtime Environment,
  Standard Edition (build 1.5.0_03-b07)
  Java HotSpot(TM) Client VM
  (build 1.5.0_04-b05, mixed mode, sharing)

Z:\>
```

Note that Windows uses the backslash character \ as a path delimiter while Unix and Linux use a forward slash /. Java applications recognize the slash / as a delimiter even if these applications are run on a Java VM in a Windows environment.

To further check the JDK installation, it is useful to compile and run a simple Java application. We follow tradition and create a HelloWorldApp test program. Create a project directory named hello and create two subdirectories named src and classes. Create a subdirectory named hello in src. Use a simple text editor to create HelloWorldApp.java code file shown in Listing 1.1 and save this file in the hello subdirectory. Note how the directory we have created matches the package declaration at the beginning of the code.

Listing 1.1 HelloWorldApp tests the JDK installation.

```
package hello;
public class HelloWorldApp {
  public static void main(String[] args) {
    System.out.println("Hello World");
  }
}
```

Compile the HelloWorldApp program by executing the javac command within a console after using the cd command to navigate to the project file.

```
cd \hello_project
javac  -d classes/ src/hello/HelloWorldApp.java
```

Entering the javac -help command explains the various command line options including the -d option used above to specify where the compiler should place its output. The specified directory is the root of an output directory hierarchy that matches the package names. The compiler creates directories in this hierarchy automatically as needed.

If you have not made any typing mistakes (Java is case sensitive), the compiler silently processes the HelloWorldApp.java file and produces the HelloWorldApp.class file in the classes/hello directory. You can now run the HelloWorldApp program using the java command.

```
java -classpath classes/ hello/HelloWorldApp
```

Again, entering the java -help command explains the various command line options such as -classpath.

1.6 ■ PACKAGES

Because of the large number of Java source files, both Sun Microsystems and the Open Source Physics project organize code using Java packages. A Java package is a collection of related files located in a single directory. Code is given privileged access to other code within the same package because it is assumed that a package is designed and written by a programmer or by a small group of programmers who understand the code's interdependencies. Access to a class's data and methods from outside the package is far more restrictive and is only granted if the programmer has added either the protected or public keywords to a variable or method definition.

We often refer to a group of packages as a library and we refer to a group of files designed for a specific task as a framework. For example, the Open Source Physics core library contains a 3D drawing framework, and this framework uses files from multiple packages such as `ElementBox.java` in the simple 3D package and `Transformation.java` in the numerics package as shown in the following code fragments:

```
// import a concrete implementation of OSP 3D
import org.opensourcephysics.display3d.simple3d.*;
import org.opensourcephysics.numerics.*;
// create a 3D world
DrawingPanel3D panel = new DrawingPanel3D();
DrawingFrame3D frame = new DrawingFrame3D(panel);
// place an Element into the 3D world
Element box = new ElementBox();
Transformation transformation =
    Matrix3DTransformation.rotationX(Math.PI/6);
box.setTransformation(transformation);
panel.addElement(box);
```

The Open Source Physics library is organized in directories (packages) starting at the `org.opensourcephysics` root directory. Examples referred to in this Guide can be found in the `org.opensourcephysics.manual` directory. Open Source Physics programs written by students and faculty at Davidson College can be found in the `org.opensourcephysics.davidson` package. Statistical and thermal physics programs can be found in the `org.opensourcephysics.clark` package. These programs were written at Clark University and Kalamazoo College.

Directories in the core Open Source Physics library are shown in Table 1.3. The compiled core library is available in a Java archive `osp.jar` on the CD and from the Open Source Physics website. Source code for the core library is available in the `osp_core.zip` file. Additional libraries, such as high precision differential equation solvers by Andrew Lvovitch Gusev and Yuri B. Senichenkov at Saint Petersburg Polytechnic University, Russia, are available from the OSP website.

Although Java does not formally recognize the concept of a *subpackage* we often refer to subdirectories as such. Because it is awkward to refer to the code within the `org.opensourcephysics.numerics` package using the fully qualified package name, we often refer to a package using a name such as the *numerics* package when the complete path can be inferred from the context.

To compile and run an Open Source Physics program, first download the OSP core library archive `osp_core.zip` and place it in a project directory (folder) such as `osp_project`. Unpack the archive and notice the directory called `src`, which in turn contains a directory called `org`, which in turn contains a directory called `opensourcephysics`, which in turn contains directories for the various packages. All directories except the *resources* directory contain code files. The resources directory contains noncode files such as images and text. Files with a `properties` extension contain text data such as button and menu labels to configure the user interface and operating system information such as default file paths. Placing noncode resources in a separate directory makes it easy to locate text data and to internationalize the Open Source Physics library by translating this text into other languages. Because the Java compiler does not process text files, resource files must be copied to the output directory. In the `osp_project` directory, create a new directory called `classes` which contains a directory called `org` which contains a directory called `opensourcephysics`. This directory is where the compiler will place the byte code (binary output). Copy the resources from the source `src/org/opensourcephysics/resources` to the output `classes/org/opensourcephysics/resources` directory. We are now ready to test the core Open Source Physics library by plotting the function shown in Figure 1.3.

Table 1.3   Packages (directories) within the core Open Source
Physics library.

org.opensourcephysics	
display	A drawing framework based on the `Drawable` interface and the `DrawingPanel` class.
display2d	Visualization package for two-dimensional data such as contour and surface plots.
display3d	A three-dimensional modeling framework.
controls	A framework based on the `Control` interface for building graphical user interface (GUI) components.
numerics	A numerical analysis package containing tools such as Fourier analysis and ordinary differential equation (ODE) solvers.
tools	Utility programs, such as Launcher and LaunchBuilder, that are designed to work with other OSP programs.
resources	Text files and configuration data.

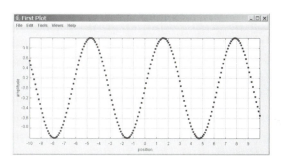

Figure 1.3   A simple Open Source Physics graph.

The project directory contains two subdirectories: `src` and `classes`. Create another subdirectory named `plot` in `src`. Use a simple text editor to create `FirstPlotApp.java` code file shown in Listing 1.2 and save this file in the `plot` subdirectory. Figure 1.4 shows the package structure for this project.

Listing 1.2   The `FirstPlotApp` tests the Open Source Physics
library installation.

```
package plot;
import org.opensourcephysics.frames.*;
import javax.swing.JFrame;

public class FirstPlotApp {
  public static void main(String[] args) {
    PlotFrame frame =
        new PlotFrame("position", "amplitude", "First Plot");
    frame.setSize(400, 400);
    for(double x = -10, dx = 0.1;x<10;x += dx) {
      frame.append(0, x, Math.sin(x));
    }
```

Figure 1.4   The package structure for the `FirstPlotApp` program. Notice how the OSP library is organized.

```
      frame.setVisible(true);
      frame.setDefaultCloseOperation(JFrame.EXIT_ON_CLOSE);
   }
}
```

Open the console (terminal) application, change directory (cd) to the project directory, and compile the project by executing the `javac` command.

```
javac  -d classes/ -sourcepath src/ src/plot/FirstPlotApp.java
```

Run the `FirstPlotApp` program by executing the `java` command.

```
java -classpath classes/plot/FirstPlot
```

A window similar to Figure 1.3 should appear if the OSP library is properly installed.

You can specify multiple code files in a single `javac` command, but this is usually not necessary. The Java compiler is very helpful and automatically searches for and compiles classes referenced by our code. Examine the files produced in the output directory tree. Because the `FirstPlotApp` example references the `PlotFrame` class, this class and all dependent classes in the OSP library are automatically compiled.

Occasionally classes are not compiled because they are not directly referenced in the code, but these classes are later needed when a program is run, resulting in a *ClassNot-FoundException* error message. If this runtime error occurs because the necessary class has not been compiled, go to the project directory and compile the entire package.

```
javac -d classes/ org/opensourcephysics/packagename/*.java
```

Examples from *Open Source Physics: A User's Guide with Examples* are distributed in the `osp_guide.zip` archive as shown in Table 1.4. The `osp_guide.zip` archive is organized in packages named by chapter number and these packages are located in the `org.opensourcephysics.manual` package. Create a new project directory with `src` and `classes` subdirectories as before and download and unpack the examples into the `src` directory. You can copy the Open Source Physics core library into this project's source

Table 1.4   Code archives (zip) files available on the CD and on the Open
Source Physics website.

`osp_core.zip`	The basic Open Source Physics library without any programs or curricular material.
`osp_csm.zip`	Examples from the third edition of *An Introduction to Computer Simulation Methods* by Gould, Tobochnik, and Christian.
`osp_guide.zip`	Examples from this Guide.
`jep.zip`	Java expression parser by Nathan Frank.
`osp_ode.zip`	High-order differential equation solvers developed by Andrew Gusev and Yuri B. Senichenkov.
`osp_jogl.zip`	A Java for Open GL (JOGL) implementation of the OSP 3D API developed by Glenn Ford.
`osp_media.zip`	Video tools developed by Doug Brown.

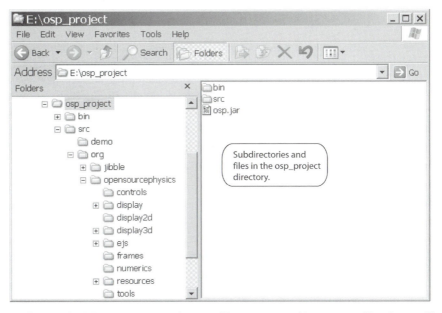

Figure 1.5   A typical directory structure for compiling programs without recompiling the core library.

directory, but this is not necessary. We now demonstrate how to compile and run programs
using already compiled byte code in the `osp.jar` file.

Open Source Physics programs are compiled and run without recompiling the core library
by using a command line parameter that specifies the *classpath* to the byte code. Download
and copy the `osp.jar` into the project directory as shown in Figure 1.5. We can now compile
programs by adding this jar file to the classpath. For example, in order to compile and run
the `SHO3DApp` program in the `ch02` code package, we open the console and navigate to the
Guide's project directory before executing the `javac` and `java` programs:[2]

---

[2]Some operating systems use a colon rather than a semicolon to delimit the search path.

```
javac -d classes -classpath osp.jar
    src/org/opensourcephysics/manual/ch02/*.java
java -classpath
    osp.jar;.\classes org.opensourcephysics.manual.ch02.SHO3DApp
```

The compiler assumes that the `osp.jar` file is in the console's current directory and that the subdirectory structure for the examples has a directory structure that matches the Java package names. Although locations of the source and output files can be overridden with command line options, it is best to keep things simple and locate everything in a single project directory when starting out.

To make it easy to identify OSP programs, we have adopted the convention that classes that contain a *main* method, and can therefore be run as a Java application, end with *App*. For example, the ubiquitous "HelloWorld" program was defined in a file named `HelloWorldApp.java`. This convention makes it easy to find example programs in the Guide by searching for files matching the pattern "*App.java."

1.7 ■ ECLIPSE

The command line examples we have shown are simple and useful as a starting point for building shell (command) scripts for compiling and running Java programs. Large development projects, however, are more efficiently managed using an integrated development environment (IDE) such as *Eclipse*, Borland's *JBuilder*, Sun's *NetBeans*, or Apple's *XTools*. Good IDEs provide color-coded syntax highlighting and syntax checking as well as access to tools such as the `jar` archive builder and the `javadoc` documentation generator. They can easily incorporate the entire OSP source code library thereby allowing debugging and single-stepping through an entire project. And most importantly, an IDE provides easy access to documentation. Highlighting a method or variable name and right-clicking (control or option clicking in some operating systems) takes the user directly to the source code and documentation for the given object or variable.

Eclipse is available for Mac OS X, Windows, Linux, and Solaris. Although it supports multiple programming languages, it is written in Java and is an excellent Java IDE. Eclipse is open source and is freely available from `http://eclipse.org`.

Although it takes time to fully master Eclipse, it is not difficult to get started. An Eclipse *project* organizes the code and resources needed to compile, run, and distribute a program. Download and install Eclipse and then select the *Create Project* item from the File menu after starting the Eclipse application as shown in Figure 1.6. Select the check box to use separate directories for the source code and the compiler output and enter the name of the project and the location of the source code and the compiler output. We typically use `src` and `classes` for the names of the the output directories.

Although it is possible to copy the Open Source Physics code library into the source directory, it is preferable to use the Import menu item to obtain the source code. Using the Eclipse menu copies the source code into the correct directory and insures that internal variables are up to date. You should press the *F5* function key to refresh the Eclipse display if you copy the source code directly.

Eclipse displays the project in a graphical interface known as the Eclipse workspace shown in Figure 1.7. Source code is accessed using the tree-like package explorer in the left-hand side of the the workspace. Expand a node in the Eclipse package explorer by

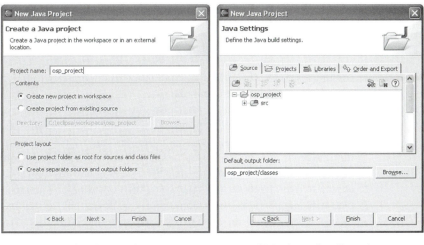

(a) Creating a project.   (b) Setting project directories.

Figure 1.6   Dialog boxes guide the user through the process of creating a project in Eclipse.

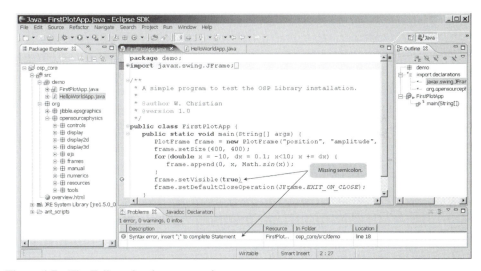

Figure 1.7   The Eclipse development environment.

double-clicking and select a class definition containing a Java application. The application's code appears in an editor pane near the center of the workspace. Run the application using either the Run menu or the Toolbar button with the white triangle inside the green circle.

When you edit code in the Eclipse editor, an asterisk appears in front of the file name in the title bar indicating that the code has not yet been saved. If the *Build Automatically* option under the Project menu is selected, the edited code is automatically compiled when the code is saved using either the File menu or the standard <control>-s keyboard shortcut. Errors, warnings, and program output are displayed in a tabbed view near the bottom of the workspace.

1.8 ■ BUILDING PROGRAMS WITH ANT

Ant is an open source alternative to generic shell scripts and a powerful complement to Java IDEs such as Eclipse. Version 3.1 of Eclipse comes with Ant 1.6.5, and there is an extensive Ant interface in Eclipse. To understand how Ant operates within Eclipse, it is useful to learn how to run Ant from the command line.

Ant is a Java application that uses Extensible Markup Language (xml) to build Java projects. Ant invokes other programs know as tasks and the most common Ant task is, of course, to compile and run Java programs using the Java compiler `javac` and the Java runtime environment `java`. But there are other tasks such as creating documentation using `JavaDoc`, creating Java archive (jar) files, signing jar files with security keys, and deploying applications to Web servers. Because Ant does this and much more and because Ant is platform independent, we recommend that developers distribute an Ant build file with their projects even if they use other development environments for their day-to-day work. Although it is not our intention to teach Ant programming, a basic build file shown in Listing 1.3 is not difficult to decipher. The latest Ant distribution and online documentation is available at `http://ant.apache.org`.

Ant takes its instructions from a build file that contains one or more *targets*. Each target has a name such as `compile` or `run`. If Ant is properly installed, you execute a target by passing the target's name to Ant. For example, the command line `ant compile` executes the *compile* target in the default file `build.xml`. An Ant file that compiles and runs an example is shown in Listing 1.3. Note how we include the `osp.jar` archive in the *classpath*.

Listing 1.3   A simple Ant build file for compiling and running an OSP program. The default Ant build file is named `build.xml`.

```
<?xml version="1.0" encoding="utf-8" ?>
<project name="osp" default="all" basedir="./">
  <property name="author" value="W. Christian"/>
  <property name="classes.dir" value="./classes/"/>
  <property name="srcdir.dir" value="./src/"/>
  <property name="lib.dir" value="./lib/"/>
  <path id="run.class.path">
    <pathelement location="./classes/"/>
    <pathelement location="${lib.dir}/osp.jar"/>
  </path>

  <!--compiles the source files-->
  <target name="compile" >
    <delete dir="${classes.dir}"/>
    <mkdir dir="${classes.dir}"/>
    <javac srcdir="${srcdir.dir}"
      destdir="${classes.dir}"
      classpath="${lib.dir}/osp.jar" />
  </target>

  <!--runs the PlotterApp-->
  <target name="run" depends="compile">
    <echo>Running Program.</echo>
    <java
      classname="plot.FirstPlotApp"
      classpathref="run.class.path"
      fork="true">
```

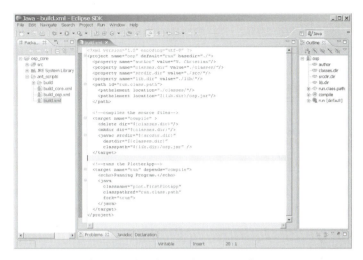

Figure 1.8 Eclipse has well-integrated and extensive support for Ant.

```
    </java>
  </target>
</project>
```

Targets within the build file contain tasks that are executed by Ant. These tasks typically invoke programs that take input and produce output. The `compile` target executes a task that invokes the Java compiler `javac` to generate class files from Java source files. The compile target shown in Listing 1.3 executes two additional (nested) tasks to delete and make the output directory.

Names, such as the locations of source and output directories, are usually defined near the beginning of an Ant file using *property* variables. These variables are referenced by enclosing the property name with a dollar sign and braces. Defining properties makes it easy to adjust build files to match the local installation. Note how the path to the `osp.jar` library is specified.

Ant tasks often depend on the successful completion of one or more previous targets. Adding the `depends="compile"` attribute to the run target insures that the Java code is compiled before the program is executed. This compilation is done automatically and it is not necessary to invoke the compile target first. The user simply types

```
ant run
```

to compile and run the `PlotterApp` program from the build file.

An Ant build file to manage the entire Open Source Physics library is available on the OSP website. The `build_osp.xml` file defines targets to compile the code and to create the documentation and the `osp.jar` archive. Entering the following command line executes a lengthy task that removes old files and produces a clean copy of the core OSP library. You should edit the build file's property values to match the directory structure and Java version on your computer.

```
ant -buildfile build\_osp.xml all
```

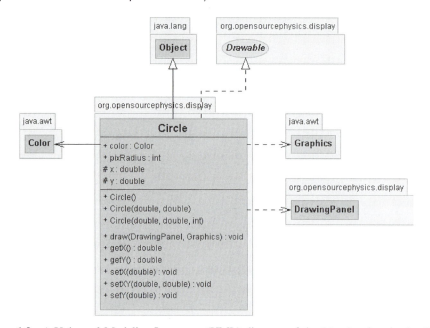

Figure 1.9 A Universal Modeling Language (UML) diagram of the `Circle` class in the display framework.

You can import the `build_osp.xml` file into an Eclipse project and execute targets from within the workspace as shown in Figure 1.8. An icon for the Ant build file appears in the package explorer and the xml code is displayed in a color-coded editor view to highlight declarations, attributes, and keywords. Eclipse catches syntax errors when you edit the build file and provides an explanation of the error. Clicking on an Ant target in the outline view executes that target and sends output to the Eclipse console view.

Other Ant IDEs exist, but the big players are behind Eclipse, and support for Ant in Eclipse is extensive and robust. The combination of Ant and Eclipse makes an ideal development environment for Java programmers.

1.9 ■ DOCUMENTATION

Although Java code should be documented using `javadoc` comments, we assume that readers can access OSP code and documentation in an IDE and have therefore removed most block comments from this Guide in order to compact the listings.[3] Complete *Java Documentation Generator* output is available on the Guide's CD and on the OSP website. This official OSP documentation is generated using a custom `javadoc` add-on *doclet* that creates Universal Modeling Language (UML) diagrams as shown in Figure 1.9.

[3]The Java 2 Software Development Kit (J2SDK) provides a Java API documentation generator known as `javadoc`. The `javadoc` program generates html-based documentation by processing the Java source code and extracting information such as comments, method names, and class variables.

1.10 ■ PROGRAMS

FirstPlotApp

`FirstPlotApp` in the `plot` package tests the Open Source Physics library by creating a plot of the sine function. See Section 1.6.

HeliumApp

The `HeliumApp` program described in Section 1.3 is discussed in Chapter 9.

HelloWorldApp

`HelloWorldApp` in the `hello` package tests the Java installation by printing a string in the console. See Section 1.5.

CHAPTER

2

A Tour of Open Source Physics

The Open Source Physics project provides high-level components for drawing, numerical analysis, and user interfaces. This chapter provides an overview of these capabilities.

2.1 ■ OBJECT-ORIENTED PROGRAMMING

Object-oriented programming (OOP) enables us to represent the real world using a computer. We are fortunate that in physics, computer objects can more nearly reproduce real-world behavior than in most other disciplines. This chapter presents an overview of some important *abstractions* in the Open Source Physics library by building a function plotter and by modeling simple phenomena using first-order differential equations. These examples make use of functions, differential equation solvers, and strip-chart plots without worrying how they are implemented. Because our programs frequently make use of graphical user interface (GUI) components such as buttons, text fields, and check boxes, we introduce *Easy Java Simulations* (Ejs) components. Ejs components can be used without the user being an expert in the Java Swing library. You do not, for example, need to know much about a Java JButton to understand that the following statement creates a button and adds it to a graphical user interface named gui.

```
gui.add ("Button", "parent=inputPanel; text=Plot; action=doPlot");
```

The button displays the text label Plot and (without additional programming) invokes the program's doPlot method when it is pressed.

2.2 ■ OSP FRAMES

We have come to expect programs to have attractive graphical user interfaces that support common tasks such as printing, disk access, and copy-paste data exchange with other applications. These features can be implemented using packages in the Java Development Kit (JDK), but it usually requires much programming and a good knowledge of the Java API. For example, adding a Print menu item requires importing eight classes in three different packages. Learning the API is essential for software developers, but even experienced programmers cut and paste this type of boilerplate code. To enable users to quickly begin writing their own programs while they are learning the core Java and the Open Source Physics APIs and in order to provide examples of how these APIs should be used, we have defined a collection of high-level objects in the frames package for routine data visualization

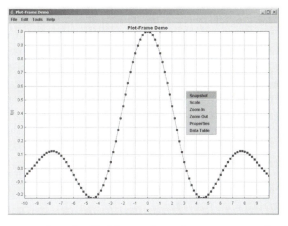

Figure 2.1   OSP frames, such as the PlotFrame shown here, are composite objects that contain a menu bar and a pop-up menu that allow users to access the frame's data.

tasks[1]. These frames inherit from the JFrame class in the Java javax.swing package and add the functionality needed to plot datasets, display vector and scalar fields, and do numerical analysis such as fast Fourier transformations.

One such frame is the PlotFrame class that plots points in one or more datasets as shown in Figure 2.1. A Dataset object contains an array of points and drawing attributes such as color and marker shapes. Data points are added directly to a PlotFrame using the append method and the frame automatically creates the necessary Dataset object as shown in Listing 2.1. The first parameter in the append method identifies the dataset. The second and third parameters specify the point's coordinates. The append method has other signatures that accept arrays of data points and data points with error estimates.

> **Listing 2.1**   The PlotFrameApp program displays an xy-plot of the $\sin x/x$ function.

```
package org.opensourcephysics.manual.ch02;
import org.opensourcephysics.frames.PlotFrame;
import javax.swing.JFrame;

public class PlotFrameApp {
  public static void main(String[] args) {
    PlotFrame frame = new PlotFrame("x", "f(x)", "Plot-Frame Demo");
    frame.setConnected(true); // connects dataset points
    frame.setXYColumnNames(0, "x","sin(x)/x"); // datatable columns
    double dx = 0.2;
    for(double x = -10;x<=10;x += dx) {
      frame.append(0, x, Math.sin(x)/x);
    }
    frame.setVisible(true);
    frame.setDefaultCloseOperation(JFrame.EXIT_ON_CLOSE);
  }
}
```

---

[1]An application programming interface (API) defines how a piece of software is used. In other words, an API specifies how a method (subroutine) is invoked (called) and what it does.

OSP frames, such as the `PlotFrame` in the example, are composite objects with a great deal of functionality. Right (control)-clicking within the display area shows a pop-up menu. The default menu allows users to create a snapshot (a gif image) of the frame's contents, set the scale, zoom, and examine the frame's internal data. The Data Table item displays data points in a Java `JTable` and the Properties item presents an xml tree view of the frame and its contents. (See Chapter 12 for an introduction to xml.)

OSP frames have a menu bar that gives additional options. A typical menu bar has a File menu that allows users to print, save, and inspect the contents of the frame; an Edit menu that allows users to cut and paste between OSP frames; and a Tools menu that contains links to other OSP programs such as data analysis tools.

2.3 ■ FUNCTION PLOTTER

We now use the `PlotFrame` introduced in Section 2.2 to build a simple function plotter with a graphical user interface. This program will be assembled using components from multiple packages and is outlined in Listing 2.2. It builds a custom user interface using components from the Ejs package and draws the plot using components from the display package. Mathematical abstractions, such as a function $f(x)$, are defined in the numerics package.

Listing 2.2 A function plotter program outline.

```
import org.opensourcephysics.display.*;
import org.opensourcephysics.ejs.control.*;
import org.opensourcephysics.numerics.*;

public class PlotterApp {
  PlottingPanel plot =
      new PlottingPanel("x", "f(x)", "y = f(x)");
  Dataset dataset = new Dataset();   // (x,y) data
  EjsControl gui;        // graphical user interface
  Function function;     // function that will be plotted

  public PlotterApp() {
    dataset.setConnected(true); // connect with lines
    dataset.setMarkerShape(Dataset.NO_MARKER);
    plot.addDrawable(dataset);   // add dataset to panel
    buildUserInterface();         // creates user interface
    doPlot();                     // plots default function
  }

  public void doPlot(){
    // code omitted for compactness
  }

  void buildUserInterface() {
    // code omitted for compactness
  }

  public static void main(String[] args) {
    new PlotterApp();  // creates program
  }
}
```

The `PlotterApp` program creates a plotting panel with Cartesian axes, a dataset to store (x, y) points, and identifiers for the user interface and the function. The function and the user interface are created later in the `doPlot` and `buildUserInterface` methods, respectively. The program's constructor sets various display parameters, creates the `PlottingPanel` object, and adds the `Dataset` object to the panel so that it can be drawn. It then builds the graphical user interface and plots a default function. The static `main` method instantiates the application.

The `doPlot` method shown below does what you would expect. It reads a string from the user interface and attempts to convert the string's characters into a mathematical function using a parser (see Section 10.4). If a user mistypes a character or enters an invalid function, the function is set to zero using a utility class in the numerics package. The program then clears the old data, reads the domain of the independent variable, and evaluates the function at a suitable number of points. Although the plot's default x-scale is $[-10, 10]$, this range can be changed by the user at runtime by right (control)-clicking within the plot.[2] Setting the scale and other routine operations are implemented in the plotting panel.

```
public void doPlot() {
  try{ // read input and parse text
    function =
        new ParsedFunction(gui.getString("fx"),"x");
  }catch(ParserException ex){ // input errors are common
    // set f(x)=0 if there is an error
    function= Util.constantFunction(0);
  }
  dataset.clear();              // removes old data
  double xmin=plot.getXMin(), xmax=plot.getXMax();
  double dx=(xmax-xmin)/200;
  for(double x=xmin; x<=xmax; x+=dx){ // loop generates data
    double y=function.evaluate(x);
    dataset.append(x,y);
  }
  plot.repaint(); // new data so repaint
}
```

Although the Open Source Physics library contains a number of simple general purpose graphical user interfaces, it also defines tools that enable us to quickly design and create a custom user interface. The `buildUserInterface` method constructs a custom user interface using the Ejs framework developed by Francisco Esquembre at the University of Murcia, Spain. The complete *Easy Java Simulations* (Ejs) program is a high-level modelling tool that builds Java applications and applets using a drag and drop metaphor. The core OSP library includes a subset of this framework. The `buildUserInterface` method in the function plotter example creates a user interface consisting of a text field and a button as shown in Figure 2.2.

Ejs builds controls without any preconditions or assumptions as to the type of object that will be controlled. The user invokes methods in the `PlotterApp` because a reference to the model `this` is passed to the `EjsControl` constructor.

```
void buildUserInterface() {
  gui = new EjsControl(this); // Ejs user interface
```

---

[2]Users with a one-button mouse should press the Control, Alt, or Option key while clicking to modify mouse actions.

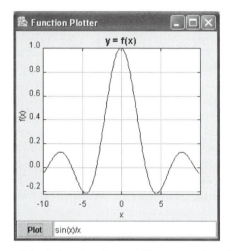

Figure 2.2   A function plotter with a custom interface.

```
gui.add("Frame", "name=controlFrame; title=Plotter;
  layout=border; exit=true; size=pack");
gui.addObject(plot, "Panel", "name=contentPanel;
  parent=controlFrame; position=center");
gui.add ("Panel", "name=inputPanel; parent=controlFrame;
  layout=hbox; position=south");
gui.add ("Button", "parent=inputPanel; text=Plot; action=doPlot;");
gui.add ("TextField", "parent=inputPanel; variable=fx;
  value=sin(x); size=125,15");
}
```

Ejs supports a wide variety of user interface components including labels, radio buttons, check boxes, and text fields. These components are created in a script-like style by invoking the control's `add` method using the following signature:

```
add(String type, String property_list);
```

The `add` method's first parameter specifies the type of object that is being created. In the plotter program we create a frame, a button, and a one-line text field. The second parameter is a semicolon delimited list of properties. These properties depend on the type of object being created. For example, the button's action property specifies the method in the plotter program that will be invoked. The text field's variable property `fx` is used later in the `doPlot` method to read the function string. Note that the Ejs package provides a convenient way to add the preexisting `plot` component to an `EjsControl` using the `addObject` method.

The function plotter program makes use of three core OSP packages. The display package contains the definition of the `Dataset` class and the `PlottingPanel` class. The numerics package defines the `Function` interface and the `ParsedFunction` class. The controls package defines the `Control` interface and the `ejs.control` package provides a concrete implementation of this interface. These packages must, of course, be imported into the program.

```
org.opensourcephysics.numerics.*; import
    org.opensourcephysics.ejs.control.*;
```

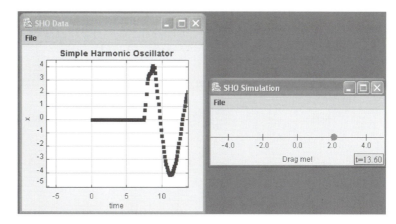

Figure 2.3   A simple harmonic oscillator simulation with a position graph.

The function plotter program can be found in the `org.opensourcephysics.manual.ch02` package.

2.4 ■ SHO

The simple harmonic oscillator (SHO) shown in Figure 2.3 occurs in many different educational contexts. Our simulation of this phenomenon requires two classes SHO and SHOApp to define and to display the evolution of the model, respectively. The SHOApp class constructs the object that draws a picture of the oscillating body, the object that plots position as a function of time, and the object that contains the dynamical equations of motion. The SHO class shown in Listing 2.3, contains the physics.

Listing 2.3   The SHO class defines the dynamical equations of motion for a damped simple harmonic oscillator.

```
package org.opensourcephysics.manual.ch02;
import org.opensourcephysics.display.InteractiveCircle;
import org.opensourcephysics.numerics.*;

public class SHO extends InteractiveCircle implements ODE {
  // initial state values = {x, v, t}
  double[] state = new double[] {0.0, 0.0, 0.0};
  double k = 1;   // spring constant
  double b = 0.2; // damping constant
  ODESolver ode_solver = new RK4(this);

  public double getTime() {
    return state[2];
  }

  public double[] getState() {
    // insure that the state matches the screen position
    state[0] = getX();
    return state;
  }
```

NMU LIBRARY

```
public void getRate(double[] state, double[] rate) {
  rate[0] = state[1]; // dx/dt = v
  double force = -k*state[0]-b*state[1];
  rate[1] = force; // dv/dt = force
  rate[2] = 1;      // dt/dt = 1
}

public void setXY(double x, double y) {
  super.setXY(x, 0); // y is always zero
  state[0] = x;
}

public void stepTime() {
  ode_solver.step();
  setX(state[0]);
}
}
```

The SHO class extends InteractiveCircle so that it can respond to mouse actions and can paint itself as a circle on the computer screen. A user just drags the circle to set the oscillator's initial position. Because mouse actions and painting are defined in the oscillator's *superclass*, we need not concern ourselves with the details here. It is sufficient to know that the oscillator inherits this capability.

The oscillator's dynamics is implemented via the ODE interface to define a system of first-order ordinary differential equations. The OSP numerics package defines a number of differential equation solvers that make use of the ODE interface. In this example, we instantiate a fourth-order Runge–Kutta solver to advance the dynamical system using variables that are stored in the state array. The spring and damping constants are parameters that do not evolve and are therefore not included in this array. Because a user of the SHO class may not know the order of the variables in the array, we implement the getTime convenience method for clarity. The ODE solver calls the oscillator's getState and getRate methods as needed when the solver's step method is invoked. In the spirit of object-oriented programming, we again do not concern ourselves with the details. We do assume that the differential equation solver has implemented the numerical method correctly.

The SHOApp class defines a concrete implementation of an AbstractAnimation by implementing the doStep method. It also creates the necessary drawing and plotting panels to show the oscillator and the generated data. Drawing and plotting follow similar paradigms. Panels are created and added to frames. In the constructor, objects such as the oscillator and the strip-chart dataset are added to the panels. The only requirement is that these objects implement the Drawable interface defined in the display package.

The startAnimation method in the application's AbstractAnimation superclass creates a Thread that invokes the doStep method approximately ten times per second. In this example, the doStep method advances the oscillator's state by stepping a differential equation. It then retrieves the position and time values and appends a data point to the strip-chart dataset. The drawing and the plot are then repainted because objects within these panels have changed state.

Listing 2.4   The SHOApp class creates the objects needed for a simple harmonic oscillator simulation.

```
package org.opensourcephysics.manual.ch02;
import org.opensourcephysics.controls.*;
import org.opensourcephysics.display.*;
```

```
import org.opensourcephysics.display.axes.XAxis;
import javax.swing.JFrame;

public class SHOApp extends AbstractAnimation {
  PlottingPanel plot =
       new PlottingPanel("time", "x", "Simple Harmonic Oscillator");
  DrawingFrame plottingFrame = new DrawingFrame("SHO Data", plot);
  DrawingPanel drawing = new InteractivePanel();
  DrawingFrame drawingFrame =
       new DrawingFrame("SHO Simulation", drawing);
  Dataset stripChart = new Stripchart(20, 10); // chart of x(t)
  SHO sho = new SHO(); // simple harmonic oscillator

  public SHOApp() {
    drawing.setPreferredMinMax(-5, 5, -1, 1);
    drawing.addDrawable(new XAxis("Drag me!"));
    drawing.addDrawable(sho);
    drawingFrame.setSize(300, 150);
    drawingFrame.setVisible(true);
    drawingFrame.setDefaultCloseOperation(JFrame.EXIT_ON_CLOSE);
    plot.addDrawable(stripChart);
    plottingFrame.setLocation(400, 300);
    plottingFrame.setVisible(true);
    plottingFrame.setDefaultCloseOperation(JFrame.EXIT_ON_CLOSE);
  }

  protected void doStep() {
    sho.stepTime();
    stripChart.append(sho.getTime(), sho.getX());
    drawing.setMessage("t="+decimalFormat.format(sho.getTime()));
    drawing.repaint();
    plot.repaint();
  }

  public static void main(String[] args) {
    new SHOApp().startAnimation();
  }
}
```

2.5 ■ THREE-DIMENSIONAL FRAMEWORK

The Java 2D drawing API is designed to represent and manipulate two-dimensional objects. The DrawingPanel and PlottingPanel classes assume this model. However, the physical world is three-dimensional, and we have therefore defined a number of high-level abstractions for manipulating 3D models. Listing 2.5 shows that it is not much more difficult to define and manipulate a three-dimensional ball than a two-dimensional ball using the display3d.simple3d package. The most significant change is that the program instantiates a DrawingPanel3D and adds Element objects to this panel.

The Element interface defines objects that can be added to a Display3DFrame as shown in Figure 2.4. This interface contains methods that allow us to work with elements in a way similar to the way that we work with physical objects. Using concrete implementations of Element in the display3d.simple3d package, we can create and manipulate objects such as arrows, cylinders, and ellipsoids in space. The OSP 3D API defines objects that

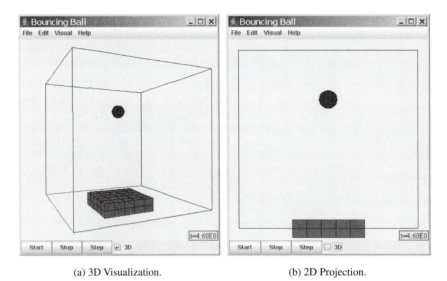

(a) 3D Visualization.   (b) 2D Projection.

Figure 2.4   Objects can be represented in three dimensions using the OSP 3D framework.

- have a position and size in space,

- have visual properties such as color,

- can change their visibility status,

- post interaction events and can have one or more interaction targets,

- and can be grouped with other elements.

Listing 2.5   Particles can be displayed in three dimensions using the simple 3D package.

```
package org.opensourcephysics.manual.ch02;
import org.opensourcephysics.controls.AbstractSimulation;
import org.opensourcephysics.display3d.simple3d.*;
import org.opensourcephysics.ejs.control.EjsControl;
import org.opensourcephysics.frames.Display3DFrame;
import java.awt.Color;

public class Ball3DApp extends AbstractSimulation {
  EjsControl gui;
  Display3DFrame frame = new Display3DFrame("3D Ball");
  Element ball = new ElementEllipsoid();
  double time = 0, dt = 0.05;
  double vz = 0;

  public Ball3DApp() {
    frame.setPreferredMinMax(-5.0, 5.0, -5.0, 5.0, 0.0, 10.0);
    ball.setZ(9);
    ball.setSizeXYZ(1, 1, 1);
```

```
      frame.addElement(ball);
      Element block = new ElementBox();
      block.setXYZ(0, 0, 0);
      block.setSizeXYZ(4, 4, 1);
      block.getStyle().setFillColor(Color.RED);
      // divide the block in subblocks
      block.getStyle().setResolution(new Resolution(5, 5, 2));
      frame.addElement(block);
      buildUserInterface();
   }

   protected void doStep() {
      time += dt;
      double z = ball.getZ()+vz*dt-4.9*dt*dt;
      vz -= 9.8*dt;
      if(vz<0&&z<1.0) {
        vz = -vz;
      }
      ball.setZ(z);
      frame.setMessage("t="+decimalFormat.format(time));
   }

   public void set3d() {
      if(gui.getBoolean("3d")) {
        frame.setProjectionMode(Camera.MODE_PERSPECTIVE);
      } else {
        frame.setProjectionMode(Camera.MODE_PLANAR_YZ);
      }
      frame.repaint();
   }

   void buildUserInterface() {
      gui = new EjsControl(this); // Ejs to build user interface
      gui.addObject(frame, "Frame",
          "name=controlFrame; title=Bouncing Ball; location=400,0;
          exit=true; size=pack");
      gui.add("Panel",
          "name=inputPanel; parent=controlFrame; layout=hbox;
          position=south");
      gui.add("Button",
          "parent=inputPanel; text=Start; action=startAnimation;");
      gui.add("Button",
          "parent=inputPanel; text=Stop; action=stopAnimation;");
      gui.add("Button",
          "parent=inputPanel; text=Step; action=stepAnimation;");
      gui.add("CheckBox",
          "parent=inputPanel; variable=3d; text=3D; selected=true;
          action=set3d;");
   }

   public static void main(String[] args) {
      new Ball3DApp();
   }
}
```

2.6 ■ PROGRAMS

The following examples are in the `org.opensourcephysics.manual.ch02` package.

Ball3DApp

`Ball3DApp` creates a bouncing ball simulation by extending `AbstractSimulation` and implementing the `doStep` method. See Section 2.5.

PlotFrameApp

`PlotFrameApp` demonstrates how to use a `PlotFrame`. See Section 2.2.

PlotterApp

`PlotterApp` reads a string from a text field, converts this string to a function, and plots the function on the interval -10 to 10. See Section 2.3.

SHOApp

`SHOApp` creates a 2D harmonic oscillator simulation by extending `AbstractAnimation` and implementing the `doStep` method. See Section 2.5.

CHAPTER
3

Frames Package

Because visualization is a very common requirement in the sciences, the Open Source Physics library defines the *frames* package (`org.opensourcephysics.frames`) that contain high-level data visualization and data analysis objects. Examples of how this package is used are located in the `org.opensourcephysics.manual.ch03` package unless otherwise stated.

3.1 ■ OVERVIEW

To enable users to quickly begin writing their own programs while they are learning the Java and the Open Source Physics APIs and in order to provide examples of how these APIs are used, we have defined a collection of high-level objects for routine visualization tasks. These objects are defined in the frames package. They extend the `DrawingFrame` superclass and add the functionality needed to plot datasets, display vector and scalar fields, and do numerical analysis such as Fourier transformations. A simple frame is shown in Figure 3.1 and Table 3.1 lists the classes in the package.

Frames in the `org.opensourcephysics.frames` package are high-level composite objects that perform methods by forwarding them to instances of other objects such as `Dataset` or `ContourPlot`. This chapter describes the classes in the frames package. Subsequent chapters in this Guide will show how these frames are composed of objects defined in other Open Source Physics packages. Users should consult the code listings and each frame's `javadoc` documentation for additional information as they gain familiarity with the Open Source Physics library.

The frames described in this chapter are designed to provide convenient general-purpose data visualization and analysis tools for *An Introduction to Computer Simulation Methods* by Gould, Tobochnik, and Christian, and these classes may not meet the needs of other Open Source Physics projects. Ease of use and not computation speed or efficient use of memory motivated the frames package API. However, it is not difficult for readers to create their own optimized and customized visualizations. Subsequent chapters describe how the core OSP classes are implemented so that computer resources can be used most efficiently.

3.2 ■ DISPLAY FRAME

The Open Source Physics library defines objects such as circles, rectangles, images, and single-line text that can be rendered on an output device such as a monitor or printer. These objects are referred to as being drawable because they implement the `Drawable` interface by

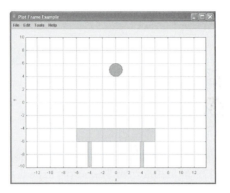

Figure 3.1 A ball bouncing off of a table can be represented using a circle and three rectangles within a DisplayFrame as shown in Listing 3.1.

Table 3.1   The frames package contains plotting and data analysis objects.

org.opensourcephyysics.frames	
ComplexPlotFrame	Displays xy-plots of complex datasets. Data points are added using the append method. Real and imaginary components can be drawn as separate curves or as a single function using color to show phase.
Complex2DFrame	Displays a complex scalar field. The complex field is defined using two-dimensional arrays for the real and imaginary components.
DisplayFrame	Draws objects such as images, circles, and rectangles.
Display3DFrame	Displays three-dimensional objects using the OSP 3D implementation in the simple 3D package.
FFTFrame	Displays the fast Fourier transform of a complex dataset.
FFT2DFrame	Displays the fast Fourier transform of a two-dimensional scalar field.
HistogramFrame	Sorts the data into bins and plots the bins as bars.
LatticeFrame	Displays a two-dimensional array of integers.
PlotFrame	Displays xy-plots.
RasterFrame	Converts a scalar field into an image.
Scalar2DFrame	Displays a scalar field defined using a two-dimensional array of double.
TableFrame	Displays a table. Elements in any row must be of the same data type, but elements in different rows can contain different data types. Rows are not required to be the same length.
Vector2DFrame	Displays a vector field. The scalar field is defined using a two-dimensional array for each vector component.

defining a `draw` method. They are drawn in the order that they are added to an OSP drawing frame. Although drawable objects can be added to any frame in the frames package, this section uses the simplest of these frames, the `DisplayFrame` class, to demonstrate how drawable objects are displayed and manipulated.

The `DisplayFrame` class defines a general-purpose frame and is shown in Figure 3.1. The frame's constructor has signatures to create a display area with and without axes. The axes labels and title are set when the frame is created, but these strings can later be changed using accessor methods such as `setXLabel`.

```
// constructor with axes
DisplayFrame frame =
    new DisplayFrame("x axis label","y axis label", "Plot Title");
// constructor without axes
DisplayFrame frame = new DisplayFrame("Plot Title");
```

The instantiated `DisplayFrame` is initially empty (blank) and is now ready to be customized by adding drawable objects. A typical use of this frame is to display a physical model. For example, a ball bouncing off of a table can be represented using a circle and three rectangles as shown in Listing 3.1. Note how the axes change as the frame is resized.

Listing 3.1 `BallAndTableApp` creates a visualization of a ball and a table.

```
package org.opensourcephysics.manual.ch03;
import org.opensourcephysics.display.DrawableShape;
import org.opensourcephysics.frames.DisplayFrame;
import java.awt.Color;
import javax.swing.JFrame;

public class BallAndTableApp {
  public static void main(String[] args) {
    DisplayFrame frame =
        new DisplayFrame("x", "y", "Plot Frame Example");
    DrawableShape circle = DrawableShape.createCircle(0.0, 5.0, 2);
    circle.setMarkerColor(new Color(128, 128, 255), Color.BLUE);
    frame.addDrawable(circle);
    DrawableShape rectangle =
        DrawableShape.createRectangle(-4, -8.0, 0.5, 4);
    frame.addDrawable(rectangle); // left leg
    rectangle = DrawableShape.createRectangle(4, -8.0, 0.5, 4);
    frame.addDrawable(rectangle); // right leg
    rectangle = DrawableShape.createRectangle(0.0, -5.0, 12, 2);
    frame.addDrawable(rectangle); // table top
    frame.setVisible(true);
    frame.setDefaultCloseOperation(JFrame.EXIT_ON_CLOSE);
  }
}
```

Some, but not all, drawable objects can respond to mouse events and are called *interactive* objects. The `InteractiveShape` class in the display package defines interactive objects that can be moved by clicking and dragging. The `BoundedShape` class defines full-featured objects that can respond to additional mouse actions so that they can be rotated and resized. Run the program in Listing 3.2 to see the capabilities of these interactive objects. Note that you must double-click a `BoundedShape` to change its properties, whereas you can simply click-drag an `InteractiveShape`. It is usually not a good idea to mix `BoundedShape` and

InteractiveShape objects in a drawing panel because users will not know if they should click-drag an object or double-click to select and then click-drag.

Listing 3.2   InteractionApp creates objects that can be manipulated using a mouse.

```
package org.opensourcephysics.manual.ch03;
import org.opensourcephysics.display.*;
import org.opensourcephysics.frames.DisplayFrame;

public class InteractionApp {
  public static void main(String[] args) {
    DisplayFrame frame = new DisplayFrame("x", "y", "Drawable Shapes");
    frame.addDrawable(DrawableShape.createCircle(0.0, 0, 4));
    frame.addDrawable(InteractiveShape.createCircle(-5.0, -5.0, 4));
    BoundedShape circle =
        BoundedShape.createBoundedCircle(-5.0, 5.0, 4);
    circle.setWidthDrag(true);
    circle.setHeightDrag(true);
    frame.addDrawable(circle);
    BoundedShape rectangle =
        BoundedShape.createBoundedRectangle(0, -2, 4, 8);
    rectangle.setRotateDrag(true);
    frame.addDrawable(rectangle);
    BoundedShape arrow =
        BoundedShape.createBoundedArrow(-4, -8.5, 8, 0);
    arrow.setRotateDrag(true);
    frame.addDrawable(arrow);
    frame.setVisible(true);
    frame.setDefaultCloseOperation(javax.swing.JFrame.EXIT_ON_CLOSE);
  }
}
```

Right (control)-clicking within the display area shows a pop-up menu. The *Properties* menu item allows a user to examine the properties of objects in the frame. Other menu items enable a user to control the display area's scale.

Table 3.2 gives an overview of some simple predefined geometric shapes that can be added to a DrawingFrame. These are just samples. Almost any geometric shape can be created using Java 2D as described in Chapter 4. Composite drawables such as contour plot, vector field plot, and datasets are described in subsequent chapters. Programs should, of course, use the simplest drawable object that implements the required functionality.

3.3 ■ PLOT FRAME

The PlotFrame class is designed to create *xy*-plots by adding points to one or more *datasets* as shown in Figure 3.2. A Dataset is an object that contains an array of points and drawing attributes such as color and marker shapes. The details of dataset creation and management are hidden inside the frame. Data points are added directly to a PlotFrame using the append method. The frame creates Dataset objects as needed and appends points to the appropriate dataset.

```
PlotFrame frame = new PlotFrame("x", "y", "Plot Frame Test");
frame.append(0, 0.2, 3.1);  // point (x,y)=(0.2, 3.1)
// xy-arrays
frame.append(1, new double[] {-2,-1,0}, new double[] {4,5,6});
```

35

Table 3.2 The Open Source Physics library defines `Drawable` objects with varying capabilities.

---

org.opensourcephysics.display.Drawable

AWT drawables	Objects, such as `Circle`, that use the original pixel-based drawing mechanism introduced in the Java 1.0 *Abstract Windows Toolkit* (AWT). Size properties are usually specified in pixel units.
Swing drawables	Objects, such as `DrawableShape`, that use the Java 2D API introduced in Java 1.2 to transform themselves from world to pixel coordinates. Properties are specified in world units.
Interactive drawables	Objects, such as `InteractiveShape` or `BoundedShape`, that support mouse actions to adjust their properties.

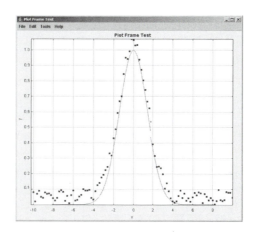

Figure 3.2 A `PlotFrame` that displays a spectral line and data with random noise.

The first parameter in the frame's `append` method is an index that identifies the dataset. The second and third parameters specify coordinates. In the code fragment, dataset zero contains a single data point and dataset one contains three data points. Note that the `append` method has signatures that accept points, point arrays, and points with error estimates.

The `PlotFrame` class creates datasets with rectangular data point *markers* having different colors. A dataset's marker shape, size, and color, as well as a boolean that determines if markers should be connected with line segments, can be set using accessor methods. Listing 3.3 uses two datasets to simulate a Gaussian spectral line with random noise.

Listing 3.3 `PlotFrameApp` displays two datasets.

```
package org.opensourcephysics.manual.ch03;
import org.opensourcephysics.display.Dataset;
import org.opensourcephysics.frames.PlotFrame;

public class PlotFrameApp {
  public static void main(String[] args) {
    PlotFrame frame = new PlotFrame("$\\Delta$f",
        "intensity", "Gaussian Lineshape");
    frame.setConnected(0, true);
```

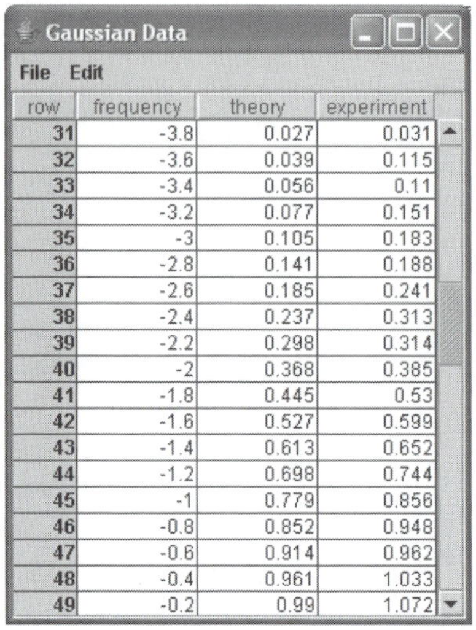

Gaussian Data

File   Edit

row	frequency	theory	experiment
31	-3.8	0.027	0.031
32	-3.6	0.039	0.115
33	-3.4	0.056	0.11
34	-3.2	0.077	0.151
35	-3	0.105	0.183
36	-2.8	0.141	0.188
37	-2.6	0.185	0.241
38	-2.4	0.237	0.313
39	-2.2	0.298	0.314
40	-2	0.368	0.385
41	-1.8	0.445	0.53
42	-1.6	0.527	0.599
43	-1.4	0.613	0.652
44	-1.2	0.698	0.744
45	-1	0.779	0.856
46	-0.8	0.852	0.948
47	-0.6	0.914	0.962
48	-0.4	0.961	1.033
49	-0.2	0.99	1.072

Figure 3.3   A PlotFrame can display its data in a table.

```
frame.setMarkerShape(0, Dataset.NO_MARKER);
for(double x = -10;x<10;x += 0.2) {
  double y = Math.exp(-x*x/4);
  frame.append(0, x, y);                      // datum
  frame.append(1, x, y+0.1*Math.random()); // datum + noise
}
frame.setVisible(true);
frame.setDefaultCloseOperation(javax.swing.JFrame.EXIT_ON_CLOSE);
frame.setXPointsLinked(true); // default is true
frame.setXYColumnNames(0, "frequency", "theory");
frame.setXYColumnNames(1, "frequency", "experiment");
frame.setRowNumberVisible(true);
  }
}
```

It is often convenient to examine data in a table and the PlotFrame class implements this feature using a table as shown in Figure 3.3. A data table can be created using the frame's file menu or using a right (control) mouse click to show a pop-up menu within the display area. Because it is common for points in multiple datasets to have the same independent variable, a PlotFrame assumes that x-coordinates in multiple datasets are linked and suppresses all but the first column of x-values in a data table. This behavior can be changed using the setXPointsLinked accessor method. In addition, the table's column names can be set as shown in Listing 3.3.

3.4 ■ SCALAR AND VECTOR FIELDS

Imagine a plate that is heated at an interior point and cooled along its edges. In principle, the temperature of this plate can be measured at every point. A scalar quantity, such as temperature, pressure, or light intensity, that is defined throughout a region of space is known as a *scalar field*. Other physical quantities, such as the wind velocity or the force on

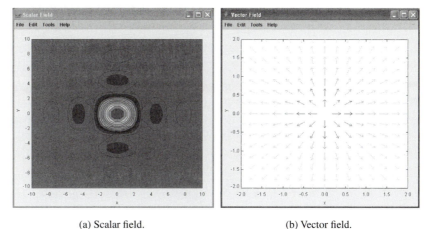

(a) Scalar field.   (b) Vector field.

Figure 3.4   Visualizations of two-dimensional fields.

a particle near Earth, require that a vector be measured at every point and therefore define a *vector field*. The Open Source Physics library contains tools that help us visualize two-dimensional scalar and vector fields as shown in Figure 3.4. These low-level components are defined in the 2D display package. The frames package defines high-level `Scalar2DFrame` and `Vector2DFrame` components that are not as flexible but are easier to use.

The `Scalar2DFrame` class allows us to view 2D scalar fields using representations such as contour plots and 3D surface plots. Listing 3.4 shows a `Scalar2DFrame` being used to visualize the diffraction pattern from a rectangular aperture. We plot the square root of the light intensity to favor the regions of low intensity.

Listing 3.4   A scalar field test program.

```
package org.opensourcephysics.manual.ch03;
import org.opensourcephysics.frames.Scalar2DFrame;
import javax.swing.JFrame;

public class Scalar2DFrameApp {
  public static void main(String[] args) {
    Scalar2DFrame frame = new Scalar2DFrame("x", "y", "Scalar Field");
    // generate sample data
    double[][] data = new double[32][32];
    frame.setAll(data, -10, 10, -10, 10); // initialize field and scale
    for(int i = 0, nx = data.length;i<nx;i++) {
      double x = frame.indexToX(i);
      double ax = (x==0) ? 1 : Math.abs(Math.sin(x)/x);
      for(int j = 0, ny = data[0].length;j<ny;j++) {
        double y = frame.indexToY(j);
        double ay = (x==0) ? 1 : Math.abs(Math.sin(y)/y);
        data[i][j] = ax*ay; // square root of intensity
      }
    }
    frame.setAll(data); // new values
    frame.setVisible(true);
    frame.setDefaultCloseOperation(JFrame.EXIT_ON_CLOSE);
  }
}
```

Table 3.3 `Scalar2DFrame` visualizations are available from the frame's Views menu.

org.opensourcephysics.frames.Scalar2DFrame	
Contour Plot	Draws contour lines and colors the area between contours.
Grayscale Plot	Renders a 2D scalar field as a grayscale image.
Grid Plot	Renders a scalar field using multicolored rectangles.
Interpolated Plot	Renders a scalar field by coloring pixels using values that are interpolated between grid points.
Surface Plot	Renders a scalar field by drawing a 3D surface whose height is proportional to the field's value.

Run the scalar field example and note the various types of visualizations available under the frame's Views menu. Many of the visualizations shown in Table 3.3 produce useful representations even if the grid is small. Some visualizations may look better using a smaller grid size.

The frames package contains the `Vector2DFrame` class for displaying two-dimensional vector fields. To use this class we instantiate a multidimensional array to store vector components. The first array index is zero or one in order to specify the x or y vector component, respectively. The second array index iterates over the x-coordinate and the third array index iterates over the y-coordinate. As for scalar fields, the vectors are set by passing the data array to the frame using the `setAll` method. The program in Listing 3.5 demonstrates how a `Vector2DFrame` is used by displaying the electric field due to a unit charge located at the origin.

Listing 3.5   A vector field test program.

```
package org.opensourcephysics.manual.ch03;
import org.opensourcephysics.frames.Vector2DFrame;
import javax.swing.JFrame;

public class VectorFrameApp {
  public static void main(String[] args) {
    Vector2DFrame frame = new Vector2DFrame("x", "y", "Vector Field");
    double a = 2; // world units for vector field
    int nx = 15, ny = 15;
    // generate the data
    double[][][] data = new double[2][nx][ny]; // vector field
    frame.setAll(data, -a, a, -a, a); // initialize field and scale
    for(int i = 0;i<nx;i++) {
      double x = frame.indexToX(i);
      for(int j = 0;j<ny;j++) {
        double y = frame.indexToY(j);
        double r2 = x*x+y*y;                   // distance squared
        double r3 = Math.sqrt(r2)*r2;          // distance cubed
        data[0][i][j] = (r2==0) ? 0 : x/r3; // x-component
        data[1][i][j] = (r2==0) ? 0 : y/r3; // y-component
      }
    }
    frame.setAll(data); // vector field displays new data
    frame.setVisible(true);
    frame.setDefaultCloseOperation(JFrame.EXIT_ON_CLOSE);
  }
}
```

The arrows representing the vectors have a fixed length that is chosen to match the grid spacing. The arrow's color represents the field's magnitude. The frame's Legend item in the Tools menu shows this mapping.

3.5 ■ COMPLEX FUNCTIONS

Complex functions are essential in many areas of physics such as quantum mechanics, and the frames package contains classes for displaying and analyzing these functions. Listing 3.6 uses a ComplexPlotFrame to display a one-dimensional complex wave function.

Listing 3.6   ComplexPlotFrameApp displays a one-dimensional quantum wave packet with a phase modulation.

```
package org.opensourcephysics.manual.ch03;
import org.opensourcephysics.frames.ComplexPlotFrame;
import javax.swing.JFrame;

public class ComplexPlotFrameApp {
  public static void main(String[] args) {
    ComplexPlotFrame frame =
        new ComplexPlotFrame("x", "Psi(x)","Complex Function");
    int n = 128;
    double xmin = -Math.PI, xmax = Math.PI;
    double x = xmin, dx = (xmax-xmin)/n;
    double[] xdata = new double[n];
    // real and imaginary values alternate in zdata
    double[] zdata = new double[2*n];
    int mode = 4;
    for(int i = 0;i<n;i++) {
      double a = Math.exp(-x*x/4); // wave function amplitude
      // function is e^(-x*x/4)e^(i*mode*x) where x=[-pi,pi)
      zdata[2*i] = a*Math.cos(mode*x);
      zdata[2*i+1] = a*Math.sin(mode*x);
      xdata[i] = x;
      x += dx;
    }
    frame.append(xdata, zdata);
    frame.setVisible(true);
    frame.setDefaultCloseOperation(JFrame.EXIT_ON_CLOSE);
  }
}
```

Figure 3.5 shows two representations of a one-dimensional complex (quantum) wave function. The real and imaginary representation displays the real and imaginary parts of the wave function $\Psi(x)$ by drawing two curves. The amplitude and phase representation uses height to show wave function magnitude and color to show phase. Other wave function visualizations can be selected at runtime using the Views menu or programmatically using convert methods such as convertToPostView and convertToReImView. The Views menu also allows the user to display a data table and a legend to examine the wave function's values and to display the color to phase relationship, respectively.

The Complex2DFrame class is designed to display two-dimensional complex wave functions (see Figure 3.6). This class is designed to plot a two-dimensional array containing the field's real and imaginary components. The Complex2DFrameApp example is available on the CD.

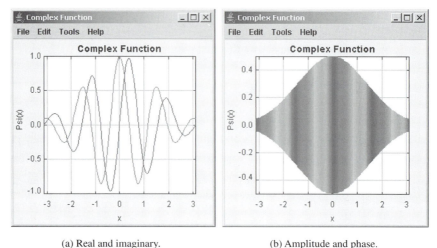

(a) Real and imaginary.   (b) Amplitude and phase.

Figure 3.5   Two representations of complex wave functions.

3.6 ■ FOURIER ANALYSIS

An arbitrary periodic function $f(t)$ can be approximated as a series of sinusoidal functions. In other words, a function $f(t)$ of period T can be approximated as a sum of N sine and cosine functions:

$$f(t) = \frac{1}{2}a_0 + \sum_{k=1}^{N}(a_k \cos \omega_k t + b_k \sin \omega_k t), \tag{3.1}$$

where

$$\omega_k = k\omega_0 \quad \text{and} \quad \omega_0 = 2\pi/T. \tag{3.2}$$

The quantity ω_0 is the fundamental frequency and the sum is called a Fourier series. Because it is mathematically convenient to work with complex numbers, the sine and cosine functions of the same frequency are often combined into a single complex exponential using Euler's formula[1]:

$$e^{i w_k t} = \cos \omega_k t + i \sin \omega_k t. \tag{3.3}$$

By using (3.3) we can express an arbitrary periodic complex function $f(t)$ as

$$f(t) = \sum_{k=-N/2}^{N/2} c_k e^{i\omega_k t}, \tag{3.4}$$

where the expansion coefficients c_k are complex numbers.

The process of approximating a function by computing the expansion coefficients c_k is called *Fourier analysis*, and the most efficient algorithm for performing this computation

[1]Euler's formula is a special case of De Moivre's formula $(\cos x + i \sin x)^n = \cos nx + i \sin nx$ that preceded it.

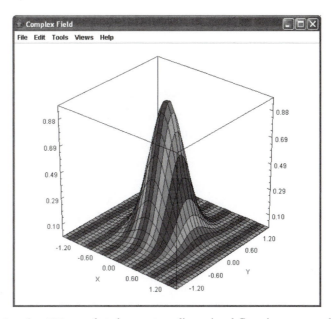

Figure 3.6   A Complex2DFrame that shows a two-dimensional Gaussian wave packet with a phase modulation.

is known as the Fast Fourier Transformation (FFT). Computing Fourier coefficients using one of the many publicly available FFT implementations is straightforward but may require a fair amount of bookkeeping. To simplify the process, we have defined the FFTFrame class in the frames package to perform a one-dimensional complex FFT and display the coefficients. This utility class accepts either data arrays or functions as input parameters to the doFFT method. The code shown in Listing 3.7 transforms an input array containing the complex exponential e^{imt}. The mode number m determines the Fourier coefficient.

Listing 3.7   FFTFrameApp displays the coefficients of the function $e^{2\pi mt}$ where m is the harmonic (mode).

```
package org.opensourcephysics.manual.ch03;
import org.opensourcephysics.frames.FFTFrame;
import javax.swing.JFrame;

public class FFTFrameApp {
  public static void main(String[] args) {
    FFTFrame frame =
        new FFTFrame("omega", "amplitude", "FFT Frame Test");
    int n = 16; // number of data points
    double tmin = 0, tmax = 2*Math.PI;
    double t = tmin, delta = (tmax-tmin)/n;
    double[] data = new double[2*n];
    frame.setDomainType(FFTFrame.OMEGA);
    int mode = 2; // function is e^(i*mode*x) where x=[0,2pi]
    for(int i = 0;i<n;i++) {
      data[2*i] = Math.cos(mode*t);
      data[2*i+1] = Math.sin(mode*t);
      t += delta;
    }
```

```
        frame.doFFT(data, tmin, tmax);
        frame.setVisible(true);
        frame.setDefaultCloseOperation(JFrame.EXIT_ON_CLOSE);
    }
}
```

The extension of the ideas of Fourier analysis to two dimensions is simple and direct. If we assume a complex function of two variables $f(x, y)$, then a two-dimensional series is constructed using harmonics of both variables. The expansion functions are the products of one-dimensional functions $e^{i(xq_x+yq_y)}$ and the Fourier series is written as a sum of these harmonics:

$$f(x, y) = \sum_{n=-N/2}^{N/2} \sum_{m=-M/2}^{M/2} c_{n,m}\, e^{iq_n x} e^{iq_m y}, \tag{3.5}$$

where

$$q_n = \frac{2\pi n}{X} \quad \text{and} \quad q_m = 2\pi \frac{m}{Y}. \tag{3.6}$$

The function $f(x, y)$ is assumed to be periodic in both x and y with periods X and Y, respectively.

Because of the large number of coefficients $c_{n,m}$, the discrete two-dimensional Fourier transform is also implemented using the FFT algorithm. FFT2DFrameApp shows how to compute and display a 2D FFT using the FFT2DFrame class.

The Fourier analysis implementation in the OSP library is based on FFT routines contributed to the GNU Scientific Library (GSL) by Brian Gough and adapted to Java by Bruce Miller at NIST. We initialize our data array to conform to the GSL API using a one-dimensional array such that rows follow sequentially. This ordering is known as row-major format. Because the input function is assumed to be complex, the array has dimension $2N_x N_y$, where N_x and N_y are the number of grid points in the x and y direction, respectively. One disadvantage of using row-major format is that the number of rows must be specified as an additional parameter in order to interpret the data. The FFT2DFrame object transforms and displays the data when the doFFT method is invoked. Note that the doFFT method is invoked with a position space scale and that this scale is used to compute the spatial frequencies of the Fourier components.

Listing 3.8   FFT2DFrameApp displays the coefficients for the two-dimensional Fourier analysis of the complex function $e^{i2\pi(nx+my)}$.

```
package org.opensourcephysics.manual.ch03;
import org.opensourcephysics.frames.FFT2DFrame;
import javax.swing.JFrame;

public class FFT2DFrameApp {
  public static void main(String[] args) {
    FFT2DFrame frame = new FFT2DFrame("k_x", "k_y", "2D FFT");
    double xmin = 0, xmax = 2*Math.PI, ymin = 0, ymax = 2*Math.PI;
    // generate data on a grid of size (nx,ny)
    int nx = 5, ny = 5;
    int xMode = -1, yMode = -1;
    // field stored in row-major format with array size 2*nx*ny
    double[] zdata = new double[2*nx*ny];
    // function is e^(i*xmode*x)e^(i*ymode*y) where x and y domains
    // are [0,2pi]
```

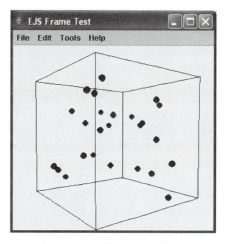

Figure 3.7 A Display3DFrame displaying particles at random locations.

```
double y = 0, yDelta = 2*Math.PI/ny;
for(int iy = 0;iy<ny;iy++) {
  // offset to beginning of a row; row length is nx
  int offset = 2*iy *nx;
  double x = 0, xDelta = 2*Math.PI/nx;
  for(int ix = 0;ix<nx;ix++) {
    // real part
    zdata[offset+2*ix] =
        Math.cos(xMode*x)*Math.cos(yMode*y)-
        Math.sin(xMode*x)*Math.sin(yMode*y);
    // imaginary part
    zdata[offset+2*ix+1] =
        Math.sin(xMode*x)*Math.cos(yMode*y)+
        Math.cos(xMode*x)*Math.sin(yMode*y);
    x += xDelta;
  }
  y += yDelta;
}
frame.doFFT(zdata, nx, xmin, xmax, ymin, ymax);
frame.setVisible(true);
frame.setDefaultCloseOperation(JFrame.EXIT_ON_CLOSE);
  }
}
```

3.7 ■ DISPLAY 3D FRAME

There are a number of APIs available for three-dimensional visualizations using Java. Although Sun has developed *Java 3D*, this library is currently not included in the standard Java runtime environment. The *Java bindings for OpenGL* (JOGL) library is a popular alternative because it is based on the OpenGL language. Because the basic OSP library should not rely on other 3D libraries to be downloaded and installed on a client computer and because we want a three-dimensional visualization framework designed for physics simulations, we have developed our own framework that relies only on the standard Java API (see Figure 3.7). This 3D framework uses packages (subdirectories) in the OSP `display3d` directory.

The OSP 3D API is defined in the `org.opensourcephysics.display3d.core` package. The interfaces defined in this package allow us to develop multiple implementations using various graphics libraries. The OSP 3D API is similar to the OSP 2D API except that 3D objects are referred to as `Elements` rather than `Drawables`. Three-dimensional drawable objects are created and added to a 3D container using the `addElement` method as shown in Listing 3.9.

The `Display3DFrame` class in the frames package uses concrete implementations of 3D `Elements` defined in the `org.opensourcephysics.display3d.simple3d` package. This package uses only the standard Java library and is well suited for displaying small objects that do not intersect. It avoids advanced rendering techniques by decomposing large objects into smaller polygons. You may notice artifacts at the polygon intersections, but large objects can be decomposed into smaller pieces and the results are almost always satisfactory for simple physical models such as collections of particles and vectors.

Listing 3.9   `Display3DFrameApp` displays twenty-five interactive particles using the OSP 3D framework.

```
package org.opensourcephysics.manual.ch03;
import org.opensourcephysics.display3d.simple3d.*;
import org.opensourcephysics.frames.Display3DFrame;
import javax.swing.JFrame;

public class Display3DFrameApp {
  public static void main(String[] args) {
    Display3DFrame frame = new Display3DFrame("Random Circles");
    for(int i = 0;i<25;i++) {
      Element ball = new ElementCircle();
      ball.setSizeXYZ(0.1, 0.1, 0.1);
      ball.setXYZ(-1.0+2*Math.random(), -1.0+2*Math.random(),
                  -1.0+2*Math.random());
      frame.addElement(ball);
    }
    frame.setVisible(true);
    frame.setDefaultCloseOperation(JFrame.EXIT_ON_CLOSE);
  }
}
```

We will not study the simple3D implementation in detail here but will describe only key OSP 3D concepts (see Chapter 11). `Display3DFrameApp` creates a simple three-dimensional visualization. Run the program and drag the mouse within the panel. Line and perspective details are hidden from the user as the viewing (camera) position is changed.

Figure 3.8 shows another example, a 3D visualization of an electromagnetic wave. The wave is constructed using arrows as shown in Listing 3.10.

Listing 3.10   A 3D visualization of an electromagnetic wave is constructed using arrows.

```
package org.opensourcephysics.manual.ch03;
import org.opensourcephysics.display3d.simple3d.*;
import org.opensourcephysics.frames.Display3DFrame;
import javax.swing.JFrame;

public class Wave3DApp {
```

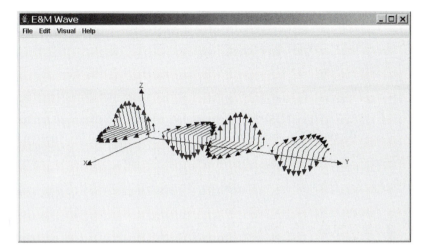

Figure 3.8 A 3D visualization of an electromagnetic wave.

```java
public static void main(String[] args) {
  // Data for the wave
  int n = 48; // number of arrows
  double time = 0.0, dt = 0.1, E0 = 100.0, Vy = 40.0, B0 = E0;
  double period = 5.0, omega = 2.0*Math.PI/period, k = omega/Vy;
  double[] y = new double[n];
  Display3DFrame frame = new Display3DFrame("E&M Wave");
  frame.setPreferredMinMax(-100, 100, 0, 400, -100, 100);
  frame.setDecorationType(VisualizationHints.DECORATION_AXES);
  ElementArrow[] fieldE = new ElementArrow[n];
  ElementArrow[] fieldB = new ElementArrow[n];
  for(int i = 0;i<n;i++) {
    y[i] = i*400.0/n;
    fieldE[i] = new ElementArrow();
    fieldE[i].getStyle().setFillColor(java.awt.Color.RED);
    fieldE[i].setXYZ(0, y[i], 0);
    fieldE[i].setSizeXYZ(0, 0, 0);
    frame.addElement(fieldE[i]);
    fieldB[i] = new ElementArrow();
    fieldB[i].getStyle().setFillColor(java.awt.Color.BLUE);
    fieldB[i].setXYZ(0, y[i], 0);
    fieldB[i].setSizeXYZ(0, 0, 0);
    frame.addElement(fieldB[i]);
  }
  frame.setSquareAspect(false);
  frame.setVisible(true);
  frame.setDefaultCloseOperation(JFrame.EXIT_ON_CLOSE);
  while(true) { // animate until the program exits
    try {
      Thread.sleep(100);
    } catch(InterruptedException ex) {}
    time += dt;
    for(int i = 0;i<n;i++) {
      fieldE[i].setSizeZ(E0*Math.sin(k*(y[i]-Vy*time)));
```

```
                fieldB[i].setSizeX(B0*Math.sin(k*(y[i]-Vy*time)));
            }
            frame.render();
        }
    }
}
```

Additional OSP 3D implementations that use advanced rendering packages such as Java 3D and JOGL are under development. All that will be required for a program to use these more sophisticated libraries is to install the library and import the corresponding OSP 3D package.

3.8 ■ TABLES

The TableFrame class is designed to show a table without the overhead of creating a plot. Data are appended one row at a time using the appendRow method.

```
TableFrame tableFrame = new TableFrame("Root Table");
tableFrame.appendRow(new String[]{"one","two"});
tableFrame.appendRow(new int[]{2,3,4,5});
```

Each row is a one-dimensional array of double, integer, string, or byte data, but these arrays need not be the same length. Listing 3.11 uses a TableFrame to create a table of square and cube roots. Users can copy data from this table into the system clipboard by highlighting a block of cells and then using the keyboard copy command.

> **Listing 3.11**   TableFrameApp demonstrates how to create
> a table of numerical values.

```
package org.opensourcephysics.manual.ch03;
import org.opensourcephysics.frames.TableFrame;

public class TableFrameApp {
  public static void main(String[] args) {
    TableFrame tableFrame = new TableFrame("Root Table");
    tableFrame.setRowNumberVisible(false);
    tableFrame.setColumnNames(0, "x");
    tableFrame.setColumnNames(1, "square root");
    tableFrame.setColumnNames(2, "cube root");
    for(int i = 0;i<10;i++) {
      try {
        tableFrame.appendRow(new double[] {i, Math.sqrt(i),
            Math.pow(i, 1.0/3)});
      } catch(Exception ex) {}
    }
    tableFrame.setVisible(true);
    tableFrame.setDefaultCloseOperation(
        javax.swing.JFrame.EXIT_ON_CLOSE);
  }
}
```

3.9 ■ PROGRAMS

The following examples are in the `org.opensourcephysics.manual.ch03` package.

BallAndTableApp

`BallAndTableApp` creates a visualization of a ball and a table using a `DisplayFrame`. See Section 3.2.

Complex2DFrameApp

`Complex2DFrameApp` demonstrates how to plot a 2D complex scalar field by displaying a two-dimensional Gaussian wave function with a momentum boost. See Section 3.5.

ComplexPlotFrameApp

`Complex2DFrameApp` demonstrates how to plot a complex function by displaying a one-dimensional Gaussian wave function with a momentum boost. See Section 3.5.

Display3DFrameApp

`Display3DFrameApp` demonstrates how to create a simple 3D view by displaying 25 particles in a 3D box. See Section 3.7.

FFT2DFrameApp

`FFT2DFrameApp` tests the `FFT2DFrame` class by taking the two-dimensional Fast Fourier Transform (FFT) of a harmonic function. See Section 3.6.

FFTFrameApp

`FFTFrameApp` tests the `FFTFrame` class by taking the one-dimensional Fast Fourier Transform (FFT) of a harmonic function. See Section 3.6.

InteractionApp

`InteractionApp` creates three bounded shapes and places them in a `DisplayFrame`. Double-click on a shape to select it and then drag the hot spots. See Section 3.2.

PlotFrameApp

`PlotFrameApp` simulates a best fit to a Gaussian spectral line using a `PlotFrame` with two datasets. These datasets uses different point markers and drawing styles. See Section 3.3.

Scalar2DFrameApp

`Scalar2DFrameApp` tests the `Scalar2DFrame` class by plotting the field $f(x, y) = xy$. See Section 3.4.

TableFrameApp

`TableFrameApp` demonstrates how to create and use a `TableFrame`. See Section 3.8.

VectorFrameApp

`VectorFrameApp` plots a $1/r^2$ vector field using a `Vector2DFrame`. See Section 3.4.

Wave3DApp

`Wave3DApp` displays a visualization of a transverse traveling wave. See Section 3.7.

CHAPTER

4

Drawing

The display package (org.opensourcephysics.display) defines the Open Source Physics drawing framework. Examples of how this package is used to create simple drawings are located in the org.opensourcephysics.manual.ch04 package unless otherwise stated.

4.1 ■ OVERVIEW

One of the attractions of Java is that device and platform independent graphics is incorporated directly into the language as shown in Figure 4.1. Lines, ovals, rectangles, images, and text can be drawn with just a few statements. The position of each shape is specified using an integer coordinate system that has its origin located at the upper left-hand corner of the device—the computer display or printer paper—and whose positive *y* direction is defined to be down. Although we can use Java's graphics capabilities to produce visualizations and animations, creating even a simple graph in such a coordinate system can require a fair amount of programming. A scale must be established, axes need to be drawn, and data needs to be transformed. An additional complication arises due to the fact that a graph must be able to redraw itself whenever an application is exposed or a window resized. But data visualization is a fairly common operation. We have developed a drawing framework that not only scales data but also can be used for animations and other visualizations. Listing 4.1 uses this framework to create a rectangle and a circle in a plot frame.

Run DrawingApp and do the following:

1. Click-drag within the drawing and notice the coordinate display in the lower left-hand corner.
2. Double-click on a shape to select it and drag the hotspots to change the shape's properties.
3. Right-click within the drawing to display a pop-up menu and explore the menu items.
4. Resize the frame and observe changes in the axes.
5. Explore the menu items in the frame's menu bar.

The drawing framework is defined in the org.opensourcephysics.display package. As the above example shows, the display package defines a drawing API with a high level of abstraction. This chapter examines the details of how the OSP 2D drawing API is constructed.

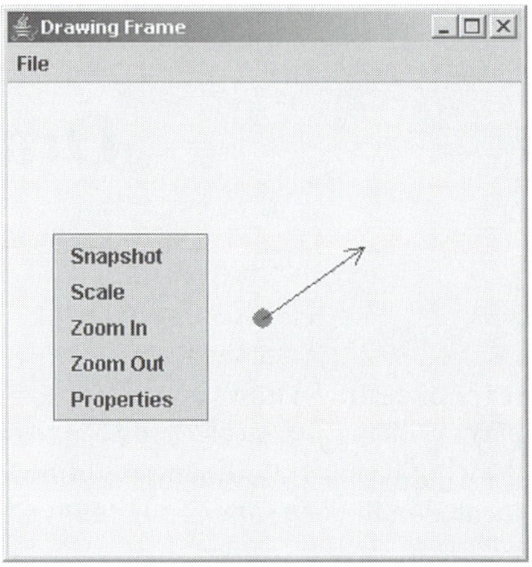

Figure 4.1   A drawing panel within a drawing frame. The drawing panel in the figure contains a drawable circle and a drawable arrow. The panel's pop-up menu is activated by right (control)-clicking within the panel.

Listing 4.1   A drawing with a drawable shape.

```
package org.opensourcephysics.manual.ch04;
import org.opensourcephysics.display.BoundedShape;
import org.opensourcephysics.frames.PlotFrame;
import javax.swing.JFrame;

public class DrawingApp {
  public static void main(String[] args) {
    PlotFrame frame = new PlotFrame("x", "y", "Drawing Demo");
    frame.setPreferredMinMax(-10, 10, -10, 10);
    BoundedShape ishape = BoundedShape.createBoundedCircle(3, 4, 5);
    ishape.setHeightDrag(true);
    ishape.setXYDrag(false);
    frame.addDrawable(ishape);
    ishape = BoundedShape.createBoundedRectangle(0, 0, 9, 3);
    frame.addDrawable(ishape);
    frame.setVisible(true);
    frame.setDefaultCloseOperation(JFrame.EXIT_ON_CLOSE);
  }
}
```

4.2 ■ DRAWABLES

The drawing framework is based on the `Drawable` interface and the `DrawingPanel` class that are defined in the `org.opensourcephysics.display` package. The `Drawable` interface contains a single method `draw` that is invoked automatically from within the drawing panel's `paintComponent` method.

```
public interface Drawable {
    public void draw(DrawingPanel drawingPanel, Graphics g);
}
```

Table 4.1   Examples of drawable objects defined in the display package.

org.opensourcephysics.display	
Arrow	An arrow with the location of its tail and xy-components defined in world units.
Circle	A circle with its center defined in world units and its radius defined in pixels.
Dataset	Data points with various rendering options such as color and marker styles.
FunctionDrawer	Draws a graph by evaluating a function at every pixel on the drawing panel's x-ordinate.
Trail	An array of connected data points.

Drawable objects, that is, objects that implement this interface, are instantiated and then added to a drawing panel where they will draw themselves in the order that they are added. Listing 4.2 shows a program that creates a drawing panel containing a circle and an arrow. Other drawable objects are listed in Table 4.1.

Listing 4.2   Drawable objects in a DrawingPanel.

```
import org.opensourcephysics.display.*;
import javax.swing.JFrame;

public class CircleAndArrowApp {
   public static void main(String[] args) {
       // create a drawing frame and a drawing panel
       DrawingPanel panel = new DrawingPanel();
       panel.setPreferredMinMax(-10,10,-10,10);
       DrawingFrame frame = new DrawingFrame(panel);
       panel.setSquareAspect(false);
       Drawable circle = new Circle(0, 0);   // create circle
       panel.addDrawable(circle);            // add to panel
       Drawable arrow = new Arrow(0, 0,4,3); // create arrow
       panel.addDrawable(arrow);             // add to panel
       frame.show();
       frame.setDefaultCloseOperation(JFrame.EXIT_ON_CLOSE);
   }
}
```

Drawables are easy to create using either Java AWT or Java 2D drawing techniques.[1] For example, the drawable circle defined in the display package has a center at (x, y) in world coordinates and a radius in pixels (see Listing 4.3). It implements Drawable using a standard Java graphics context Graphics.

Better drawing paradigms than those in the Java 1.0 AWT API are now common. This API only supports solid single-pixel lines, limited fonts, and no rotation or scaling. The Java2D API introduced in Java 1.2 is not restricted to pixel coordinates nor is it restricted to solid single-pixel lines. Much of its flexibility is proved by affine transformations as described in Section 4.9. *Java Advanced Imaging* (JAI) is available from Sun as an add-

[1]The Java Abstract Windows Toolkit (AWT) was introduced in Java 1.0 and implements primitive pixel-based drawing methods.

Listing 4.3   A drawable circle with fixed pixel radius.

```
import java.awt.*;
public class Circle implements Drawable {
    int  pixRadius = 6;
    Color color = Color.red;
    double x = 0;
    double y = 0;

    public Circle(double _x, double _y) {
        x = _x;
        y = _y;
    }

    public void draw(DrawingPanel panel, Graphics g) {
        int xpix = panel.xToPix(x) - pixRadius;
        int ypix = panel.yToPix(y) - pixRadius;
        g.setColor(color);
        g.fillOval(xpix, ypix, 2*pixRadius, 2*pixRadius);
    }
}
```

on package that provides extended imaging processing capabilities beyond those found in Java 2D.

The Open Source Physics library contains numerous other components for data visualization including a contour component, a wire mesh component, and a vector field component as described in Chapter 8. These components all implement the Drawable interface so that it is possible, for example, to show the motion of a circle superimposed on potential energy contours by adding a contour plot and a circle to a drawing panel.

4.3 ■ DRAWING PANEL AND FRAME

The DrawingPanel class is a subclass of JPanel as shown in Figure 4.2. Its primary purpose is to define a system of *world coordinates* that allow us to specify the location and size of objects in units other than pixel units. The DrawingFrame class is a subclass of OSPFrame which is a subclass of JFrame. A DrawingFrame is designed to display a single drawing panel in the center of its content pane. This frame also has a default menu including a Print option.

The InteractivePanel class is a subclass of DrawingPanel that gathers mouse and keyboard actions and passes them to a listener. The PlottingPanel class is a subclass of InteractivePanel that adds axes. Because plotting data is an important application of drawing, the full capabilities of this class are presented in Chapter 6. We will use the PlottingPanel class in various examples in this chapter when it is convenient to display a scale.

Drawable objects do not "belong" to any panel and can be rendered in more than one panel. These objects can represent themselves in different panels because the draw method obtains a reference to the drawing panel and to the graphics context. Drawable objects can, for example, simultaneously render themselves in a zoom and an extended view.

Swing components paint themselves using the paintComponent method and the drawing panel's paintComponent method begins by reading the panel's pixel width and height. The

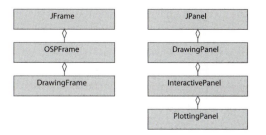

Figure 4.2 `DrawingFrame` and `DrawingPanel` inherit from the Swing `JFrame` and `JPanel` components, respectively.

Table 4.2 Common drawing panel methods.

org.opensourcephysics.display.DrawingPanel	
`getAspectRatio`	Gets the ratio of x pixels per unit to y pixels per unit.
`getPixelTransform`	Gets the Java 2D affine transformation that converts from world coordinates to pixel coordinates.
`pixToX`	Converts horizontal pixel values to world coordinates.
`pixToY`	Converts vertical pixel values to world coordinates.
`setPreferredMinMax`	Sets the preferred coordinate scale.
`xToPix`	Converts x world coordinates to horizontal pixel values.
`yToPix`	Converts y world coordinates to vertical pixel values.

panel then calculates the optimum scale using preferred values of x_{min}, x_{max}, y_{min}, and y_{max}. Finally, it calculates the world-to-pixel coordinate transformation. If an aspect ratio of unity is selected, the pixels per unit are equal along both axes, and the minimum and maximum values for each axis are calculated to be at least as large as the preferred values. Examples of common drawing panel methods are shown in Table 4.2.

When a drawing panel is repainted, it iterates through all drawable objects invoking each object's `draw` method. Each drawable object has access to the panel's coordinate transformation because the `draw` method is passed a reference to the drawing panel. It is, of course, possible to combine drawable objects to build compound objects such as graphs (see Chapter 6). The display package defines all the components necessary to produce a plot including axes and a `Dataset` class that stores and renders xy-data points.

4.4 ■ INSPECTORS

It is often convenient to change a drawing panel's properties while a program is running by right-clicking to activate the panel's pop-up menu. We can, for example, *zoom in* and *zoom out* for axes that are not autoscaled. If an axis is autoscaled, then that axis scale is set by the objects within the panel. The pop-up menu also allows a user to display the drawing panel's *inspector*.

Drawing panel properties can be examined using an inspector. The default drawing panel inspector is shown in Figure 4.3. If the inspector option has been enabled, it can be displayed by right (control)-clicking within the panel.

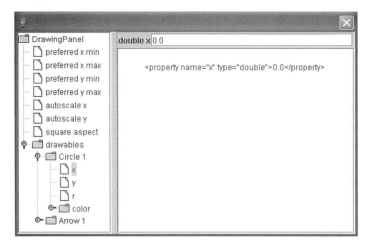

Figure 4.3   The default drawing panel inspector.

```
PlottingPanel plottingPanel = new PlottingPanel ("x", "y", "Title");
plottingPanel.enableInspector(true);
plottingPanel.showInspector();
```

The default inspector uses an *extensible markup language* (xml) based syntax and this syntax may not be appropriate for nontechnical users (see Chapter 12). It is, however, very useful for developers because it allows them to inspect and sometimes edit an object's properties while a program is running. It is straightforward to create a user friendly inspector for distribution to less technical audiences. We replace the default inspector with a custom `Window` component using the `setCustomInspector` method.

```
PlottingPanel plottingPanel = new PlottingPanel ("x", "y", "Title");
// window is an instance of JWindow
plottingPanel.setCustomInspector(window);
```

4.5 ■ MESSAGE BOXES

Drawing panels borrow the `GlassPane` concept from the Java `JRootPane` class. The drawing panel's glass pane is a transparent component that sits on top of the panel. It is designed to display Swing components without interfering with the drawing panel. We use the glass pane to layout yellow message boxes that display coordinate values when the mouse is dragged and that display custom messages.

The default glass pane contains four message boxes located in the drawing panel's four corners. Messages can be placed into these boxes using the `setMessage` method.

```
double t = 0, en = 10;
PlottingPanel plottingPanel = new PlottingPanel("x","y","title");
plottingPanel.setMessage("time = "+t); // lower right is default
plottingPanel.setMessage("energy = "+en, DrawingPanel.TOP_LEFT);
```

Message boxes are located at the four corners and the default message box is located in the lower right corner.

Users can add their own custom components to the glass pane by obtaining a reference from the drawing panel. Listing 4.4 shows how this is done. Note that the glass pane uses a custom OSPLayout manager that supports corner and center placement of components.

Listing 4.4 Messages are drawn using a glass pane that sits on top of the drawing panel.

```
package org.opensourcephysics.manual.ch04;
import org.opensourcephysics.display.*;

public class MessageApp {
  public static void main(String[] args) {
    // create a drawing frame and a drawing panel
    DrawingPanel panel = new DrawingPanel();
    DrawingFrame frame = new DrawingFrame(panel);
    panel.setMessage("time="+0);
    panel.setMessage("energy="+0, 2); // top right
    MessagePanel messagePanel = new MessagePanel();
    panel.getGlassPanel().add(messagePanel, OSPLayout.CENTERED);
    frame.setVisible(true);
    frame.setDefaultCloseOperation(javax.swing.JFrame.EXIT_ON_CLOSE);
  }
}

class MessagePanel extends javax.swing.JPanel {
  MessagePanel() {
    setPreferredSize(new java.awt.Dimension(100, 100));
    this.setBackground(java.awt.Color.RED);
  }
}
```

4.6 ■ SCALE

Computers typically use an integer coordinate system based on pixels to access locations on a device. The origin $(0, 0)$ is located in the upper left-hand corner, the horizontal x-coordinate increases toward the right, and the vertical y-coordinate increases toward the bottom. We refer to these coordinates as *pixel coordinates*. Physical models use *world coordinates* in whatever units are appropriate. For example, the distance from Earth to Sun is defined to be one astronomical unit (AU) and this scale is often used when modeling the solar system. In order to make it easy for programs to create visualizations, the position of a physical object is almost always computed in world coordinates. In a computer simulation, a model is solved in whatever units are appropriate and the location of drawable objects representing the physics are then set using world coordinates.

A DrawingPanel contains a transformation that is used to scale and locate drawable objets before drawing their visual representation on the output device. The default scale along each axis is –10 to 10, but these limits can be changed a number of ways. The most direct approach is to specify the minimum and maximum values along the x (horizontal) and y (vertical) axes. The following code fragment sets the x-axis to $[0, 20]$ and the y-axis to $[-10, 10]$.

```
//setPreferredMinMax signature is (xmin, xmax, ymin, ymax)
drawingPanel.setPreferredMinMax(0, 20, -10, 10);
drawingPanel.setSquareAspect(true); // default is true
```

These minimum and maximum values specify a preferred range. The true range can differ because the panel may not be square and the computed pixels per unit in the *x* direction may differ from the pixels per unit in the *y* direction. The algorithm adjusts the scale's minimum and maximum values to insure that pixels per unit are the same in both directions. This algorithm also guarantees that the true range contains the preferred range. Setting the square aspect property to false allows the panel to honor the preferred minimum and maximum values. A nonsquare aspect will, of course, distort the appearance of geometric shapes.

The number of screen pixels per world unit can also be set explicitly using the `setPixelsPerUnit` method. The following code fragment sets the panel's scale to be exactly 10 pixels per unit in both the *x* and *y* directions.

```
drawingPanel.setPixelsPerUnit(true,10,10);
```

The center of the display area is set to the average of the preferred minimum and maximum values, and true minimum and maximum values are then computed using the window size and the given pixels per unit values. In other words, setting the pixels per unit causes the size of the visible "world" to decrease and increase in direct proportion to the window size.

A third scaling option computes a drawing panel's maximum and minimum values using data supplied by the drawable objects themselves. For example, `Dataset` objects keep track of their minimum and maximum *x*- and *y*-values as data are appended, and these values can be used to insure that all points are displayed. Drawable objects can provide these minimum and maximum values to a panel by implementing the `Measurable` interface as described in the next section. If autoscaling is enabled and if measurable objects are added to the panel, then the drawing panel computes a scale using data supplied by the drawable objects themselves.

```
drawingPanel.setAutoscaleX(true);
drawingPanel.setAutoscaleY(true);
```

Although it is sometimes convenient to allow objects to adjust a panel's scale, this method can lead to inappropriate values if the *xy*-range is small. We might, for example, wish to insure that the minimum *y*-value be no greater than zero and the maximum *y*-value be no smaller than ten even if every point has a *y*-value of 5. This behavior can be obtained by setting floor and ceiling limits.

```
drawingPanel.setAutoscaleY(true);
drawingPanel.limitAutoscaleY(0,10);   // sets floor and ceiling
```

Limits allow the panel to autoscale the axis if the measurable object's minimum and maximum values exceed the given interval [0, 10], but they insure that the axis always contains the interval.

4.7 ■ MEASURABLE

If the drawing panel's `autoscaleX` or `autoscaleY` properties are set to true, the panel reads the minimum and maximum values of its measurable objects at the beginning of the drawing cycle and sets the scale appropriately. The `Measurable` interface in the display package provides this functionality and is defined as follows:

```
public interface Measurable extends Drawable {
  public double   getXMin();
  public double   getXMax();
  public double   getYMin();
  public double   getYMax();
  public boolean isMeasured();  // true if the measure is valid
}
```

Note that the isMeasured method should return false if the object, such as an empty dataset, does not have meaningful minimum and maximum values to report.

The measurable interface is very flexible. We can, for example, define a MeasuredCircle class that reports the center of the circle as its minimum and maximum values.

```
public class MeasuredCircle extends Circle implements Measurable {
    boolean enableMeasure = true;
    // x and y are defined in the circle superclass
    public MeasuredCircle (double x, double y) { super (x, y); }

    public boolean isMeasured () { return enableMeasure; }
    public double   getXMin () { return x; }
    public double   getXMax () { return x; }
    public double   getYMin () { return y; }
    public double   getYMax () { return y; }
}
```

If an ensemble of measured circles is created and these circles are added to a drawing panel, then the drawing panel will always rescale itself so that at least the center of every circle is visible.

Another simple but instructive example of Measurable uses Dataset to produce a strip-chart recorder by overriding get minimum value so that a fixed *x*-interval is obtained.

```
public Stripchart extends Dataset {
  double range = 10;
  public double getXMin(){ return xmax-range; }
}
```

Because Dataset already implements Measurable, additional methods need not be defined. A more robust strip chart would trim excess data so as to limit the size of the internal data arrays. See the Stripchart class in the display package for the code for such a class.

4.8 ■ INTERACTIVE

It is often desirable for a user to interact with a program using a mouse, which can be done using an interactive drawing panel InteractivePanel in combination with the Interactive interface and the InteractiveMouseHandler interface. The following main method shows how easy it is to create a circle and a rectangle that can be positioned within a panel by clicking and dragging using the mouse.

```
import org.opensourcephysics.display.*;

public class InteractiveShapeApp {
  public static void main(String[] args) {
    DrawingPanel panel = new PlottingPanel("x","y","Interactive Demo");
    DrawingFrame frame = new DrawingFrame(panel);
    panel.setPreferredMinMax(-10,10,-10,10);
```

```
    // creates two interactive drawables and adds them to the panel
    panel.addDrawable(new InteractiveCircle(0, 0));
    InteractiveShape ishape=InteractiveShape.createRectangle(3,4,2,2);
    panel.addDrawable(ishape);
    frame.setVisible(true);
    frame.setDefaultCloseOperation(javax.swing.JFrame.EXIT_ON_CLOSE);
  }
}
```

InteractivePanel is a subclass of DrawingPanel that collects mouse events. Because the PlottingPanel class is a subclass of InteractivePanel, it too supports the interactions. The interactive panel collects incoming events but only stores the last event. It notifies a mouse handler that an action has occurred. The handler's handleMouseAction method has the option of responding to this action. Because the interactive drawing panel's default handler is the panel itself, simple programs are not required to handle the mouse action. Interactive objects that are added to an InteractivePanel or a subclass are draggable without additional programming.

Interactive Mouse Handler

The InteractiveMouseHandler interface enables objects to define their own interactive mouse actions.

```
public interface InteractiveMouseHandler {
  public void handleMouseAction(InteractivePanel panel,
      MouseEvent evt);
}
```

Interactive mouse handlers register their interest in events using an interactive panel's setInteractiveMouseHandler method. The object then receives all interactive mouse actions.

```
InteractivePanel interactivePanel = new InteractivePanel();
interactivePanel.setInteractiveMouseHandler(interactiveMouseHandler);
```

Mouse handlers determine the type of action using the method in the InteractivePanel class. A handler will often invoke the panel's default action method in order to drag objects. The following handler is typical. The method uses the panel's handler to drag the interactive object and then uses a switch statement to display the position of the mouse action.

```
public void handleMouseAction(InteractivePanel panel, MouseEvent evt) {
  panel.handleMouseAction(panel,evt);  //drag the objects in the panel
  // add custom actions
  switch(panel.getMouseAction() ) {
    case InteractivePanel.MOUSE_PRESSED:
        double x = panel.getMouseX();
        double y = panel.getMouseY();
        System.out.println("Mouse pressed at x="+x+" y="+y);
        break;
    case InteractivePanel.MOUSE_RELEASED:
        double x = panel.getMouseX();
        double y = panel.getMouseY();
        System.out.println("Mouse released at x="+x+" y="+y);
        break;
  } // end of switch
}
```

Interactive Interface

Interactive objects implement the `Interactive` interface to get and set an object's position, to enable dragging, and to determine the interactive "hotspots." The `findInteractive` method determines which hotspot contains the given coordinates. The interactive object compares the mouse coordinates to one or more locations and returns a reference to an object that can be used to drag, resize, or otherwise modify the selected object. If there is only one hotspot, it is likely that the object will return a reference to itself, `this`. If the coordinates are not within a hotspot, the method should return `null`.

```
public interface Interactive extends Measurable {
  public Interactive findInteractive(
      DrawingPanel panel, int xpix, int ypix);
  public void     setEnabled(boolean enabled);
  public boolean isEnabled();
  public void     setXY(double x, double y);
  public void     setX(double x);
  public void     setY(double y);
  public double   getX();
  public double   getY();
}
```

Listing 4.5 defines an interactive circle by extending the `MeasuredCircle` class which extends the `Circle` class introduced in Section 4.2.

Listing 4.5   Interactive circle class.

```
import java.awt.*;

public class InteractiveCircle extends MeasuredCircle
    implements Interactive {
  boolean enableInteraction = true;

  public InteractiveCircle(double x, double y) {
    super(x, y);
  }

  public void setEnabled(boolean _enableInteraction) {
    enableInteraction = _enableInteraction;
  }

  public boolean isEnabled() {
    return enableInteraction;
  }

  // Set and get methods for x and y are defined in superclass
  public void setXY(double _x, double _y) {
    x = _x;
    y = _y;
  }

  public Interactive findInteractive(
      DrawingPanel panel, int _xpix, int _ypix) {
    if(!enableInteraction) {
      return null;
    }
    int xpix = panel.xToPix(x); // convert the center x to pixels
```

```
    int ypix = panel.yToPix(y); // convert the center y to pixels
    if((Math.abs(xpix-_xpix)<pixRadius)
        &&(Math.abs(ypix-_ypix)<pixRadius)) {
      return this;
    } else {
      return null;
    }
  }
}
```

The `InteractiveHandlerApp` program on the CD demonstrates how to handle mouse actions from an interactive panel by creating an interactive circle. Click-dragging within the panel moves interactive objects and causes a message to appear.

In summary, a program must do the following to use interactive objects within an interactive panel.

1. Instantiate an interactive object and add it to an interactive panel.
2. Implement the interactive mouse handler in an appropriate class.
3. Instantiate the interactive mouse handler and register this handler with the interactive panel.

4.9 ■ AFFINE TRANSFORMATIONS

Affine transformations are powerful tools that can be used to quickly transform objects including points, shapes, and text. An affine transformation is a transformation in which parallel lines are still parallel after being transformed. Two examples of such transformations (written in matrix form) are rotation by an angle θ

$$\begin{bmatrix} x' \\ y' \\ 1 \end{bmatrix} = \begin{bmatrix} \cos\theta & -\sin\theta & 0 \\ \sin\theta & \cos\theta & 0 \\ 0 & 0 & 1 \end{bmatrix} \cdot \begin{bmatrix} x \\ y \\ 1 \end{bmatrix}, \tag{4.1}$$

and translation along the x- and y-axes by a, b

$$\begin{bmatrix} x' \\ y' \\ 1 \end{bmatrix} = \begin{bmatrix} 1 & 0 & a \\ 0 & 1 & b \\ 0 & 0 & 1 \end{bmatrix} \cdot \begin{bmatrix} x \\ y \\ 1 \end{bmatrix}. \tag{4.2}$$

The most general affine transformation is a combination of scaling, translation, and rotation. It can be written as

$$\begin{bmatrix} x' \\ y' \\ 1 \end{bmatrix} = \begin{bmatrix} m_{00} & m_{01} & m_{02} \\ m_{10} & m_{11} & m_{12} \\ 0 & 0 & 1 \end{bmatrix} \cdot \begin{bmatrix} x \\ y \\ 1 \end{bmatrix}. \tag{4.3}$$

The `AffineTransform` class the `java.awt.geom` package defines affine transformations. Instances of this class are constructed as follows:

```
AffineTransform(double m00, double m10,
                double m01, double m11,
                double m02, double m12);
```

Note that the `AffineTransform` class also defines convenience methods for constructing pure rotations, pure translations, and pure shears. The usual rules of matrix arithmetic apply when these transformations are combined.[2]

An affine transformation can be used to change a drawing panel's graphics context before drawing takes place. For example, we can define a draw method that renders an 80 pixel by 160 pixel rectangle at 30 degrees from a standard position by concatenating a rotation with the current graphics context's transformation.

```
public void draw (DrawingPanel panel, Graphics g) {
  Graphics2D g2 = (Graphics2D) g;
  AffineTransform oldAT=g2.getTransform();
  AffineTransform at=g2.getTransform();
  at.concatenate(AffineTransform.getRotateInstance(Math.PI/6,140,230));
  // rectangle with center at (140, 230)
  g2.drawRect(100,150,80,160);
  g2.setTransform(oldAT);
}
```

This code fragment demonstrates a number of important features. First, because affine transformations uses the Java 2 API, the graphics context `Graphics g` must be cast to a Java 2D graphics context `Graphics2D g2`. Second, although the graphics context that is passed to a drawable object is in pixel units, it may have already been transformed by the Java VM. Consequently, we save the original context before drawing and restore the original context after drawing. Third, because of the rules of matrix multiplication, we concatenate the rotation with the existing transformation. Note that the `getRotateInstance` method in the Java geometry package has two signatures. We have used a rotation instance that allows us to specify a rotation about the rectangle's center, (140, 230).

Although transforming the entire graphics context is sometimes useful, there are side effects. Scaling changes the location of every point except the origin. In other words, applying a scale instance not only changes the width and height of a rectangle but also changes its location. In addition, scaling changes line thicknesses and text sizes. Text labels may, for example, become unreadable at high magnifications. It is usually better to instantiate a Java `Shape` using world coordinates and then use affine transformations to transform this shape into pixel coordinates. This technique is used in the next section to define OSP drawable shapes.

4.10 ■ SHAPES

The Java 2D API defines a set of classes that can be used to create high-quality graphics using composition, image processing, anti-aliasing, and text layout. We use a subset of the Java 2D API in the Open Source Physics display package. The `Shape` interface is one of the most useful.

The `java.awt.Shape` interface is a cornerstone of the Java 2D API, and we make it easy to incorporate objects that implement this interface into drawing panels. A shape represents a geometric object that has an outline and an interior. Concrete representations defined in the `java.awt.geom` package include `Rectangle2D`, `Ellipse2D`, `Line2D`, and `GeneralPath`. A typical shape is constructed by passing geometric information to a constructor. A Java 2D

[2]See the online Java 2D documentation from Sun Microsystems and *Java 2D Graphics* by Jonathan Knudsen (O'Reilly 1999) for a more complete discussion of affine transformations.

Table 4.3 Static (factory) methods in the `DrawableShape` class and `InteractiveShape` class create simple geometric shapes.

---

org.opensourcephysics.display.DrawableShape
org.opensourcephysics.display.InteractiveShape

`createArrow`	Creates an arrow with the specified center, horizontal component, and vertical component.
`createCircle`	Creates a circle with the given center and radius.
`createEllipse`	Creates an ellipse with the given center, width, and height.
`createRectangle`	Creates a rectangle with the given center, width, and height.
`createSquare`	Creates a circle with the given center and width.

---

rectangle, for example, is constructed by passing upper left-hand corner (x, y) coordinates, width w, and height h.

```
// Shape is defined in the Java 2D API
Shape rectangle = new Rectangle2D.Double(x,y,w,h);
```

Because a `DrawingPanel` object contains an affine world-to-pixel transformation, we can easily transform Java 2D shapes before they are rendered within a drawing panel. The `DrawableShape` class uses the world-to-pixel transformation to render shapes whose position and size have been specified using world coordinates. Note that the location of a Java 2D shape is usually the top left corner of the object, but we can offset the shape's origin by half the width and half the height to position it using the geometric center. In the following code fragment we instantiate a Java 2D `Rectangle` with a width of three and a height of one with its top left corner at $(-1.5, -0.5)$. The (x, y) position of the drawable shape's center is $(-5, -4)$ in world coordinates.

```
// DrawableShape draws Java 2D shapes in a DrawingPanel
DrawableShape rect =
    new DrawableShape(new Rectangle2D.Double(-1.5,-0.5,3,1),-5,-4);
drawingPanel.addDrawable(rect);
```

Having to first create a Java 2D shape is awkward. In order to facilitate the creation of simple geometric shapes, the `DrawableShape` class and `InteractiveShape` class contain static (factory) methods to create common shapes, such as circles and rectangles, whose xy-locations are set to the geometric center (see Table 4.3). A program that creates a rectangle with a width of four and a height of five centered at $(-3, -4)$ is shown in Listing 4.6.

Listing 4.6 `DrawableShapeApp` tests the `DrawableShape` class by creating and manipulating shapes.

```
package org.opensourcephysics.manual.ch04;
import org.opensourcephysics.display.*;
import java.awt.Color;

public class DrawableShapeApp {
  public static void main(String[] args) {
    PlottingPanel panel = new PlottingPanel("x", "y", null);
    panel.setSquareAspect(true);
    DrawingFrame frame = new DrawingFrame(panel);
    DrawableShape rectangle =
        DrawableShape.createRectangle(-3, -4, 4, 5);
```

```
    rectangle.setTheta(Math.PI/4);
    Color fillColor = new Color(255, 128, 128, 128);
    Color edgeColor = new Color(255, 0, 0, 255);
    rectangle.setMarkerColor(fillColor, edgeColor);
    panel.addDrawable(rectangle);
    DrawableShape circle = DrawableShape.createCircle(3, 4, 6);
    panel.addDrawable(circle);
    frame.setDefaultCloseOperation(javax.swing.JFrame.EXIT_ON_CLOSE);
    frame.setVisible(true);
  }
}
```

The position of a shape can be changed after it is created using accessor methods such as setX, setY, and setXY. These methods move the drawable shape's origin. The setTheta sets the orientation.

Another approach is to transform the shape using an affine transformation. The following code fragment illustrates the difference between transforming a DrawableShape and transforming an internal Java 2D Shape.

```
DrawableShape rectangleOne = DrawableShape.createRectangle(0,0,2,3);
DrawableShape rectangleTwo = DrawableShape.createRectangle(0,0,2,3);
rectangleOne.setXY(6,8);
rectangleTwo.transform(AffineTransform.getTranslateInstance(6,8));
System.out.println(rectangleOne.getX()); // prints 6
System.out.println(rectangleTwo.getX()); // prints 0
rectangleOne.setTheta(Math.PI/4);  // rotates about (6,8)
rectangleTwo.setTheta(Math.PI/4);  // rotates about (0,0)
```

Both rectangles instantiated with center at $(0, 0)$ and have a width of 2 and height of 3. These rectangles are then translated so that they are drawn with centers at $(6, 8)$. The last transformation will, however, produce dramatically different results because the rotation is performed about the drawable shape's origin. The first drawable shape has its origin shifted, whereas the second rectangle has shifted the center of the internal Java 2D object. The transform method applies an affine transformation, such as a shear and a rotation, directly to the internal geometry, whereas set commands such as setXY are applied to an object that positions the shape.

Java shapes are not restricted to simple geometric objects. The GeneralPath class in the java.awt.geom package defines a shape using line segments and Bézier curves. The code fragment shown below uses this class to define a triangle using a sequence of moveTo and lineTo methods. The triangle is converted to a drawable and added to a drawing panel.

```
// GeneralPath is defined in java.awt.geom package and implements Shape
GeneralPath path = new GeneralPath();
path.moveTo(3,0);
path.lineTo(0,3);
path.lineTo(0,-3);
path.closePath();
DrawableShape triangle = new DrawableShape(path,0,0);
drawingPanel.addDrawable(triangle);
```

A shape's border is a path and this path can be accessed using the getPathIterator method in the Shape interface. A rectangle's perimeter, for example, consists of a series of four line segments and each iteration will return the next segment endpoint. This paradigm allows us to compute path integrals as shown in Listing 4.7. Note that because we have flattened the shape's path, only straight line segments are returned by the iterator.[3]

Listing 4.7   A path integral can be calculated using an `Iterator`.

```
package org.opensourcephysics.manual.ch04;
import org.opensourcephysics.display.InteractiveShape;
import java.awt.geom.PathIterator;

public class PathIntegralApp {
  public static void main(String[] args) {
    InteractiveShape circle = InteractiveShape.createCircle(-2, 2, 1);
    // get an iterator with line flatness less than 0.001
    PathIterator it = circle.getShape().getPathIterator(null, 0.001);
    double[] coord = new double[6];
    double sum = 0, x1 = 0, y1 = 0, xstart = 0, ystart = 0;
    while(!it.isDone()) {
      switch(it.currentSegment(coord)) {
      case PathIterator.SEG_LINETO:
        sum += Math.sqrt((x1-coord[0])*(x1-coord[0])
            +(y1-coord[1])*(y1-coord[1]));
        x1 = coord[0];
        y1 = coord[1];
        break;
      case PathIterator.SEG_MOVETO:
        x1 = coord[0];
        y1 = coord[1];
        xstart = x1; // start of the path
        ystart = y1;
        break;
      case PathIterator.SEG_CLOSE:
        sum += Math.sqrt((x1-xstart)*(x1-xstart)
            +(y1-ystart)*(y1-ystart));
        xstart = x1;
        ystart = y1;
        break;
      default:
        System.out.println("Segment Type not supported. Type="
            +it.currentSegment(coord));
      }
      it.next();
    }
    System.out.println("path integral="+sum);
  }
}
```

4.11 ■ INTERACTIVE SHAPES

The display package defines two types of interactive shapes, `InteractiveShape` and `BoundedShape`. An `InteractiveShape` is similar to a `DrawableShape` except that its xy-position can be adjusted by click-dragging within an interactive panel. A `BoundedShape` can be dragged, rotated, and resized.

---

[3]If the flatten parameter is omitted when getting the path iterator, then a shape may return quadratic or cubic line segments and a more sophisticated integration algorithm must be employed.

Listing 4.8 The `InteractiveShapeApp` program tests the InteractiveShape class.

```
package org.opensourcephysics.manual.ch04;
import org.opensourcephysics.display.*;
import javax.swing.JFrame;

public class InteractiveShapeApp {
  public static void main(String[] args) {
    PlottingPanel panel =
        new PlottingPanel("x", "y", "Interactive Demo");
    panel.setPreferredMinMax(1, 10, 1, 10);
    DrawingFrame frame = new DrawingFrame(panel);
    // create interactive shapes and add them to the panel
    InteractiveShape ishape =
        InteractiveShape.createRectangle(3, 4, 2, 2);
    panel.addDrawable(ishape);
    InteractiveShape arrow = InteractiveShape.createArrow(3, 4, 1, 5);
    panel.addDrawable(arrow);
    frame.setVisible(true);
    frame.setDefaultCloseOperation(JFrame.EXIT_ON_CLOSE);
  }
}
```

Because it can be computationally expensive (and visually confusing) to perform these mouse actions if a panel contains multiple bounded shapes, the user must first double-click a `BoundedShape` object to select it. Selecting a shape highlights it by showing a light-blue bounding box and one or more *hotspots*. Click-dragging the hotspot changes an object's properties. Note that only the xy-drag interaction is enabled when a `BoundedShape` is instantiated. Other interactions are explicitly enabled using accessor methods such as `setRotateDrag` and `setWidthDrag`.

Listing 4.9 BoundedShapeApp tests the BoundedShape class.

```
package org.opensourcephysics.manual.ch04;
import org.opensourcephysics.display.*;

public class BoundedShapeApp {
  public static void main(String[] args) {
    PlottingPanel panel =
        new PlottingPanel("x", "y", "Bounded Shape Demo");
    DrawingFrame frame = new DrawingFrame(panel);
    panel.setPreferredMinMax(-10, 10, -10, 10);
    BoundedShape bShape =
        BoundedShape.createBoundedRectangle(3, 4, 5, 6);
    bShape.setRotateDrag(true);
    panel.addDrawable(bShape);
    bShape = BoundedShape.createBoundedEllipse(-3, -4, 6, 6);
    bShape.setHeightDrag(true);
    bShape.setWidthDrag(true);
    panel.addDrawable(bShape);
    frame.setVisible(true);
    frame.setDefaultCloseOperation(javax.swing.JFrame.EXIT_ON_CLOSE);
  }
}
```

Although shapes usually specify their xy-location using world coordinates, it is sometimes convenient to use pixels to specify their size. It would, for example, be awkward if text, images, or dataset point markers changed their appearance when the coordinate scale changed. The `InteractiveShape` class and `BoundedShape` class implement both pixel and world size units. The following code fragment creates a 40×80 pixel rectangle at location $(2, 3)$ in world coordinates.

```
InteractiveShape ishape=InteractiveShape.createRectangle(2,3,40,80);
ishape.setPixelSized(true);
```

4.12 ■ TEXT

Java 1.4 renders text through software by forming characters pixel by pixel. We can do better if the characters are always the same, in the same font, at the same size, and in the same color. We render the text as an image and then rotate and scale the image using affine transformations. The `DrawableTextLine` class implements this drawing technique. In addition, we have added processing that allows us to incorporate Greek characters and to change the size and location of text to produce superscripts and subscripts using a syntax that mimics the TeX markup language. Subscripts begin with an underscore _ and are enclosed in braces { }. Superscripts begin with a caret ˆ and are also enclosed in braces.

Listing 4.10   OSP supports special characters using a syntax that mimics the TeX markup language.

```
package org.opensourcephysics.manual.ch04;
import org.opensourcephysics.display.*;
import org.opensourcephysics.frames.*;

public class GreekCharApp {
  public static void main(String[] args) {
    PlotFrame frame =
        new PlotFrame("$\\theta$","$\\Psi$_{$\\theta$}",
                        "Special Characters: $\\alpha$ to $\\Omega$");
    String inputStr = "Wave function $\\Psi$_{$\\theta$}";
    DrawableTextLine textLine = new DrawableTextLine(inputStr, -8, 0);
    frame.addDrawable(textLine);
    textLine.setFontSize(22);
    textLine.setFontStyle(java.awt.Font.BOLD);
    frame.setVisible(true);
    frame.setDefaultCloseOperation(javax.swing.JFrame.EXIT_ON_CLOSE);
  }
}
```

The Greek characters shown in Figure 4.4 are specified using a double backslash followed by the character name enclosed within matching $ delimiters. We use two backslash characters because a single backslash is the Java escape sequence and has special meaning. The `DrawableTextLine` class, axis labels, and frame titles support the commonly used Greek characters shown in Table 4.4. Additional *unicode* character mappings are defined in the `GUIUtil` class in the display package. As in TeX, if the Greek character name begins with a capital letter, the capital Greek symbol is used.

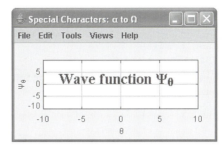

Figure 4.4 The `DrawableTextLine` class supports special characters, superscripts, and subscripts using a TEX-like markup language.

4.13 ■ MEASURED IMAGES AND SNAPSHOTS

Images can be added to a drawing panel using the `MeasuredImage` class. These images can either be computed by the program or loaded as a resource from a disk drive or URL using the `ResourceLoader`.

Image pixels are data just as array elements are data. In fact, it is possible to access image pixels using the Java `BufferedImage` class in the `java.awt.image` package. The `MeasuredImage` class in the display package accepts a `BufferedImage` and assigns a scale that converts image pixels to and from a drawing panel's world coordinates. Listing 4.11 shows how the `MeasuredImage` class is used to display an image that is computed within the program. Because `MeasuredImage` implements the `Measured` interface, the drawing panel sets its coordinate scale to match that given in the `MeasuredImage` constructor if the panel's autoscale option is enabled.

Table 4.4 Greek characters are defined in the `GUIUtil` class in the display package.

	Lowercase			Uppercase	
α	\alpha	μ	\mu	Γ	\Gamma
β	\beta	ν	\nu	Δ	\Delta
γ	\gamma	ξ	\xi	Θ	\Theta
δ	\delta	π	\pi	Π	\Pi
ϵ	\epsilon	ρ	\rho	Σ	\Sigma
ζ	\zeta	σ	\sigma	Φ	\Phi
η	\eta	τ	\tau	Ψ	\Psi
θ	\theta	ϕ	\phi	Ω	\Omega
ι	\iota	χ	\chi	Ξ	\Xi
κ	\kappa	ψ	\psi		
λ	\lambda	ω	\omega		

Listing 4.11   Measured image test program.

```
package org.opensourcephysics.manual.ch04;
import org.opensourcephysics.display.*;
import java.awt.*;
import java.awt.image.BufferedImage;

public class MeasuredImageApp {
  static final int SIZE = 20;

  public static void main(String[] args) {
    PlottingPanel panel =
        new PlottingPanel("x", "y", "Measured Image");
    DrawingFrame frame = new DrawingFrame(panel);
    panel.setPreferredMinMax(-2, 2, -2, 2);
    BufferedImage image =
        new BufferedImage(SIZE, SIZE, BufferedImage.TYPE_INT_ARGB);
    Graphics g = image.getGraphics();
    g.setColor(Color.RED);
    g.fillRect(0, 0, SIZE, SIZE);
    g.dispose();
    int color = 0xFF00FF00; // opaque and green in ARGB color space
    for(int i = 0;i<SIZE;i++) {
      image.setRGB(i, 0, color);
      image.setRGB(i, i, color);
      image.setRGB(i, SIZE-1, color);
      image.setRGB(i, SIZE-i-1, color);
    }
    MeasuredImage mi = new MeasuredImage(image, -1, 1, -1, 1);
    panel.addDrawable(mi);
    frame.setDefaultCloseOperation(javax.swing.JFrame.EXIT_ON_CLOSE);
    frame.setVisible(true);
    frame.setSize(500, 400);
  }
}
```

The drawing panel uses a measured image to capture a *snapshot* of its visual representation. The snapshot method invokes the getRenderedImage method and places a copy of this image into a new frame. Listing 4.12 shows how this method is implemented. This snapshot method can be invoked from within a program or by using the drawing panel's Snapshot menu item.

Listing 4.12   The drawing panel's snapshot method.

```
public void snapshot() {
  DrawingPanel panel = new DrawingPanel();
  DrawingFrame frame = new DrawingFrame(panel);
  frame.setDefaultCloseOperation(WindowConstants.DISPOSE_ON_CLOSE);
  frame.setKeepHidden(false);
  panel.setSquareAspect(false);
  int w = (isVisible())
          ? getWidth()
          : getPreferredSize().width;
  int h = (isVisible())
          ? getHeight()
          : getPreferredSize().height;
  if((w==0)||(h==0)) {
    return;
  }
```

```
BufferedImage snapimage =
    new BufferedImage(w, h, BufferedImage.TYPE_INT_ARGB);
render(snapimage);
MeasuredImage mi = new MeasuredImage(snapimage, pixToX(0),
    pixToX(w), pixToY(h), pixToY(0));
panel.addDrawable(mi);
panel.setPreferredMinMax(pixToX(0), pixToX(w), pixToY(h), pixToY(0));
panel.setPreferredSize(new Dimension(w, h));
frame.setTitle("Snapshot");
frame.pack();
frame.setVisible(true);
}
```

Java *resources* are ancillary files that are used within a program. They may be loaded from the disk or the Internet and can be almost anything including text files such as html pages or binary files such as images or sound. Because Java uses different loading mechanisms, depending on whether the program is running as an application or as an applet, we have created the ResourceLoader class in the tools package. Listing 4.13 uses the ResourceLoader to load an image containing a Mercator projection of Earth. Note that the corners of the MeasuredImage are assigned a minimum and maximum value in order to set the image size in the drawing panel.

Listing 4.13   Image resource loader test program.

```
package org.opensourcephysics.manual.ch04;
import org.opensourcephysics.display.*;
import org.opensourcephysics.tools.ResourceLoader;
import java.awt.image.BufferedImage;
import javax.swing.JFrame;

public class LoadImageApp {
  public static void main(String[] args) {
    PlottingPanel panel = new PlottingPanel("x", "y", "Load Image");
    DrawingFrame frame = new DrawingFrame(panel);
    String s = "org/opensourcephysics/manual/ch04/earthmap.gif";
    BufferedImage earth = ResourceLoader.getBufferedImage(s);
    MeasuredImage mi = new MeasuredImage(earth, -180, 180, -90, 90);
    panel.addDrawable(mi);
    frame.setDefaultCloseOperation(JFrame.EXIT_ON_CLOSE);
    frame.setVisible(true);
    frame.setSize(500, 400);
  }
}
```

If the program is an application, the ResourceLoader searches for the resource in the specified directory; if the program is an applet, the program searches for the resource starting at the applet's codebase. If the resource cannot be found in these locations, the loader searches in the jar file from which the program was loaded. The ResourceLoader caches resources in memory so that a requested file is only loaded once.

4.14 ■ EFFECTIVE JAVA

The book *Effective Java* by Joshua Bloch provides a useful guide to how the Java language is best used in practice. Three of his fifty-seven rules help us to decide when to use inheritance and when to use interfaces.

- Prefer interfaces to abstract classes.

- Favor composition over inheritance.

- Design and document for inheritance or else prohibit it.

It is generally considered safe to use inheritance within a package where all the classes are under the control of the same programmers. Carefully designed and documented class hierarchies, such as the user interface library distributed by Sun, can easily be subclassed. We have done so in designing the drawing panel, the interactive panel, and the plotting panel in the display package. The `DrawingPanel` adds world coordinates, coordinate transformations, and the ability to manage drawable objects to the basic panel class, the `InteractivePanel` adds a framework for repositioning objects using the mouse, and the `PlottingPanel` adds axes, titles, and logarithmic scales.

Inheritance is a powerful way to achieve code reuse, but it breaks encapsulation. A subclass all too often depends on the implementation details of its superclass if for no other reason than it invokes the superclass constructor. Because of the hierarchy shown on Figure 4.2, changing the world-to-pixel coordinate transformation in the drawing panel might break logarithmic axes in the plotting panel.

In addition to allowing and encouraging encapsulation, interfaces are easier to use because any class can be retrofitted to implement an interface. For example, after we solve for the trajectory of a baseball using a differential equation, we can create a visualization by implementing the `Drawable` interface anywhere within a `baseball` object's class hierarchy. Adding a `baseball` to a drawing panel (and redrawing the panel containing the `baseball` as the solution is generated) produces an animation as shown in Chapter 7. The `Drawable` interface makes it possible for the `baseball` object to display itself in a drawing panel. Similarly, the `Measurable` interface makes it possible for an object to rescale a drawing panel, and the `Interactive` interface makes it possible to reposition an object in a drawing panel. In subsequent chapters we introduce additional interfaces, including the `Control` interface to store and retrieve values from a graphical user interface and the `ODE` interface to solve differential equations.

It is reasonable to declare variables using an interface rather than a class. For example, rather than defining a variable to be an interactive circle, we can define a variable that references an `Interactive` object if only the methods in the interface are required.

```
// ball at location x = 20, y = 0
Interactive baseball = new InteractiveCircle(20,0);
```

The one disadvantage to interfaces is that it is easier to evolve a class hierarchy or an abstract class than it is to evolve an interface. If we add a new method to the particle class named `launch` that gives particles an initial velocity, all existing subclasses will be able to perform this method. If we were to add the `launch` method to the `Drawable` interface, we would have to add this method to all classes that implement this interface. This example is contrived, but the problem is real. If other programmers use our particle class, they will probably not notice if the new `launch` method exists. But if `launch` is added to the `Drawable` interface, any code that uses this interface will cease to work.

The safest approach to object design is to use a technique know as *composition*. A new class is defined that implements an interface, such as `Drawable`. This class contains private references to other objects that implement the necessary methods. Methods in the new class implement interfaces by invoking methods in the private variables. Defining a method in an object that invokes a method in another object is known as *forwarding*. Although this technique may not always be possible or efficient, it decouples a class from all implementation details.[4]

4.15 ■ PROGRAMS

The following examples are in the `org.opensourcephysics.manual.ch04` package.

AffineTestApp

`AffineTestApp` demonstrates how to transform 2D shapes using the 2D `AffineTransform` class. See Section 4.9.

BoundedShapeApp

`BoundedShapeApp` tests the `BoundedShape` class by creating a rectangle and an ellipse. Click on a bounded shape to select it and then drag a hotspot. See Section 4.11.

CircleAndArrowApp

`CircleAndArrowApp` tests the `DrawingPanel` class by creating two drawable objects. This program is described in Section 4.3.

DrawableShapeApp

`DrawableShapeApp` demonstrates how to create, translate, and rotate a `DrawableShape`. This program is described in Section 4.10.

DrawingApp

`DrawingApp` demonstrates the OSP drawing framework. See Section 4.1.

GreekCharApp

`GreekCharApp` demonstrates Greek characters and other drawing features such as superscripts and subscripts. See Section 4.12.

InteractiveHandlerApp

`InteractiveHandlerApp` demonstrates how to handle mouse actions from an interactive drawing panel. Pressing the mouse within the panel causes a message to appear. This program is described in Section 4.8.

InteractiveShapeApp

`InteractiveShapeApp` tests the `InteractiveShape` class by creating a circle and a square. Users can drag these shapes within the panel. This program is described in Section 4.8.

[4]See *Effective Java* by Joshua Bloch, Addison–Wesley (2001).

LoadImageApp

LoadImageApp loads an image as a resource and adds it to a drawing panel using the MeasuredImage class. See Section 4.13.

MeasuredImageApp

MeasuredImageApp creates an image, assigns a measure in world units, and places the image in a plotting panel. This program is described in Section 4.13.

MessageApp

MessageApp tests message boxes and adds a component to the glass pane. See Section 4.5.

PathIntegralApp

PathIntegralApp demonstrates how to calculate a path integral using an iterator. This program is described in Section 4.10.

StripChartApp

StripChartApp demonstrates how to use the Measurable interface to create a strip chart. This program is described in Section 4.7.

CHAPTER

5

Controls and Threads

The Open Source Physics library contains packages for creating graphical user interfaces and for storing data. The *controls* package (`org.opensourcephysics.controls`) defines the API and implements a number of simple controls. Custom user interfaces are constructed using components in the *Ejs* package (`org.opensourcephysics.ejs`). Examples of how these packages are used are located in the `org.opensourcephysics.manual.ch05` package unless otherwise stated.

5.1 ■ OVERVIEW

Physics models require parameters (such as mass) that are held constant during a calculation and initial values of state variables (such as position and velocity) that evolve during a calculation. These parameters and state variables are usually stored in a model using `int` and `double` identifiers. However, users cannot access these variables directly. Users interact with a graphical user interface that collects data (usually from a keyboard) and passes this data to the model in response to an action such as a button click. This user interface object is an example of a *control*.

A control is an object that stores data and invokes methods in another object. Controls often have a graphical user interface such as shown in Figure 5.1, whereas controlled objects are often complex models whose job it is to carry out a computation or simulation. In principle, any object can have a control. In practice, the most important control is the control that starts, stops, and initializes a program.

The `OSPControlApp` program shown in Listing 5.1 creates an `OSPControl` object. Because the control must store a reference to the model in order for the Get button to invoke methods in the model, we pass a reference to `this` to the control's constructor. The program's `reset` method stores default values for four variables in the control. These variables can be edited and are read when the user clicks the Get button. Run the program and edit the input values. Note the behavior of the control if incorrect data are typed into the table and the button is pressed.

Listing 5.1 The `OSPControlApp` program creates the user interface shown in Figure 5.1. It reads and displays values in response to button clicks.

```
package org.opensourcephysics.manual.ch05;
import org.opensourcephysics.controls.*;

public class OSPControlApp {
   OSPControl control = new OSPControl(this); // creates the control
```

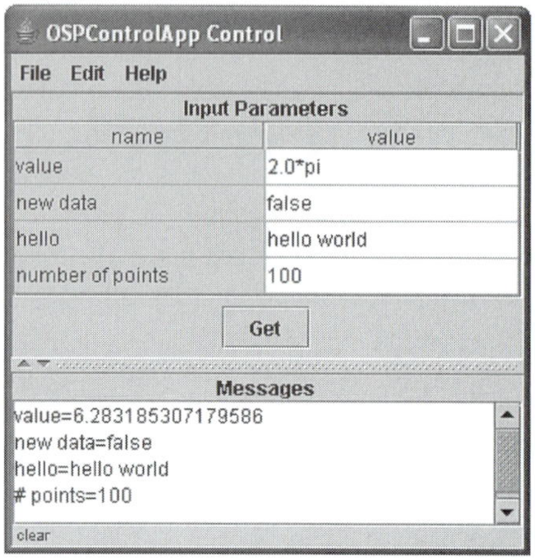

Figure 5.1   An `OSPControl` stores variables using a table.

```
public OSPControlApp() {
  control.addButton("getValues", "Get",
       "Gets values from the control.");
  reset(); // stores default values
}

public void getValues() {
  double val = control.getDouble("value");
  control.println("value="+val);
  boolean newData = control.getBoolean("new data");
  control.println("new data="+newData);
  String str = control.getString("hello");
  control.println("hello="+str);
  int n = control.getInt("number of points");
  control.println("# points="+n);
}

public void reset() {
  control.setValue("value", "2.0*pi");
  control.setValue("new data", false);
  control.setValue("hello", "hello world");
  control.setValue("number of points", 100);
}

public static void main(String[] args) {
  new OSPControlApp();
}
}
```

5.2 ■ CONTROL INTERFACE

The Open Source Physics library contains a number of predefined controls and one general purpose control framework known as Ejs as shown in Figure 5.2. These controls implement

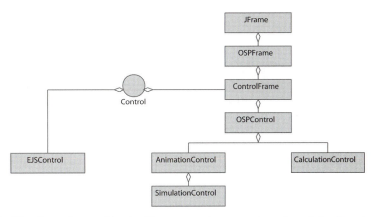

Figure 5.2 The Open Source Physics library contains a number of predefined controls and one general purpose control framework known as Ejs.

the `Control` interface. The purpose of this interface is to enable the two-way communication needed to store and retrieve values from an object using *set* and *get* methods.

The `Control` interface may seem complex, but the concept is very simple. There is a `setValue` method with signatures to store primitive data types and generic objects. There are five accessor methods to retrieve primitive data types and objects from the control. There are four methods to display and clear messages in a console-like text area. And there are four utility methods for specialized applications.

Listing 5.2 The `Control` interface in the
`org.opensourcephysics.controls` package.

```
public interface Control {
  // stores values in the control
  public void setValue(String name, Object val);
  public void setValue(String name, double val);
  public void setValue(String name, int val);
  public void setValue(String name, boolean val);

  // retrieves values from the control
  public int getInt(String name);
  public double getDouble(String name);
  public String getString(String name);
  public boolean getBoolean(String name);
  public boolean getObject(String name);

  // shows messages in the control's console
  public void println(String s);
  public void println();
  public void print(String s);
  public void clearMessages();

  // utility methods
  public void clearValues();            // clears name-value pairs
  public void calculationDone(String message);
  public void setLockValues(boolean lock);   // locks interface
  public java.util.Collection getPropertyNames(); // gets names
}
```

The superclass for the standard OSP controls is the `OSPControl` class. A program uses the control's `setValue` method to store data by passing the name of the variable and an initial value. The control then displays these name-value pairs using a table that can be edited. The user enters new values into the table and presses the Enter (Return) key. Note that the value field's background becomes yellow until the data has been entered. The new value can then be read by invoking an accessor method such as `getDouble` or `getInteger`.

OSP controls can store primitive data types, objects, and arrays. For example, if we want the user interface to provide access to the mass m, the initial position of a particle x_0, and the initial velocity of a particle $\mathbf{v} = (2, 3)$, we execute the following statements:

```
control.setValue("x0", 2.0);              // stores primitive data type
control.setValue("y0", "sqrt(2.0)");      // stores arithmetic expression
control.setValue("mass", new Double(3));// restricts input
control.setValue("v", new double[]{2.0,3.0}); // stores array
```

The control now holds a copy of these variables and displays them in a user interface. The user edits the values in the control, but these changes are not reflected in the model until they are read from the control. Note that the control is able to store mathematical expressions as strings and to convert them to numbers. The user can type almost anything if the value contains a `String` or a primitive data type, but if the value is wrapped in an object, such as `Double` or `Integer`, then the user cannot enter any other data type. The user cannot, for example, enter an arithmetic expression for the `mass` in the control.

In addition to arrays, compound objects can be stored in a control if the object has an xml loader. (Many objects in the OSP library do. See Chapter 12.) The control table displays the name of the object on a pale-green background. Double-clicking on the cell shows an inspector window that displays the object's properties.

```
control.setValue("Circle",new InteractiveCircle());
```

Objects must be cast to the correct type when they are read from the control.

```
InteractiveCircle circle = (InteractiveCircle)
    control.getObject("Circle");
```

Reading values from a control is often initiated by an action such as a button click. Buttons are added to standard OSP controls by passing a method name, a button title, and a description (hint) to the `addButton` method.

```
control.addButton("calculate", "Calculate", "Does a calculation.");
```

The button's action method invokes the `calculate` method which does something like the following:

```
public void calculate () {
  double x = control.getDouble("x0");
  double y = control.getDouble("y0");
  double m = control.getDouble("mass");
  double[] v = (double[]) control.getDouble("v");
  speed = Math.sqrt(v[0]*v[0]+v[1]*v[1]);
  // continue the calculation
}
```

The `OSPControl` class is able to store and retrieve data in multiple formats. Storing a floating point number and retrieving an integer or a string is allowed. So is storing an arithmetic expression and retrieving an integer or a double. Character strings such as `pi` or

`2*sqrt(3)` are evaluated to produce numbers when the `getDouble` method is invoked. If the user mistypes the expression, the conversion will fail. This failure is indicated in the control by turning the input field's background pale red.

As the calculation progresses, the model can modify the control's copy of the data by invoking the `setValue` method. The control will reflect this change, but constantly updating the control may demand considerable processing power. If multiple values are being accessed using set or get methods, performance can be improved by locking and unlocking the control, which is done as follows:

```
control.setLockValues(true);            // locks user interface
control.setValue("x0", 2.0);            // primitive data type
control.setValue("y0", "sqrt(2.0)");    // arithmetic expression
control.setValue("mass", new Double(3)); // restricted input
control.setValue("v", new double[]{2.0,3.0});
control.setLockValues(false);           // updates user interface
```

A locked control stores and returns its current values but is not required to update its graphical user interface until it is unlocked.

Because user actions can occur at any time, synchronization errors may result when the control sends data to a model as explained in Section 5.4. It is the programmer's responsibility to insure that actions can safely modify the model's data. The best solution is to read values after all computation has stopped, but these design considerations are entirely up to the programmer.

5.3 ■ CALCULATIONS

The `CalculationControl` shown in Figure 5.3 is designed to perform a short computation. This control inherits the `OSPControl` user interface and adds two buttons that invoke methods in a `Calculation`. Because we want this control to interact with many different programs, we define an interface in Listing 5.3 with a one-to-one correspondence between the control's buttons and the calculations's methods.

Listing 5.3   The `Calculation` interface.

```
public interface Calculation {
  // button actions
  public void calculate ();
  public void resetCalculation ();

  // register the control with the model
  public void setControl (Control control);
}
```

Models that use a `CalculationControl` implement the `Calculation` interface. The Calculate and Reset buttons shown in Figure 5.3 invoke the corresponding methods in the `Calculation` object.

A `CalculationControl` must contain a reference to its `Calculation` because the control expects to invoke certain methods. On the other hand, the `Calculation` must contain a reference to its `Control` because it will set and read variables in the control. This two-way communication is established when the control and calculation are created in the application's `main` method.

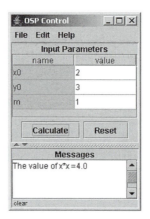

Figure 5.3 A simple control for use with a `Calculation`. The control stores three variables and shows the result of a calculation.

```
public static void main (String[] args) {
  Calculation calculation = new CalculationApp();
  Control control = new CalculationControl(calculation);
  calculation.setControl(control);
}
```

Note the use of Java interfaces. The application's `main` method starts by creating an object named `calculation` that implements the `Calculation` interface. This object (model) contains the physics. Next, the application creates a `Control` and passes the calculation to the control. Finally, the `setControl` method establishes communication from the calculation to the control so that the calculation can access the control's variables. Because the process of creating a control and a calculation occurs again and again, we have defined the following `static` method in the `CalculationControl` class to automate the process of establishing control-model communication.

```
public static void main (String[] args) {
  CalculationControl.createApp(new CalculationApp());
}
```

The entire control-model communication mechanism can be seen in action by running the short test program shown in Listing 5.4.

Listing 5.4 A calculation test program.

```
package org.opensourcephysics.manual.ch05;
import org.opensourcephysics.controls.*;

public class CalculationApp implements Calculation {
  private Control control; // the control
  private double x0;       // a parameter

  public void setControl(Control control) {
    this.control = control;
    resetCalculation();
  }
```

```
public void calculate() {
  x0 = control.getDouble("x0"); // read a parameter value
  // do some calculations here
  control.println("The value of x*x ="+(x0*x0));
}

public void resetCalculation() {
  control.clearMessages();
  control.setValue("x0", 2); // set a parameter value
  control.setValue("y0", 3); // set a parameter value
  control.setValue("m", 1);  // set a parameter value
}

public static void main(String[] args) {
  CalculationControl.createApp(new CalculationApp());
}
}
```

The Calculation interface has an important limitation—it is unable to respond to the click of a button while the calculation is being performed. This lack of response is not a problem if the computation is short. However, if the computation takes a long time, the user might assume that the program has crashed or is an infinite loop. The best way of overcoming this problem is to use a button to spawn a Thread as described in the next section.

5.4 ■ THREADS

Traditional programs start by executing a line of code, then another line, then another. There may be times when the program jumps from one block of code to another, but the program always moves from one statement to the next and never enters a state where statements are executed independently. Java is not like that. Java is almost always trying to do more than one thing at a time and these calculations are independent. In particular, the graphical user interface should be independent of the calculation. Otherwise, the interface will freeze while a calculation is being performed. This problem can be overcome by using a powerful feature of modern computer languages known as a *thread*.

A Java Thread is an object that executes statements within a program. Every application begins by executing statements within the main method from the *main thread*. This thread can be paused using the sleep method as shown in Listing 5.5. The main thread *dies* when the main method returns. The main thread cannot be restarted, but other threads can take over and continue to execute statements within the program. The most important of these is the *event dispatch thread* which invokes methods in response to keyboard and mouse events.

Listing 5.5   The main thread executes statements in the main method.

```
package org.opensourcephysics.manual.ch05;
import org.opensourcephysics.display.*;
import org.opensourcephysics.frames.*;

public class MainThreadApp {
  public static void main(String[] args) {
```

```
DisplayFrame frame = new DisplayFrame("x", "y", "Main Thread App");
InteractiveShape rect =
    InteractiveShape.createRectangle(4, 3, 8, 10);
double theta = 0, dtheta = 0.1;
frame.addDrawable(rect);
frame.setVisible(true);
frame.setDefaultCloseOperation(javax.swing.JFrame.EXIT_ON_CLOSE);
while(true) {
  rect.setTheta(theta);
  theta += dtheta; // increment theta
  frame.repaint(); // repaint with new theta
  try {             // sleep for 100 milliseconds
    Thread.sleep(100);
  } catch(InterruptedException ex) {}
 }
}
}
```

Having multiple threads is similar to having multiple applications insofar as the operating system causes one thread to run and then another. But threads are different from multiple applications; threads are independent calculations (not simultaneous calculations) that access the same data. A typical application has one thread to handle user actions and another thread to run the computation. For example, a graph may be plotting data that is currently being computed, while the user is using the mouse to drag an object within the graph. Because threads share the same data, we need to be careful to make sure that they work in such a way that neither thread corrupts data that is being used by the other.

A thread invokes a method within an object that implements the Runnable interface. This interface consists of a single method, the run method, and the thread executes the code within this method. It is usually an error for a program to invoke the run method directly. You should instead create a thread.

Listing 5.6   The Runnable interface.

```
public interface Runnable {
  public void run();
}
```

Listing 5.7 defines a Runnable object called RotationApp that contains a rectangular shape, a drawing panel, and a thread. The thread invokes the run method automatically soon after it starts. The object's run method is invoked by the thread and enters an infinite loop that increments the rectangle's orientation. The run method terminates when the window is closed and the application dies. Note that the program's main method creates two RotationApp objects and that these objects rotate independently.

Listing 5.7   A program with two animation threads. Each animation consists of a rectangle rotating within a drawing panel.

```
package org.opensourcephysics.manual.ch05;
import org.opensourcephysics.display.*;

public class RotationApp implements Runnable {
  DrawingPanel panel = new PlottingPanel("x", "y", "Rotating Shape");
  DrawingFrame frame = new DrawingFrame(panel);
  InteractiveShape ishape =
      InteractiveShape.createRectangle(2, 1, 2, 1);
  double theta = 0, dtheta = 0.1;
```

```
  Thread thread = new Thread(this);

  public RotationApp(double dtheta) {
    this.dtheta = dtheta;
    panel.setPreferredMinMax(-5, 5, -5, 5);
    panel.addDrawable(ishape);
    frame.setDefaultCloseOperation(javax.swing.JFrame.EXIT_ON_CLOSE);
    frame.setVisible(true);
    thread.start();
  }

  public void run() {
    while(true) {
      theta += dtheta;
      ishape.setTheta(theta);
      panel.repaint();
      try {
        Thread.sleep(100); // wait 100 millisecond
      } catch(InterruptedException ie) {}
    }
  }

  public static void main(String[] args) {
    new RotationApp(0.1); // create an animation
    new RotationApp(0.2); // create an animation
  }
}
```

A thread should not monopolize the computer's processor without giving other threads a chance to run. This is done by invoking the sleep method from time to time within the run method. The argument passed to sleep is the number of milliseconds that the thread should wait before continuing the calculation. A good value for computational speed is 10 milliseconds. This interval gives the operating system (event dispatch thread) the opportunity to process keyboard and mouse events. Because there is a tradeoff between flicker free animation and the time available to compute the physics, a good value for simulations is 100 milliseconds or 1/10 of a second.[1] Note that Java requires that the sleep method be enclosed in a try block to properly handle interrupt exceptions.

It is important to know how to end the run method, thereby stopping the computation and killing the thread. One way to do so is to have the run method loop until some condition is satisfied. Listing 5.8 contains a run method that tests to see if it is being executed by the proper thread or if the thread identifier named thread has been set to null. Note that the thread identifier has been declared to be volatile. The Volatile keyword guarantees that any thread that reads a variable will see the most recently written value.[2]

---

[1] Ideally the animation thread's sleep time should be no more than the reciprocal of the monitor's video refresh rate. This is usually not attainable for all but the simplest simulations because a typical refresh rate is 72 Hz.

[2] Although the Java language guarantees that reading and writing a 32 bit variable is atomic, it does not guarantee that a value written by one thread will immediately be seen by another unless that variable is declared to be volatile. A thread is allowed to copy a variable from main memory to cache. Other threads will also copy the variable from main memory and will not see changes made to the first copy unless the variable is declared to be volatile. See *The Java Language Specification*, 2nd ed., by Gosling, Steele, and Bracha. Unless you are running on a multiprocessor Java VM, you are unlikely to observe this effect.

Listing 5.8   A Runnable object with its own thread.

```
package org.opensourcephysics.manual.ch05;
public class RunComputation implements Runnable {
  private volatile Thread thread;

  public synchronized void startRunning() {
    if(thread!=null) {
      return; // thread is already running
    }
    thread = new Thread(this);
    thread.start(); // start the thread
  }

  public synchronized void stopRunning() {
    Thread tempThread = thread; // temporary reference
    thread = null; // signal the thread to die
    // return if thread already stopped; cannot join with
    // current thread
    if(tempThread==null||tempThread==Thread.currentThread()) {
      return;
    }
    try {
      tempThread.interrupt(); // get out of sleep state
      // wait up to one second for the thread to die
      tempThread.join(1000);
    } catch(InterruptedException e) {}
  }

  public void run() {
    while(thread==Thread.currentThread()) {
      // add code for computation here
      System.out.println("still running");
      try {
        Thread.sleep(100);
      } catch(InterruptedException ie) {}
    }
  }
}
```

Any of the following actions causes the while loop in Listing 5.8 to terminate and the thread to die.

1. The thread identifier has been set to null. This is the correct way to signal the run method to stop, as is done in the stopAnimation method in our example.

2. A new thread has been created but the run method is executing the previous thread. Because only one calculation should be running at a time, we exit the loop. The old calculation thread will die and the new thread will soon start and take over the run method. This condition should not occur in our code because the thread identifier is checked in the startAnimation method.

3. The run method was invoked directly rather than by a calculation thread. This is usually an error and the run method should exit.

Note that because the currentThread() method is static, we invoke it directly without creating an instance variable.

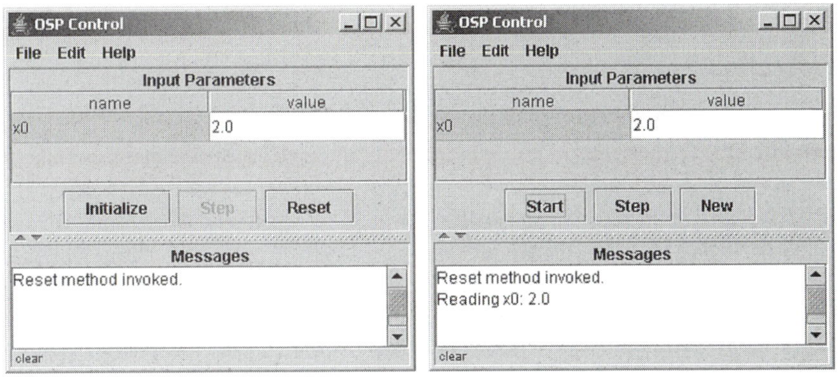

(a) Edit Mode.   (b) Run Mode.

Figure 5.4   A simple control for use with an `Animation`.

There is an important point about stopping the thread that needs to be mentioned. Setting the calculation thread to null does not immediately stop the calculation. In fact, the thread is probably not even executing when the `stopAnimation` method is invoked because the Java virtual machine is busy processing a user action. The program should, however, wait for the thread to die before returning from the `stopAnimation` method. This is accomplished using the `join` method. Notice also that the `stopAnimation` method invokes the `interrupt` method in order to awake the thread if it happens to be in a sleep state.

Finally, we mention two Sun classes, `java.util.Timer` and `javax.swing.Timer`, that can save you from programming your own thread.[3] However, you still need to be careful when using timers because invoking a timer's `stop` method does not instantly halt the timer's action method.

5.5 ■ ANIMATIONS

The `AnimationControl` class shown in Figure 5.4 is designed to start and stop threads. It implements the control interface and is similar to the `CalculationControl` discussed in Section 5.3. This control has three buttons that invoke methods in a class that implements the `Animation` interface. The labels on the buttons change depending on the control's mode so that all five of the action methods shown in Listing 5.9 can be invoked. The sixth method, `setControl`, is called when the program is instantiated from within the `main` method so that the model knows which control is being used to store and retrieve parameter values. The model's `setControl` method usually invokes other methods within the model, such as `resetAnimation`, in order to initialize the model and the user interface. If this initialization is not done, the control's parameter input table will be empty when it first appears onscreen.

Listing 5.9   The `Animation` interface.

```
public interface Animation extends Runnable {
    public void startAnimation();
```

[3]There are also thread libraries available on the Internet. See, for example, Doug Lea's `util.concurrent` package and the book *Concurrent Programming in Java: Design Principles and Patterns*, 2nd ed., by the same author.

```
public void stopAnimation();
public void initializeAnimation();
public void resetAnimation();
public void stepAnimation();
// register the control with the model
public void setControl(Control control);
}
```

Starting an animation is a two-step process. When an animation control is created, it is ready to load new parameter values or edit existing parameter values. Clicking the control's Initialize button puts the control into run mode and invokes the animation's `initializeAnimation` method. The model uses this method to read parameter values from the control and send them to other objects as needed. Run mode is designed to start and stop threads and to single-step the animation. The label on the Start button changes to Stop and the single Step button is disabled when an animation thread is running. In order to avoid synchronization problems, we do not allow users to edit parameter values while the program is running. Users can stop the animation, edit parameters, and press Start to continue. It is up to the programmer to read parameter values from the control when needed. It might, in fact, not be desirable to completely reinitialize the model when continuing an animation.

Custom buttons are disabled within an animation control when the animation thread is running. This paradigm may seem constraining at first, but it avoids synchronization problems. It insures that input parameters are clearly stated and that these parameters are not changed arbitrarily while the program is running.

In order to make it easy to create an animation, we have defined the `AbstractAnimation` class. This class implements the `Animation` interface and has one abstract method named `doStep`. The class contains an animation thread and this thread invokes the `doStep` method at regular intervals. The `doStep` method is also invoked from within the `stepAnimation` method. Examine the `AbstractAnimation` and note that we have insured that only one animation thread is running by stopping the current thread and waiting for a *join* to occur whenever the `stopAnimation` method is invoked.

Concrete animations define the `doStep` method to evolve the model. The test program shown in Listing 5.10 creates a "toy" animation by implementing this method.

Listing 5.10   An `AnimationControl` test program.

```
package org.opensourcephysics.manual.ch05;
import org.opensourcephysics.controls.*;

public class AnimationApp extends AbstractAnimation {
  public void initializeAnimation() {
    double x0 = control.getDouble("x0"); // read a parameter value
    control.println("initializeAnimation method invoked.");
    control.println("reading x0: "+x0);
  }

  public void stopAnimation() {
    control.println("stopAnimation method invoked");
    super.stopAnimation();
  }

  public void stepAnimation() {
```

```
    control.println("step method invoked");
    super.stepAnimation();
  }

  public void startAnimation() {
    control.println("startAnimation method invoked");
    super.startAnimation();
  }

  public void resetAnimation() {
    control.println("resetAnimation method invoked");
    super.resetAnimation();
    control.setValue("x0", 2.0); // initialize a parameter value
  }

  public void doStep() {
    control.println("a step");
  }

  public static void main(String[] args) {
    AnimationControl.createApp(new AnimationApp());
  }
}
```

The AbstractAnimation class is very defensive. It always stops the animation before the program is reset, and it checks to determine if the animation is running before creating another thread. We could omit these checks if the control is designed to insure that initialization, starting, stopping, stepping, and resetting always occur in the proper sequence. For example, the AnimationControl is constructed so that it is not possible to invoke initializeAnimation or resetAnimation while in run mode. However, it is easy to create custom controls that do not enforce this paradigm as shown in Section 5.7.

5.6 ■ SIMULATIONS

The animation framework provides the mechanism for stopping and starting a thread that performs a repetitive computation. The model is responsible for determining every aspect of the computation. The AbstractSimulation and SimulationControl classes in the controls package are based on the AbstractAnimation and AnimationControl classes, respectively, but include code that performs routine programming chores such as repainting windows after every computation.

A Simulation is a time dependent model of a physical system. The simulation begins with a set of initial conditions that determine the dynamical behavior of the model and generates data in the form of tables and plots as the model evolves. A SimulationControl is designed to control this type of a model. Frames are automatically repainted and data are automatically cleared when the Initialize and Reset buttons are clicked in a SimulationControl. Listing 5.11 demonstrates how the simulation framework is used by showing a two-dimensional trajectory. The program plots $x(t)$ and $y(t)$ using sinusoidal functions with angular frequencies ω_1 and ω_2, respectively. The resulting patterns are known as Lissajous figures and are shown in Figure 5.5.

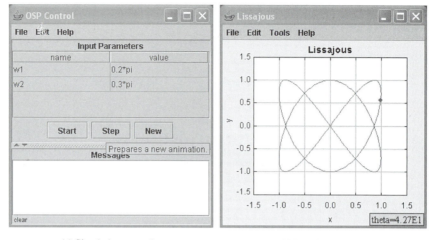

(a) Simulation control.   (b) Lissajous figure.

Figure 5.5   A graphical user interface controls the Lissajous simulation.

Listing 5.11   `LissajousApp` demonstrates how to create a simulation by drawing Lissajous figures.

```
package org.opensourcephysics.manual.ch05;
import org.opensourcephysics.controls.*;
import org.opensourcephysics.display.*;
import org.opensourcephysics.frames.PlotFrame;

public class LissajousApp extends AbstractSimulation {
  PlotFrame frame = new PlotFrame("x", "y", "Lissajous");
  double time = 0, dt = 0.1;
  double w1, w2;
  Circle circle = new Circle(0, 0, 3);

  public void reset() {
    time = 0.0;
    frame.setPreferredMinMax(-1.5, 1.5, -1.5, 1.5);
    frame.setConnected(true);
    frame.setMarkerShape(0, Dataset.NO_MARKER);
    frame.addDrawable(circle);
    control.setValue("omega 1", "0.2*pi");
    control.setValue("omega 2", "0.3*pi");
    initialize();
  }

  public void initialize() {
    w1 = control.getDouble("omega 1");
    w2 = control.getDouble("omega 2");
    time = 0.0;
    frame.append(0, 0, 0);
    circle.setXY(0, 0);
    frame.setMessage("theta="+decimalFormat.format(time), 1);
  }
```

```
  protected void doStep() {
    time += dt;
    double x = Math.sin(w1*time), y = Math.sin(w2*time);
    frame.append(0, x, y);
    circle.setXY(x, y);
    frame.setMessage("theta="+decimalFormat.format(time), 1);
  }

  public static void main(String[] args) {
    SimulationControl.createApp(new LissajousApp());
  }
}
```

LissajousApp creates a SimulationControl in its main method, and the program uses the control's setValue and getValue accessor methods to save and retrieve parameters from this control. Note how the angular velocity is expressed as arithmetic expressions and how these strings are automatically converted to numbers by the getDouble method.

The AbstractSimulation automatically performs a number of common programming chores. It makes all DrawingFrames visible when the Initialize button is clicked. It repaints frames whose animated property is true after every call to doStep.

```
frame.setAnimated(true); // true by default within frames package
```

Simulations must explicitly invoke a drawing method such as repaint or render if a frame is not animated as described in Chapter 7.

Some frames such as PlotFrame store data; these data are usually cleared when a simulation is reset or initialized. The SimulationControl clears a frame's data when the Initialize button is pressed if the *autoclear* property is true.

```
frame.setAutoclear(true); // true by default within frames package
```

Although we usually clear data when initialing a simulation, we might set a plot frame's autoclear property to false if we wish to compare trajectories with different parameters. Data are always cleared when the Reset button is pressed because this action recreates the program's initial state.

Deciding what should be cleared from a frame depends on the situation. A projectile simulation might use a circle to show the particle's xy-position, and it would be inconvenient if this circle was automatically removed during initialization. The particle is always present. A molecular dynamics simulation, on the other hand, may create N circles and add them to a frame after reading the number of particles. Existing circles should be removed when this program is initialized. In general, frames will automatically clear data stored within the frame's drawable objects but will not remove the drawable objects themselves. Drawable objects can, however, be removed from a frame using any of the following methods:

```
Circle circle = new Circle();         // creates a drawable circle
frame.addDrawable(circle);            // adds circle to frame
// examples of how to remove drawables
frame.clearDrawables();               // removes all drawable objets
frame.removeDrawable(circle);         // removes a single object
frame.removeObjectsOfClass(Circle.class); // removes all circles
```

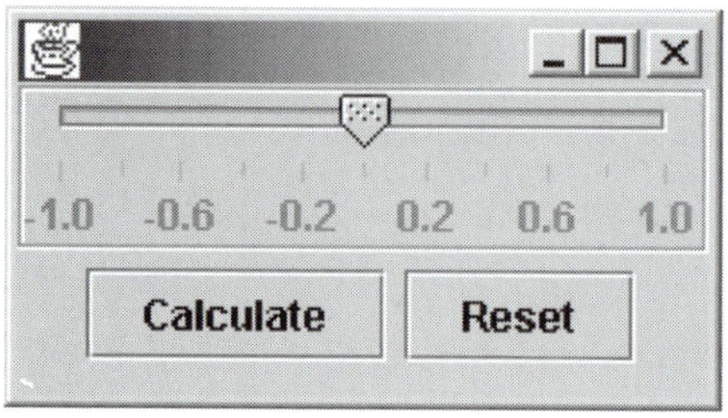

Figure 5.6   A custom Ejs control that can control any object.

5.7 ■ EJS CONTROLS

Easy Java Simulations (Ejs), is a high-level modeling tool, written by Francisco Esquembre at the University of Murcia, Spain, that builds Java applications and applets using a drag and drop graphical user interface. We have included some Ejs components in the core Open Source Physics library in the Ejs package in order to enable programmers to quickly build custom user interfaces. This section provides an overview of the Ejs package components. A more complete description is given in the Ejs technical manual available on the OSP website. A description of the entire Easy Java Simulations project, including the complete modeling program, is available at

http://fem.um.es/Ejs/

The Ejs package builds controls without any preconditions as to the type of object that will be controlled. That is, a controlled class need not implement any given interface. For example, the following code fragment creates the user interface shown in Figure 5.6 consisting of a slider and two buttons that invoke methods in anyObject.

Listing 5.12   An Ejs control containing a slider and two buttons.

```
GroupControl control = new GroupControl(anyObject);
control.add("Frame",
    "name=controlFrame; exit=true; size=200,90; layout=vbox");
control.add("Slider", "parent=controlFrame; minimum=-1; maximum=1;
    ticks=11; ticksFormat=0.0; variable=x0; dragaction=sliderMoved");
control.add("Panel", "name=controlPanel; parent=controlFrame;
    size=300,300; position=south; layout=flow");
control.add("Button",
    "parent=controlPanel; text=Calculate; action=calculate()");
control.add("Button",
    "parent=controlPanel; text=Reset; action=resetCalculation()");
control.update();
```

The code starts by instantiating a group control named control and passing it a reference to the object that will be controlled, anyObject. This controlled object will be referred to as the *target*. The target receives actions, such as button clicks or slider moves, from the control. The second statement begins the process of creating the control's user interface by instantiating a ControlFrame. The next statement adds a slider whose action invokes the sliderMoved method in the target. The remaining statements repeat this process by adding another panel and two buttons to the frame. The actions of the slider and the buttons are public methods within the target.

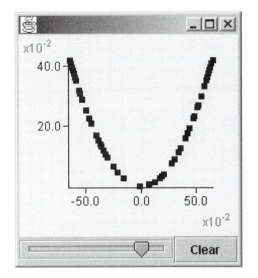

Figure 5.7 A user interface containing a plotting panel and two Ejs controls.

Ejs supports a wide variety of user interface components including labels, radio buttons, check boxes, and one-line text fields. These components are created in a script-like style by invoking the group control's `add` method with the following signature:

```
GroupControl add(String type, String property_list);
```

The `add` method's first parameter specifies the type of object that is being created. The second parameter is a semicolon delimited list of properties. Properties depend on the type of object being created but typically include the following:

- **name**: The name of the instance. This name can later be used to access the object.

- **parent**: The name of the object's container. Typical containers are panels and windows.

- **action**: The name of a method in the model that will be invoked when the user clicks, slides, or otherwise activates the control.

Listing 5.13 shows a "toy" program that creates the user interface shown in Figure 5.7 by combining two Ejs controls with a plotting panel. Moving the slider varies the control's x-parameter and plots the function $y = x^2$. Pressing the button clears the data.

> **Listing 5.13** A demonstration of how to use an Ejs control with a drawing panel.

```
package org.opensourcephysics.manual.ch05;
import org.opensourcephysics.display.*;
import org.opensourcephysics.ejs.control.GroupControl;

public class EJSSliderApp {
  DrawingPanel plottingPanel =
      new PlottingPanel("x", "y", "EJS Controls Test");
```

```
      GroupControl control;
      Dataset dataset = new Dataset();

      public EJSSliderApp() {
        control = new GroupControl(this);
        control.add("Frame",
            "name=plottingFrame; title=EJS Example; exit=true;
            size=300,300");
        control.addObject(plottingPanel, "Panel",
            "name=plottingPanel; parent=plottingFrame; position=center");
        control.add("Panel",
            "name=controlPanel; parent=plottingFrame;
            position=south; layout=hbox");
        control.add("Slider",
            "parent=controlPanel; minimum=-1; maximum=1;
            variable=x; dragaction=sliderMoved()");
        control.add("Button",
            "parent=controlPanel; text=Clear; action=clearPlot()");
        plottingPanel.addDrawable(dataset);
        control.update();
      }

      public void clearPlot() {
        dataset.clear();
        plottingPanel.repaint();
      }

      public void sliderMoved() {
        double x = control.getDouble("x");
        dataset.append(x, x*x);
        plottingPanel.repaint();
      }

      static public void main(String[] args) {
        new EJSSliderApp();
      }
    }
```

The addObject method is used to add an existing object to an Ejs control, and the add method is used to instantiate and add an object to an Ejs control. The control's update method guarantees that control elements display current values.

Although GroupControl does not implement the OSP Control interface, EjsControl and EjsControlFrame do. EjsControlFrame extends EjsControl. It contains a JFrame and implements the RootPaneContainer interface in order to function as a container for user interface elements.

Unlike our predefined controls, controls that inherit from EjsControlFrame can have an arbitrary number of user interface components and actions. In Listing 5.14 we build a control with a slider to replace a CalculationControl with a single parameter. Figure 5.7 shows a screen shot of the resulting user interface.

Listing 5.14   An Ejs control can replace a calculation control's parameter input table.

```
public class CustomCalculationFrame extends EjsControlFrame {
    public CustomCalculationFrame(Object target) {
        super(target,
            "name=controlFrame; exit=true; size=200,90; layout=vbox");
```

Figure 5.8 A custom Ejs control can replace an `AnimationControl`.

```
// creates the user interface for the control
add("Slider", "parent=controlFrame; minimum=-1; maximum=1;
    ticks=11; ticksFormat=0.0; variable=x0");
add("Panel", "name=controlPanel; parent=controlFrame;
    size=300,300; position=south; layout=flow");
add("Button",
    "parent=controlPanel; text=Calculate; action=calculate");
add("Button",
    "parent=controlPanel; text=Reset; action=resetCalculation");
update();
  }
}
```

The new custom calculation control is a direct replacement for the control in the calculation test program shown in Listing 5.4. The `main` method must, of course, instantiate the custom control.

```
public static void main (String[] args) {
  Calculation model = new CalculationTestApp();
  Control control = new CustomCalculationControl(model);
  model.setControl(control);
}
```

Simulations often benefit from the visual simplicity of a custom control that starts and stops a thread. Consider the Ising model shown in Figure 5.8. Our implementation of this model contains an animation thread and a temperature parameter. In Listing 5.15 we create a custom control with two buttons and a slider to replace the original `AnimationControl`.

Listing 5.15 A custom control for the Ising model.

```
package org.opensourcephysics.manual.ch05;
import org.opensourcephysics.ejs.control.EjsControlFrame;

public class IsingControl extends EjsControlFrame {
  public IsingControl(IsingApp model) {
    super(model,
        "name=controlFrame; title=ControlFrame; size=400,400;
        layout=border; exit=true");
```

```
addObject(model.drawingPanel, "Panel",
    "name=drawingPanel; parent=controlFrame;
    layout=border; position=center");
add("Panel",
    "name=controlPanel; parent=controlFrame;
    layout=border; position=south");
add("Panel",
    "name=sliderPanel; position=north; parent=controlPanel;
    layout=border");
add("Slider",
    "position=center; parent=sliderPanel; variable=T; minimum=0;
    maximum=5; ticks=11; ticksFormat=0; action=sliderMoved()");
add("Label",
    "position=north; parent=sliderPanel; text=Temperature;
    font=italic,12; foreground=blue; horizontalAlignment=center");
add("Panel",
    "name=buttonPanel; position=south; parent=controlPanel;
    layout=flow");
add("Button",
    "parent=buttonPanel; text=Run; action=startAnimation()");
add("Button",
    "parent=buttonPanel; text=Stop; action=stopAnimation()");
  }
}
```

Ejs controls can either instantiate standard Swing user interface components using the add method or they can accept already existing objects using the addObject method. In our example, the model already instantiated a DrawingPanel, and we want to add this panel to the control. Ejs sets the existing panel's properties and adds it to the parent because the DrawingPanel is a subclass of JPanel.

When the IsingControl slider's action method adjusts the temperature, it is the model's responsibility to insure that the temperature can safely be changed by the slider even if the animation thread is running. We do so by temporarily stopping the animation thread. The Ising model is shown in Listing 5.16. The physics-related code has been removed from the listing for clarity. The Ising control is a separate class, but it is short and could have been defined as an inner class within this model.

A final word about simplicity of design—CalculationControl and AnimationControl are designed for input/output without worrying about details such as layout managers and appearance. This approach is ideal for testing code and for teaching students how to program. Ejs provides the opportunity to create very attractive user interfaces without changing the computational model. We recommend that users develop their models first and then—if a better interface is desired—build an Ejs control to fit the model.

Listing 5.16   An outline of the Ising program.

```
package org.opensourcephysics.manual.ch05;
import org.opensourcephysics.controls.*;
import org.opensourcephysics.display.DrawingPanel;
import org.opensourcephysics.display2d.BinaryLattice;

public class IsingApp extends AbstractAnimation {
  int size = 32;
  double temperature = 2;
  DrawingPanel drawingPanel = new DrawingPanel();
```

```
  BinaryLattice lattice = new BinaryLattice(size, size);

  public IsingApp() {
    drawingPanel.addDrawable(lattice);
  }

  public void sliderMoved() {
    boolean isRunning = animationThread!=null;
    stopAnimation();
    temperature = control.getDouble("T");
    // physics has been removed for clarity
    if(isRunning) {
      startAnimation();
    }
  }

  protected void doStep() {
    // physics has been removed for clarity
    lattice.randomize();
    drawingPanel.repaint();
  }

  public static void main(String[] args) {
    IsingApp model = new IsingApp();
    Control control = new IsingControl(model);
    model.setControl(control);
  }
}
```

5.8 ■ PROGRAMS

The following examples are in the `org.opensourcephysics.manual.ch05` package.

AnimationApp

`AnimationApp` tests the `Animation` interface as described in Section 5.5. Edit this example to build new animations.

AnimationFrameApp

`AnimationFrameApp` creates an animation by extending the `AbstractAnimation` class. Edit this example to build new animations.

CalculationApp

`CalculationApp` tests the `Calculation` interface as described in Section 5.3. Edit this example to build new calculations.

CustomButtonApp

`CustomButtonApp` extends `AbstractCalculation` and creates a `CalculationControl` with a custom button.

EJSSliderApp

`EJSSliderApp` creates an Ejs control containing a button and a slider as described in Section 5.7.

IsingApp

`IsingApp` builds a custom graphical user interface for the Ising model using Ejs. See Section 5.7.

LissajousApp

`LissajousApp` demonstrates how to create a `Simulation` by animating Lissajous figures using a `PlotFrame` as described in Section 5.6.

MainThreadApp

`MainThreadApp` pauses the `main` thread using the `sleep` method to produce a simple animation. See Section 5.4.

OSPControlApp

`OSPControlApp` creates an `OSPControl`, adds a button to the control, and reads parameters from the control as described in Section 5.1.

RotationApp

`RotationApp` creates two rotating, rectangle animations. This program is described in Section 5.4.

SetValueApp

`SetValueApp` tests the `setValue` method and the `setAdjustableValue` method in `SimulationControl`.

ShapeAnimationApp

`ShapeAnimationApp` creates a rotating rectangle and implements an `AbstractAnimation`.

SimulationApp

`SimulationApp` creates a rotating square by extending an `AbstractSimulation` and implementing the `doStep` method. Because the frame is set to be animated, the `render` method is invoked automatically after every animation step.

CHAPTER

6

Plotting

The Open Source Physics library defines components that can be used to create a wide variety of xy-scatter plots including polar, semilog, and log-log plots.

6.1 ■ OVERVIEW

The easiest way to create an xy-scatter plot is to use the PlotFrame (see Figure 6.1) in the frames package. This frame contains all the necessary components to plot multiple datasets, change scale, display log-log and semilog plots, and show data in tabular form. It is, however, impossible to meet every data visualization need without duplicating the functionality of specialized programs such as *GNUPlot*, and the Export item under the File menu is available if this high-level plotting program is needed. This chapter describes how the PlotFrame and PlottingPanel classes can be used to create high-quality custom plots from within an OSP program.

6.2 ■ DATASET

A plot is created by storing data points in one or more *datasets*. A Dataset object is a Drawable object that stores and renders (x, y) data points. In order to construct a simple graph, we instantiate a dataset, add it to a drawing panel, and append points to the dataset. Typical code is shown in Listing 6.1.

> **Listing 6.1** A drawing panel containing a dataset and two axes components.

```
package org.opensourcephysics.manual.ch06;
import org.opensourcephysics.display.*;
import org.opensourcephysics.display.axes.*;
import javax.swing.JFrame;

public class DatasetApp {
  public static void main(String[] args) {
    // create a drawing frame and a drawing panel
    DrawingPanel panel = new DrawingPanel();
    DrawingFrame frame = new DrawingFrame(panel);
    panel.setSquareAspect(false);
    panel.setAutoscaleX(true);
    panel.limitAutoscaleY(-5, 15);
    // create data and append it to a dataset
    double[] x = {-2, -1, 0, 1, 2};
```

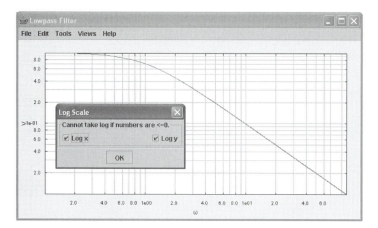

Figure 6.1   A lowpass RC filter frequency response plotted using a PlotFrame in the frames package. The log-scale dialog is available in the Views menu.

```
double[] y = {-2, 2, 6, 10, 14};
Dataset dataset = new Dataset();
dataset.setConnected(true);
dataset.setSorted(true);  // sort points as needed
dataset.append(x, y);
dataset.append(5, 11);    // a single point
dataset.append(3.5, 2.5); // this point will be sorted
// add the drawable objects to the drawing panel
panel.addDrawable(dataset);
XAxis xaxis = new XAxis("x");
panel.addDrawable(xaxis);
YAxis yaxis = new YAxis("y");
panel.addDrawable(yaxis);
panel.repaint();
frame.setVisible(true);
frame.setDefaultCloseOperation(JFrame.EXIT_ON_CLOSE);
  }
}
```

Single points and arrays of points can be added to a dataset using the append method. However, views (such as drawing panels with datasets) are not refreshed until the view is repainted. This allows a program to append multiple points to a graph without unnecessary drawing.

Data points with error estimates can also be appended to datasets using the following signature:

```
//parameters can also be double or double[]
dataset.append(x, y, err_x, err_y);
```

Error estimates are optional and can be included with only a few select points to limit the number of error bars being drawn.

Table 6.1 lists common Dataset methods. Consult the source code for a complete listing of dataset methods and more extensive documentation.

Because a dataset has many scaling and marker options, drawing large numbers of points can be slow. The setMaximumPoints method limits the number of points that will be stored.

Table 6.1   Common dataset methods.

org.opensourcephysics.display.Dataset	
`append`	Adds data to the data set as xy-coordinate pairs. The data can be either doubles or arrays of doubles. Note that data points having the nonnumerical values of `Double.NaN` or infinity are discarded.
`clear`	Removes all data.
`isMeasured`	Gets the valid-measure flag. This flag is true if the data set contains at least one valid data point.
`getXMax`	Gets the maximum x-value in the dataset. Note that this value is guaranteed to be correct only if the `isMeasured` flag is true.
`getXMin`	Similar to `getXMax`.
`getYMax`	Similar to `getXMax`.
`getYMin`	Similar to `getXMax`.
`toString`	Creates a string representation of the data with every point on a new line. The x- and y-values are separated by a tab.
`setConnected`	Connects the data points with straight lines. Default is `false`.
`setLineColor`	Sets the color of the lines connecting the points. Ignored if points are not connected. Default is `black`.
`setMarkerColor`	Sets the color of the data markers. Ignored if the data marker is set to `NO_MARKER`. Default is `black`.
`setMarkerShape`	Sets the shape of the data markers. Valid shapes are `NO_MARKER`, `CIRCLE`, `SQUARE`, `AREA`, `PIXEL`, `BAR`, and `POST` . The default is `SQUARE`.
`setMaximumPoints`	Sets the allowed number of points. The default is 16K.
`setSorted`	Sorts the dataset by increasing x-value. Default is `false`.

The default maximum number is 16K. Points at the beginning of the dataset are dropped if this number is exceeded. There are other objects that can draw many more points because they create simpler representations for each point. The `Trail` object in the display package displays a path of connected points without markers. The `DataRaster` in the 2D display package creates an image by coloring one pixel per point (see Section 7.6).

6.3 ■ DATASET MANAGER

A dataset manager is a composite object that performs many of the housekeeping chores associated with multiple datasets as shown in Figure 6.2. Datasets are accessed using an integer index, and the required dataset is created the first time the index is used. It is possible to set these (and other) options globally as well as for individual datasets. For example, the marker type, color, and connectedness of data points can be set using the following code fragment:

```
datasetManager = new DatasetManager();
// connects points in all datasets
datasetManager.setConnected(true);
// affect only dataset 3
```

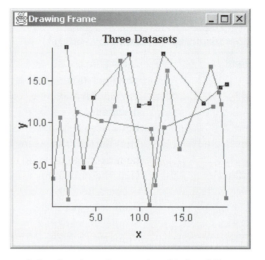

Figure 6.2   A plotting panel showing three datasets (see Listing 6.2).

```
datasetManager.setLineColor(3,Color.blue);
datasetManager.setMarkerShape(3,Dataset.CIRCLE);
```

Listing 6.2 uses a dataset manager to create three sets of random data. The first dataset has an index of 0. The manager assigns different colors and markers to each dataset.

Listing 6.2   A dataset manager containing three datasets.

```java
package org.opensourcephysics.manual.ch06;
import org.opensourcephysics.display.*;
import javax.swing.JFrame;

public class DatasetManagerApp {
  public static void main(String[] args) {
    int numberOfDatasets = 3;
    int pointsPerDataset = 10;
    DatasetManager datasetManager = new DatasetManager();
    datasetManager.setConnected(true);
    datasetManager.setSorted(true);
    for(int i = 0;i<numberOfDatasets;i++) {
      for(int j = 0;j<pointsPerDataset;j++) {
        datasetManager.append(i, Math.random()*20, Math.random()*20);
      }
    }
    PlottingPanel panel =
        new PlottingPanel("x", "y", "Three Datasets");
    panel.setAutoscaleX(true);
    panel.setAutoscaleY(true);
    panel.addDrawable(datasetManager);
    DrawingFrame frame = new DrawingFrame(panel);
    frame.setDefaultCloseOperation(JFrame.EXIT_ON_CLOSE);
    frame.setVisible(true);
  }
}
```

Table 6.2 Typical dataset manager methods.

org.opensourcephysics.display.DatasetManager	
append	Appends data to the ith dataset.
clear	Removes data from all datasets or from the ith dataset.
isMeasured	Returns true if at least one dataset in the manager has a valid point.
getXMin	Gets the minimum x-value in all datasets.
getXMax	Gets the maximum x-value in all datasets.
getYMin	Gets the minimum y-value in all datasets.
getYMax	Gets the maximum y-value in all datasets.

Many of the methods in the `Dataset` class have `DatasetManager` counterparts. Methods without an index set the default and perform the method on all existing datasets. Table 6.2 shows some common methods implemented in the manager.

6.4 ■ MARKERS

The `PlotFrame` class creates datasets with point *markers* having different drawing properties. The marker shape, size, and color as well as a boolean that determines if markers should be connected with line segments can be set using dataset accessor methods. Listing 6.3 uses two datasets with different drawing options to simulate a Gaussian spectral line with random noise. This example uses a `PlotFrame` object rather than components from the display package to emphasize that the frame's API is based on methods that are forwarded to lower-level components.

Listing 6.3 Program `GaussianPlotApp` displays two datasets.

```
package org.opensourcephysics.manual.ch06;
import org.opensourcephysics.display.Dataset;
import org.opensourcephysics.frames.PlotFrame;
import javax.swing.JFrame;

public class GaussianPlotApp {
  public static void main(String[] args) {
    PlotFrame frame = new PlotFrame("$\\Delta$ f", "intensity",
        "Gaussian Spectral Line");
    frame.setConnected(0, true);
    frame.setMarkerShape(0, Dataset.NO_MARKER);
    for(double f = -10;f<10;f += 0.2) {
      double y = Math.exp(-f*f/4);
      frame.append(0, f, y);                    // datum
      frame.append(1, f, y+0.1*Math.random()); // datum + noise
    }
    frame.setVisible(true);
    frame.setDefaultCloseOperation(JFrame.EXIT_ON_CLOSE);
  }
}
```

Table 6.3   The `Dataset` class uses named integer constants to tag common marker shapes.

Dataset Marker Shapes	
AREA	Fills the area between the connected points and the x-axis with a solid color.
BAR	A rectangle with its bottom edge on the x-axis and its top edge centered on the data point.
CIRCLE	A circle centered on the data point.
NO_MARKER	Disables marker drawing. Line segments connecting data point centers will be drawn if the dataset is connected.
PIXEL	A marker of exactly one pixel.
POST	A square centered on the data point and a line connecting the data point to the x-axis.
SQUARE	A square centered on the data point.

Marker shapes are changed by passing a named integer constant, such as `CIRCLE` or `SQUARE`, to the `setMarkerShape` method. The marker size (in pixels) can also be adjusted as shown in the following code fragment:

```
// set style for dataset 1
frame.setMarkerShape(1, Dataset.CIRCLE);
frame.setMarkerSize(1,2); //half width of marker is 2 pixels
```

Maker shapes defined in the `Dataset` class are described in Table 6.3.

Markers can be used to create a variety of graphs as shown in Figure 6.3. Color is set using the `setMarkerColor` method, and this method has a number of signatures to produce different effects. Calling this method with a single color produces a shape with a single color. It is usually more attractive to use multiple colors for the marker's interior, edge, and error bars. The color's *alpha* value can be used to define a fill color that is slightly transparent so that overlapping data points can be seen. For example, the following code creates a dataset with a pale-red semitransparent interior, red edge, and black error bars.

```
Dataset dataset = new Dataset ();
dataset.setConnected (false); // do not connect the points with lines
dataset.setMarkerShape(Dataset.SQUARE);
dataset.setMarkerColor(new
    Color(255,128,128,128),Color.red,Color.black);
```

The `BAR` marker renders each data point as a rectangle with a base centered at x and a height of y. This marker is useful for histogram-like plots. The `POST` marker is similar except that a small square is drawn centered on the data point together with a line connecting the marker to the x-axis. The line is drawn using the marker's edge color.

The `AREA` marker fills the area between the x-axis and the connected data points with a solid color.

The `PIXEL` marker draws a single pixel at the data point. It is useful for Poincaré sections. (The `DataRaster` class described in Section 7.6 is a specialized drawing component that is designed to render very large datasets quickly by coloring a single pixel for each datum.)

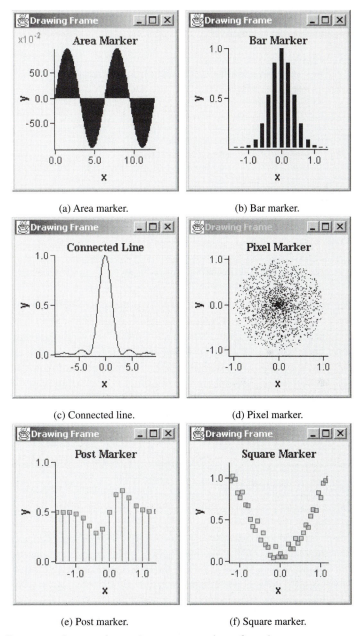

(a) Area marker.

(b) Bar marker.

(c) Connected line.

(d) Pixel marker.

(e) Post marker.

(f) Square marker.

Figure 6.3   Dataset markers can be used to create a variety of graphs.

Custom point markers can be created using the `Shape` interface in the Java 2D API. The `GeneralPath` class in Sun's geometry package defines a shape using a sequence of `moveTo` and `lineTo` methods to create an outline. The following code fragment creates a custom butterfly marker for a dataset and then sets the marker's interior color and edge color to red and black, respectively.

```
// import java.awt.geom.*;
GeneralPath marker = new GeneralPath();
marker.moveTo(3,3);
marker.lineTo(3,-3);
marker.lineTo(-3,3);
marker.lineTo(-3,-3);
marker.closePath();
frame.setCustomMarker(1, marker);
frame.setMarkerColor(1, Color.RED, Color.BLACK);
```

6.5 ■ AXES

Axes are special drawable components that implement the `DrawableAxes` interface and are drawn within a plotting panel before all other objects (see Table 6.4). An axes object always exists although it can be hidden by setting its visible property to false.

The plotting panel's default axes component `AxesType1` is based on open source code released by the University of California. `AxesType2` is similar but is based on open source code released by Leigh Bookshaw. `AxesType3` is also based on Bookshaw's code, but each axis is drawn within the plotting region and is draggable. All axes have similar APIs. When an axes component is constructed, a plotting panel is passed to the constructor. This constructor sets the panel's gutters and registers the new axes with the panel. The following code fragment is taken from the `PolarPlottingApp` program in the Chapter 2 package.

```
PlottingPanel plottingPanel = new PlottingPanel(null,null,null);
PolarAxes axes = new PolarAxes(plottingPanel);
axes.setRadialGrid(1);
axes.setThetaGrid(Math.PI/8);
plottingPanel.setTitle("PolarPlot");
plottingPanel.repaint();
```

Because axes may correspond to variables other than x and y, we have implemented mechanisms that allow us to disable the coordinate readout or to change the labels in the coordinate readout. The `setShowCoordinates` method allows us to disable the panel's coordinate display in the left-hand corner message box when the mouse is pressed. The `CoordinateStringBuilder` class is responsible for assembling the string when the mouse is pressed. The labels that are prepended to the coordinate values can be set using the `setCoordinateLabels` method.

```
plottingPanel.setShowCoordinates(true);  // default is true
plottingPanel.getCoordinateStringBuilder().
    setCoordinateLabels("t","amp");
```

Table 6.4   Concrete implementations of the `DrawableAxes` interface.

org.opensourcephysics.display.axes

`CartesianType1`	Cartesian axes that are drawn outside the plotting region.
`CartesianType2`	Alternate Cartesian axes that are drawn outside the plotting region.
`CartesianType3`	Cartesian axes that are drawn within the plotting region. Each axis is draggable.
`CustomAxes`	A basic class with an interior and a title. Drawable objects can be added to this class to produce custom axes.
`PolarType1`	Polar coordinates that remain centered in the drawing panel. The panel's minimum and maximum values are set equal to the largest preferred value.
`PolarType2`	Polar coordinates that respect the panel's preferred values and can therefore be off center.

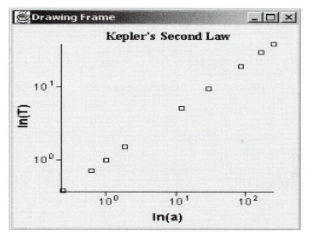

Figure 6.4   Kepler's second law demonstration. A plot of the semimajor axis of the planets in AU versus the log of the orbital period in years produces a straight line with a slope of 1.5.

6.6 ■ LOG SCALE

`PlottingPanel` contains the necessary axes and titles to produce linear, log-log, and semilog graphs. For example, Listing 6.4 shows that Kepler's second law is plausible by producing a log-log plot of the semimajor axis of the planets versus the orbital period. The data arrays `a` and `T` contain the semimajor axis of the planets in astronomical units and the orbital period in years, respectively. Note that the plot shown in Figure 6.4 automatically adjusts itself to fit the data because the autoscale option is set to true for both x- and y-axis.

Listing 6.4   Kepler's second law plot.

```
package org.opensourcephysics.manual.ch06;
import org.opensourcephysics.display.*;
```

```
public class KeplerPlotApp {
  public static void main(String[] args) {
    PlottingPanel plotPanel =
        new PlottingPanel("ln(a)", "ln(T)", "Kepler's Second Law");
    plotPanel.setLogScale(true, true);
    DrawingFrame drawingFrame = new DrawingFrame(plotPanel);
    Dataset dataset = new Dataset();
    dataset.setConnected(false);
    double[] period =
        {0.241, 0.615, 1.0, 1.88, 11.86, 29.50, 84.0, 165, 248};
    double[] a =
        {0.387, 0.723, 1.0, 1.523, 5.202, 9.539, 19.18, 30.06, 39.44};
    dataset.append(a, period);
    plotPanel.addDrawable(dataset);
    drawingFrame.setDefaultCloseOperation(
        javax.swing.JFrame.EXIT_ON_CLOSE);
    drawingFrame.setVisible(true);
    // create a data table
    DataTable dataTable = new DataTable();
    DataTableFrame tableFrame =
        new DataTableFrame("Orbital Data", dataTable);
    dataset.setXYColumnNames("T (years)", "a (AU)");
    dataTable.add(dataset);
    dataTable.setRowNumberVisible(true);
    dataTable.refreshTable();
    tableFrame.setVisible(true);
  }
}
```

Setting the log scale option causes the drawing panel to transform the data as it is being plotted and causes the axis to change how labels are rendered. If the drawing panel has minimum and maximum vales of 1 and 3, respectively, then the labels will be drawn as 10^1, 10^2, and 10^3.

6.7 ■ DATASET TABLES

Because Dataset and DatasetManager implement the TableModel interface, they can display themselves in a data table. (Section 3.8 describes how to create tables without datasets.) Dataset tables are created by instantiating a DataTable and a DataTableFrame and then adding a dataset manager to the table. The paradigm is similar to that for adding datasets to a drawing panel and is shown in Listing 6.5. Note that the code invokes refreshTable. This method is equivalent to calling a drawing panel's repaint method insofar as it forces the data table to repaint itself.

Listing 6.5   A dataset manager can be used to create a table as shown in Figure 6.5.

```
package org.opensourcephysics.manual.ch06;
import org.opensourcephysics.display.*;
import javax.swing.JFrame;

public class DataTableApp {
  public static void main(String[] args) {
    DatasetManager datasets = new DatasetManager();
    datasets.setXPointsLinked(true); // use x-values from 0th dataset
```

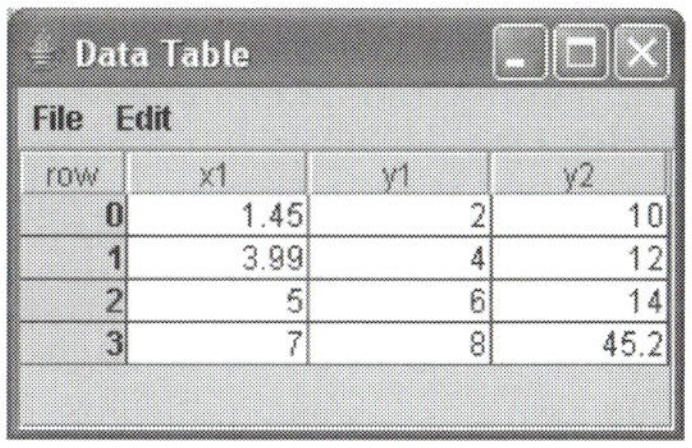

Figure 6.5   Data within a dataset manager displayed in a table.

```
        double[] xpoints = {1.45, 3.99, 5, 7};
        double[] ypoints = {2, 4, 6, 8};
        datasets.append(0, xpoints, ypoints);
        double[] ypoints2 = {10.0, 12, 14, 45.2};
        datasets.append(1, xpoints, ypoints2); // same x-values
        datasets.setXYColumnNames(0, "x1", "y1");
        datasets.setXYColumnNames(1, "x2", "y2");
        DataTable dataTable = new DataTable();
        DataTableFrame frame = new DataTableFrame("Data Table", dataTable);
        dataTable.add(datasets);
        datasets.setStride(1); // the default
        dataTable.setRowNumberVisible(true);
        dataTable.refreshTable();
        frame.setDefaultCloseOperation(JFrame.EXIT_ON_CLOSE);
        frame.setVisible(true);
    }
}
```

Each dataset produces two columns, one for x-values and another for the y-values. These columns are labeled using the setXYColumnNames method.

```
datasets.setXYColumnNames(0, "x1", "y1");
```

It is often the case that datasets have common x-values as in the example above. It is therefore convenient to suppress x-columns except those for the first dataset. This is done by setting the points-linked property.

```
datasetManager.setXPointsLinked(true); // X only for 0th dataset
```

It is also possible to show and hide individual columns and to skip rows as shown in Table 6.5. Consult the data table source code documentation for information about these data table methods.

We now show a longer example that uses multiple datasets and a table by solving the following differential equation:

$$dy/dt = -y.\qquad(6.1)$$

The analytic solution to this equation is well known, and we want to compare this solution to a numerical approximation that was generated using what is known as Euler's algorithm (see Chapter 9). We will plot the analytic and numerical solutions and we will show these values in a table with a column whose entries are their difference. Figure 6.6 shows the resulting table and graph. Managing these data is easy using a dataset manager as shown in Listing 6.6. Note that all datasets are displayed within the table, whereas only datasets 1 and 2 are shown in the plot.

Table 6.5   Common DataTable methods.

org.opensourcephysics.display.DataTable	
setRefreshDelay	Sets a short delay before table is redrawn on the screen. A delay is useful if table data changes frequently during an animation. A 500 ms delay minimizes unnecessary rendering and may reduce screen flicker.
setStride	Sets the stride in the data table. The stride allows a table to skip rows so that very long datasets can be displayed. For example, a stride of ten shows every tenth row. The default stride is one.
setXColumnVisible	Sets the visibility of the x column of the dataset in a table view.
setYColumnVisible	Sets the visibility of the y column of the dataset in a table view.
setXYColumnNames	Sets the column names when rendering the dataset in a JTable.

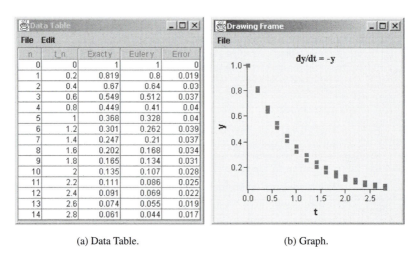

(a) Data Table.   (b) Graph.

Figure 6.6   A comparison of the analytic and Euler's method solution to a differential equation.

The first dataset keeps track of the step number and the time in the first and second columns, respectively. The remaining datasets use time for the first column, but this column is suppressed so only the time values from the first dataset are displayed in the table.

Listing 6.6   A comparison of the analytic solution and Euler solution to the exponential decay equation.

```
package org.opensourcephysics.manual.ch06;
import org.opensourcephysics.display.*;
import javax.swing.JFrame;

public class ExponentialEulerApp {
  DatasetManager datasetManager;
  DataTable dataTable;
```

```
    DataTableFrame tableFrame;
    PlottingPanel plottingPanel;
    DrawingFrame drawingFrame;
    double t0 = 0.0;              // initial value of t
    double y0 = 1.0;              // initial value of y
    double tmax = 3.0;            // t-value that stops the calculation
    double dt = 0.10;             // step size

    public ExponentialEulerApp() { // ExponentialEulerApp constructor
      // set up the dataset manager
      datasetManager = new DatasetManager();
      datasetManager.setXPointsLinked(true); // X only for 0th dataset
      datasetManager.setXYColumnNames(0, "n", "t_n");
      datasetManager.setXYColumnNames(1, "t", "Exact y");
      datasetManager.setXYColumnNames(2, "t", "Euler y");
      datasetManager.setXYColumnNames(3, "t", "Error");
      // set up table
      dataTable = new DataTable();
      tableFrame = new DataTableFrame(dataTable);
      dataTable.add(datasetManager); // add datasetManager to table
      tableFrame.setDefaultCloseOperation(JFrame.EXIT_ON_CLOSE);
      tableFrame.setVisible(true);
      // set up graph
      plottingPanel = new PlottingPanel("t", "y", "dy/dt = -y");
      drawingFrame = new DrawingFrame(plottingPanel);
      plottingPanel.addDrawable(datasetManager.getDataset(1)); // exact
      plottingPanel.addDrawable(datasetManager.getDataset(2)); // Euler
      drawingFrame.setDefaultCloseOperation(JFrame.EXIT_ON_CLOSE);
      drawingFrame.setVisible(true);
    }

    public void calculate() {
      double t = t0; // current value of x
      double y = y0; // current value of the Euler solution
      int counter = 0;
      while(t<=tmax) {
        datasetManager.append(0, counter, t); // store counter and x
        double exact = Math.exp(-t);
        datasetManager.append(1, t, exact);   // store t and exact x
        datasetManager.append(2, t, y);        // store Euler solution
        datasetManager.append(3, t, exact-y); // store Euler solution
        y = y-y*dt;                            // increase y
        t = t+dt;                              // increase t
        counter++;                             // number of steps
      }
      plottingPanel.repaint();  // repaint the graph
      dataTable.refreshTable(); // redraw the table
    }

    public static void main(String[] args) {
      ExponentialEulerApp app = new ExponentialEulerApp();
      app.calculate();
    }
}
```

6.8 ■ PROGRAMS

The following examples are in the `org.opensourcephysics.manual.ch06` package.

DatasetApp

`DatasetApp` tests the `Dataset` class. This program is described in Section 6.2. Other programs that demonstrate how to use datasets are available: `MarkerAreaApp`, `MarkerBarApp`, `MarkerCircleApp`, `MarkerConnectedApp`, `MarkerPixelApp`, and `MarkerPostApp`.

DatasetManagerApp

`DatasetManagerApp` demonstrates how to create three datasets using a dataset manager and how to append data points to these datasets. This program is described in Section 6.3.

DataTableApp

`DataTableApp` creates and displays a data table containing 3 datasets. This program is described in Section 6.7.

ExponentialEulerApp

`ExponentialEulerApp` uses a dataset manager and a data table to show the numerical and analytic solution of a simple differential equation $dy/dt = -y$. The numerical solution is obtained using Euler's method in Section 6.7.

GaussianPlotApp

`GaussianPlotApp` simulates a best fit to a Gaussian spectral line using a `PlotFrame` with two datasets as described in Section 6.4. These datasets use different drawing styles.

KeplerPlotApp

`KeplerPlotApp` demonstrates how to create a log-log plot using data from planetary orbits. The program demonstrates the validity of Kepler's second law by plotting the log of the planet's semimajor axis vs. the log of the orbital period. This program is described in Section 6.6.

LowpassFilterApp

`LowpassFilterApp` plots the frequency response for a lowpass RC filter. The output is shown in Section 6.1.

MarkerApps

Programs `MarkerAreaApp`, `MarkerBarApp`, `MarkerCircleApp`, `MarkerConnectedApp`, `MarkerPixelApp`, and `MarkerPostApp` demonstrate how to use a dataset with different marker styles as described in Section 6.4.

PolarPlottingApp

`PolarPlottingApp` and `Polar2PlottingApp` demonstrate how to create a plotting panel with polar coordinate axes as described in Section 6.5

CHAPTER

7

Animation, Images, and Buffering

Tools and techniques for producing animations are presented. Because animations use image buffers, techniques for creating and storing images are described.

7.1 ■ ANIMATION

A drawing panel paints drawable objects one at a time in the panel's `paintComponent` method as described in Chapter 4. This drawing is usually done by invoking the component's `repaint` method. Invoking `repaint` does not, however, repaint the screen directly but places a request into the program's *event queue*. The Java VM will invoke the `paintComponent` method at a later time from the *event dispatch thread* and may, in fact, collapse multiple repaint requests into a single operation. Because there is no way of knowing when painting takes place, painting may not occur at a constant rate and frames may be dropped.[1]

The `RepaintApp` program shown in Listing 7.1 attempts to produce a fast animation by drawing a rotating spiral. (The `Spiral` class is a simple implementation of `Drawable` and is not shown but is available on the CD.) The animation invokes the `sleep` method after invoking the `repaint` method to give the event dispatch thread access to system resources. The sleep time has been set to one millisecond in an attempt to produce the fastest possible frame rate.

Listing 7.1 The `repaint` method places a request into the *event queue*. The *event dispatch thread* processes these requests in the order received.

```
package org.opensourcephysics.manual.ch07;
import org.opensourcephysics.display.*;
import javax.swing.JFrame;

public class RepaintApp {
  static int sleepTime = 1;

  public static void main(String[] args) {
    DrawingPanel panel = new DrawingPanel();
    DrawingFrame frame = new DrawingFrame("Direct Drawing", panel);
    frame.setDefaultCloseOperation(JFrame.EXIT_ON_CLOSE);
    frame.setVisible(true);
    Spiral spiral = new Spiral();
```

[1]The situation is similar for other events such as a button clicks or a keyboard entry. An event is placed into the dispatch queue and this queue is processed sequentially by the event dispatch thread. Because Swing graphical user interface components are not thread safe, they should only be accessed from this thread.

```
      panel.addDrawable(spiral);
      while(true) {
        spiral.theta += 0.1;
        try {
          Thread.sleep(sleepTime); // allow the event queue to redraw
        } catch(InterruptedException ex) {}
        panel.repaint();
      }
    }
  }
```

The animation produced by RepaintApp is unsatisfying because the frame rate is not uniform and the speed of the rotation depends on the computer and the operating system. In order to produce a smooth animation, it is necessary to control the frame rate. A smooth animation paints (renders) images onto a screen faster than the human eye's response type. Film, for example, displays frames twenty times per second and video rates are even higher. Although high frame rates are possible when image data are prerecorded and when hardware is optimized to process this data, high frame rates may not be possible if both the data and the data's visualization are computed in real time. Because lowering the frame rate to ten frames per second allows more time to compute interesting physics, we usually use this setting. Execute RepaintApp with a sleep time of 100 milliseconds and note the improvement in the animation. It is usually better to have a rate that produces smooth motion at all times than to run as fast as possible for several frames and then slow to a crawl because another thread is playing catch up.

The AbstractAnimation class in the controls package is designed to produce animations with a constant frame rate. This class implements the Animation interface and the Runnable interface so that it can work with both the AnimationControl class and the Thread class. The startAnimation method in the AbstractAnimation class spawns a Thread named animationThread, and this thread is passed a reference to the Runnable instance. The thread invokes the animation's run method which is shown in Listing 7.2. The doStep method is abstract and must be implemented in a concrete subclass.

Listing 7.2   The run method in the AbstractAnimation class determines the frame rate.

```
// AbstractAnimation method
public void run(){
  long sleepTime = delayTime;
  while (animationThread==Thread.currentThread()){
    long currentTime = System.currentTimeMillis();
    doStep();
    //adjusts sleep time for a more uniform animation rate
    sleepTime = Math.max(1,delayTime -
        (System.currentTimeMillis()-currentTime));
    try{
      Thread.sleep(sleepTime);
    }
    catch (InterruptedException ie){}
  }
}
```

The run method in AbstractAnimation reads the system time before and after invoking the doStep method, and the difference in these times is used to set the sleep time. Although the System.currentTimeMillis method only has a time resolution of 10 milliseconds, this coarse resolution is adequate for our simulations.[2]

The DirectAnimationApp class shown in Listing 7.3 extends AbstractAnimation and implements the doStep method. This method sets the spiral's starting angle and invokes the panel's repaint method.

Painting objects directly onto a panel works fine in the case of static graphical user interfaces or if the object being drawn is simple. But this approach is not well suited for animation if the object is complex or if the state of the object changes while it is being drawn. Data may be corrupted because another thread modifies values while the object is being painted. Although it is possible to avoid data corruption by making copies of data and by using synchronization techniques, it is simpler to use a single Thread. This single-thread design pattern is used in the next section to synchronize both the physics computation and the data visualization.

Listing 7.3 DirectAnimationApp demonstrates how to create an animation by extending AbstractAnimation.

```
package org.opensourcephysics.manual.ch07;
import org.opensourcephysics.controls.*;
import org.opensourcephysics.display.*;

public class DirectAnimationApp extends AbstractAnimation {
  DrawingPanel panel = new DrawingPanel();
  DrawingFrame frame = new DrawingFrame("Direct Drawing", panel);
  Spiral spiral = new Spiral();
  double dtheta = 0.1;
  int counter = 0;
  long startTime = 0;

  public DirectAnimationApp() {
    panel.setPreferredMinMax(-5, 5, -5, 5);
    panel.addDrawable(spiral);
    frame.setVisible(true);
  }

  public void startAnimation() {
    delayTime = control.getInt("delay time (ms)");
    super.startAnimation();
    counter = 0;
    startTime = System.currentTimeMillis();
  }

  public void stopAnimation() {
    float rate =
        counter /(float) (System.currentTimeMillis()-startTime);
    super.stopAnimation();
    panel.repaint();
    control.println("frames pr second="+1000*rate);
    control.println("ms per frame="+(1.0/rate));
  }

  protected void doStep() {
```

[2]The resolution can be improved using the System.nanoTime method introduced in Java version 1.5 (JDK 5.0).

```
      spiral.theta += dtheta;
      panel.repaint();
      counter++;
    }

    public void resetAnimation() {
      control.setValue("delay time (ms)", 100);
    }

    public static void main(String[] args) {
      AnimationControl.createApp(new DirectAnimationApp());
    }
  }
```

7.2 ■ BUFFERED PANELS

In order to create a smooth animation at a constant frame rate without onscreen flashing or tearing, it is common to draw onto a hidden image known as a *buffer* and then copy the buffer to the screen all at once. Swing components already have a built-in buffer, and this buffer is used automatically in graphical user interfaces without having to worry about the buffer's details. Swing components are not, however, thread safe, and it is necessary to create our own buffers if we wish to render frames from an animation thread.

Setting a drawing panel's buffered option to true creates two image buffers. We refer to one of these images as the offscreen buffer and second as the working buffer. If a panel is buffered, the panel's `paintComponent` method does not recreate the drawing. Its only job is to copy the offscreen buffer onto the screen. The actual drawing is done by painting drawable objects onto the working buffer using the `render` method. After rendering is complete, the working buffer and the display buffer are switched and the new image is ready to be copied to the screen. Data integrity is insured because the physics computation and the rendering are done sequentially in the animation thread.

Listing 7.4 enables buffering by setting the drawing panel's buffered option in the program's constructor. The buffer's content is drawn by invoking the panel's `render` method in the animation's `doStep` method.

Listing 7.4   A buffered animation.

```
package org.opensourcephysics.manual.ch07;
import org.opensourcephysics.controls.*;
import org.opensourcephysics.display.*;

public class BufferedAnimationApp extends AbstractAnimation {
  DrawingPanel panel = new DrawingPanel();
  DrawingFrame frame = new DrawingFrame("Buffered Drawing", panel);
  Spiral spiral = new Spiral();
  double dtheta = 0.1;
  int counter = 0;
  long startTime = 0;

  public BufferedAnimationApp() {
    panel.setPreferredMinMax(-5, 5, -5, 5);
```

```
      panel.addDrawable(spiral);
      panel.setBuffered(true); // creates offscreen buffer
      frame.setVisible(true);
    }

  public void startAnimation() {
    panel.setIgnoreRepaint(true);
    delayTime = control.getInt("delay time (ms)");
    super.startAnimation();
    counter = 0;
    startTime = System.currentTimeMillis();
  }

  public void stopAnimation() {
    float rate =
        counter/(float) (System.currentTimeMillis()-startTime);
    super.stopAnimation();
    panel.setIgnoreRepaint(false);
    panel.repaint();
    control.println("frames per second="+1000*rate);
    control.println("ms per frame="+(1.0/rate));
  }

  protected void doStep() {
    spiral.theta += dtheta;
    panel.render();
    counter++;
  }

  public void resetAnimation() {
    control.setValue("delay time (ms)", 100);
  }

  public static void main(String[] args) {
    AnimationControl.createApp(new BufferedAnimationApp());
  }
}
```

The `BufferedAnimationApp` class differs from the `DirectAnimationApp` class in another way. Because we know that the drawing will be updated every 1/10 second from within the animation thread, we disable paint messages received from the operating system. This is done within the `startAnimation` method by setting the ignore repaint flag to true.

```
panel.setIgnoreRepaint(true);   // invoked within startAnimation
```

The `startAnimation` method in `BufferedAnimationApp` must, of course, call the superclass `startAnimation` method to start the animation thread. The order of the methods is reversed in the `stopAnimation` method. The superclass `stopAnimation` method is first invoked to stop the thread and then the `ignoreRepaint` flag is set to false.

The drawing panel's `render` method returns the current offscreen buffer, but this image may be invalid if the frame containing the image is hidden or iconified. In this case, the `render` method returns the image that was rendered when the frame was last visible. Because the properties of the internal image buffer depend on the panel's current onscreen state, it is safer to pass an image to the `render` method if an image is needed for further processing.

```
BufferedImage img = new BufferedImage(width, height,
    BufferedImage.TYPE_INT_RGB);
panel.render(img);   // draws into the image
```

Saving images is a convenient way to record the time development of a model or to compare results obtained with different input parameters. See Chapter 13 for examples of how to record a video of an animation.

The following guidelines summarize how to use a drawing panel's offscreen image buffers to produce an animation.

1. Invoke the drawing panel's `setBuffered(boolean buffered)` method to enable buffering.
2. Disable repainting from the operating system when an animation thread is running and enable repainting when the animation is stopped.
3. Invoke the panel's `render` method whenever the panel's data are changed. Calling `render` is the preferred method for animation because the rendering is done in the calling method's thread.

Buffering has one disadvantage. It increases an applications's consumption of display memory. An 800 × 600 window with a color depth of 32 bits per pixel allocates 1.9 MByte of memory. This is usually not a problem on modern operating systems, but you should consider using smaller fixed-size windows if the application requires numerous windows.

7.3 ■ BUFFER STRATEGY

You may notice an occasional ripple on your computer monitor when an animation is not synchronized to the monitor's refresh rate. This ripple is known as tearing and can be eliminated by copying the offscreen buffer to the video memory card while the electronics are being reset between refresh cycles. The `BufferStrategy` class introduced in Java version 1.4 takes care of these details.

Sun's `BufferStrategy` class creates and manages offscreen buffers and copies these buffers to the monitor without tearing. A `BufferStrategy` instance is easy to create as shown in the following code fragment:

```
JFrame frame = new JFrame();
frame.createBufferStrategy(2); // creates two buffers
BufferStrategy strategy = frame.getBufferStrategy();
```

In order to use a `BufferStrategy`, we obtain a graphics context from the strategy, draw the frame's contents, and tell the strategy to show the next buffer.

```
Graphics g = strategy.getDrawGraphics();
// insert code to draw here
strategy.show();
```

Sun claims that the `BufferStrategy` class optimizes the creation of image buffers because it creates these buffers in hardware-accelerated video memory, and we have therefore implemented a `BufferStrategy` drawing method in the `OSPFrame` class. Invoking the `bufferStrategyShow` method in `OSPFrame` draws the frame's contents. The `BufferStrategyApp` class demonstrates this animation technique, but it not shown here because it is similar to our previous examples. The only change that is needed is that the doStep method invokes bufferStrategyShow in the drawing frame, rather than the render method.

```
protected void doStep() {
  spiral.theta += dtheta;
  frame.bufferStrategyShow();
  counter++;
}
```

Although the BufferStrategy class is easy to use and gives smooth animation, we have found that it does not give the fastest frame rate. This may be because the code suspends execution until the image buffer is copied between video refreshes, or it may be due to a poor optimization strategy. The BufferStrategy class also fails using the Java 1.4 VM on dual monitor systems if the frame straddles both monitors or if the animation frame is dragged from one monitor to another. The BufferStrategy class is, however, the animation mechanism recommended by Sun and we have therefore implemented this mechanism in the OSP library. The BufferStrategyApp on the CD shows how this implementation is used.

7.4 ■ SIMULATION ARCHITECTURE

Typical physics simulations have a common architecture that is independent of the system being modeled. A simulation gathers input and starts an animation thread that periodically updates a model's state and renders views of the model's data. The SimulationControl class and AbstractSimulation class are designed to support this architecture with a minimum amount of programming.

Because simulations usually change data, the AbstractSimulation class automatically updates its drawing frames after invoking the doStep method if the frame's animated property is true. The simulation framework also assumes that data accumulated in graphs and other views should be cleared when the program is initialized or reset. The SimulationControl automatically clears data from a frame if the frame's autoclear property is set. Both the animated and autoclear properties are disabled in the OSPFrame class but are enabled for subclasses defined in the frames package. These properties can be set by invoking the following methods:

```
// frame is an instance of OSPFrame
frame.setAnimated(true);
frame.setAutoclear(true);
```

An AbstractAnimation creates an animation thread in its startAnimation method and stops this thread in the stopAnimation method. Subclasses of AbstractAnimation that override these methods should start and stop this thread by calling their superclass implementations. Concrete realizations of AbstractAnimation should also invoke superclass implementations of initializeAnimation and resetAnimation when overriding these methods. The AbstractSimulation class provides convenience methods named start, stop, initialize, and reset that are invoked when the control's buttons are pressed. Because these methods are guaranteed to be empty in the superclass, concrete subclasses need not invoke these overridden methods in the superclass. The start and stop methods are invoked before a thread is started and after a thread is stopped, respectively.

One additional feature that we have found useful is the ability to execute a block of code in conjunction with the doStep method. The startRunning and stopRunning methods are invoked before an animation thread starts and after it stops, respectively. They are also invoked when an animation is single stepped using the SimulationControl Step button but are not invoked during the animation.

Listing 7.5 demonstrates the OSP simulation framework by again implementing a rotating spiral. Note the statements that have been removed by comparing this program to Listing 7.4.

Listing 7.5   A simulation automatically updates its views after invoking the doStep method.

```
package org.opensourcephysics.manual.ch07;
import org.opensourcephysics.controls.*;
import org.opensourcephysics.frames.*;

public class RenderApp extends AbstractSimulation {
  DisplayFrame frame = new DisplayFrame("Direct Drawing");
  Spiral spiral = new Spiral();
  double dtheta = 0.1;
  int counter = 0;
  long startTime = 0;

  public RenderApp() {
    frame.setPreferredMinMax(-5, 5, -5, 5);
    frame.addDrawable(spiral);
    frame.setVisible(true);
  }

  public void startRunning() {
    delayTime = control.getInt("delay time (ms)");
    counter = 0;
    startTime = System.currentTimeMillis();
  }

  public void stopRunning() {
    float rate =
        counter /(float) (System.currentTimeMillis()-startTime);
    control.println("frames per second="+1000*rate);
    control.println("ms per frame="+(1.0/rate));
  }

  public void reset() {
    control.setAdjustableValue("delay time (ms)", 100);
  }

  protected void doStep() {
    spiral.theta += dtheta;
    counter++;
  }

  public static void main(String[] args) {
    SimulationControl.createApp(new RenderApp());
  }
}
```

7.5 ■ DRAWABLE BUFFER

It is inefficient to redraw a panel at every time step if the appearance of most objects within the drawing do not change during an animation. A better approach is to draw a static background image, copy this image into the drawing panel, and then draw only those objects that change during the animation. A DrawableBuffer is a drawable object that creates and manages a background image that is the same size as the drawing panel that contains the buffer.

A DrawableBuffer acts in many respects like a buffered drawing panel. Drawable objects are added to the buffer and the buffer stores references to these objects for later rendering. Because the visual representation of the DrawableBuffer is an image, it is the programmer's responsibility to redraw the image if the objects change their state. This can be done by invoking the buffer's updateImage(DrawingPanel drawingPanel) method. You can also set the buffer's invalid image flag and a new image will be rendered when the panel containing the DrawableBuffer is repainted. This is done using the buffer's invalidateImage method.

Because drawing panels assume that drawable buffers contain only fixed objects, there is a limitation when they are used with interactive objects. A drawing panel can only locate the buffer and not the objects in the buffer. One way around this limitation is to add an invisible proxy component to the interactive drawing panel. This proxy intercepts mouse events and forwards them to objects within the buffer.

A second limitation is that a drawable buffer should be added to only one drawing panel because the offscreen image uses the panel's dimension to set the image size. (A drawable buffer can be added to more than one drawing panel if these panels have the same dimension. In this case, the same offscreen image will be copied to multiple panels.)

It is usually not necessary to use both a DrawableBuffer and a buffered drawing panel. (This would be "triple-buffering.") A DrawableBuffer is most effective when there are a small number of animated objects that can be drawn directly on a static background. See Section 8.5 for an example of how a DrawableBuffer is used to draw a particle on a background contour plot.

7.6 ■ DATA RASTER

Drawable components can implement their own buffering. The DataRaster component, for example, is designed to display tens of thousands of data points by coloring one image pixel for every point. Because it would be inefficient to redraw every point when a new point is added, a DataRaster creates an image of the drawing panel and colors only the pixel where the datum is located. When the drawing panel is repainted, this image is copied to the screen. If the size or the scale of the drawing panel is changed, then the image is recreated from the raw data.

Listing 7.6 uses a data raster to display the logistic equation return map. The logistic equation was first proposed by Robert May as a simple model of population dynamics. This equation can be written as a one-dimensional difference equation that transforms the population in one generation x into a succeeding generation x_{n+1}. Because the population is scaled so that the maximum value is 1, the domain of x falls on the interval [0, 1]:

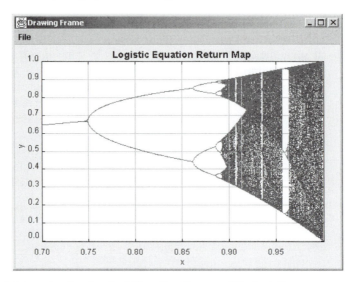

Figure 7.1   A data raster showing the bifurcation diagram of the logistic map.

$$x_{n+1} = 4rx_n(1 - x_n) \quad \text{(logistic equation)}. \tag{7.1}$$

The behavior of the logistic equation depends on the value of the growth parameter r. If the growth parameter is less than a critical value $r \approx 0.75$, then x approaches a stable fixed value. Above this value for r, the behavior of x begins to change as shown in Figure 7.1. At first the population begins to oscillate between two values. If r increases further, then x oscillates between four values, then eight values. This doubling ends when $r > 0.8924864\ldots$ after which almost any x-value is possible.

The bifurcation diagram in Figure 7.1 shows the behavior of the logistic equation by plotting successive values of x after the initial transient behavior is discarded. Listing 7.6 calculates and discards 400 iterations. It then plots 200 iterations of x in red followed by another 200 iterations in green. This process is repeated 6000 times as the growth parameter is incremented, thereby producing a data raster that contains 1.2 million data points.

Listing 7.6   The logistic equation demonstrates how very large datasets
can be plotted using a data raster.

```
package org.opensourcephysics.manual.ch07;
import org.opensourcephysics.display.*;
import org.opensourcephysics.display2d.DataRaster;
import java.awt.Color;
import javax.swing.JFrame;

public class LogisticApp {
  public static void main(String[] args) {
    PlottingPanel panel =
        new PlottingPanel("x", "y", "Logistic Equation Return Map");
    DrawingFrame frame = new DrawingFrame(panel);
    DataRaster dataRaster = new DataRaster(panel, 0.7, 1, 0, 1);
    panel.addDrawable(dataRaster);
    int nplot = 400;
```

```
   // first dataset is dark red
   dataRaster.setColor(0, new Color(128, 0, 0, 128));
   // second dataset is dark green
   dataRaster.setColor(1, new Color(0, 128, 0, 128));
   for(double r = 0.7;r<1.0;r += 0.0005) {
     double x = 0.5;                       // starting value for x
     for(int i = 1;i<=nplot;i++) { // nplot values not plotted
       x = 4*r*x*(1-x);
     }
     for(int i = 1;i<=nplot/2;i++) {
       x = 4*r*x*(1-x);
       // show x-values for current value of r
       dataRaster.append(0, r, x);
     }
     for(int i = 1;i<=nplot/2;i++) {
       x = 4*r*x*(1-x);
       dataRaster.append(1, r, x); // note different data
     }
   }
   frame.setDefaultCloseOperation(JFrame.EXIT_ON_CLOSE);
   frame.setVisible(true);
 }
}
```

7.7 ■ PROGRAMS

The following examples are in the `org.opensourcephysics.manual.ch07` package.

BufferedAnimationApp

`BufferedAnimationApp` tests the frame rate using drawing panel buffering as described in Section 7.2.

BufferStrategyApp

`BufferStrategyApp` tests the frame rate using a Java `BufferStrategy` for drawing. See Section 7.3.

DataRasterApp

`DataRasterApp` tests the `DataRaster` class by creating 100,000 random data points. Half the points are shown in red and half in green. See Section 7.6.

DirectAnimationApp

`DirectAnimationApp` tests the frame rate and the animation quality by invoking the `repaint` method. The drawing is rendered from within the event dispatch thread when the component is painted as described in Section 7.1.

LogisticApp

`LogisticApp` tests the `DataRaster` class by creating a return map for the logistic function. This program is described in Section 7.6.

RenderApp

RenderApp tests the frame rate using the render method. An `AbstractSimulation` renders the drawing within the animation thread after invoking the `doStep` method as described in Section 7.4.

RepaintApp

`RepaintApp` demonstrates erratic drawing using repaint requests as described in Section 7.1.

CHAPTER

8

Visualization of Two-Dimensional Fields

Visualization tools for two-dimensional scalar fields, vector fields, and complex fields are defined in the *2D display* package (org.opensourcephysics.display2d). Sample programs are located in the manual's ch08 package unless otherwise stated.

8.1 ■ OVERVIEW

Imagine a sheet of metal that is heated at an interior point and cooled along its edges. In principle, the temperature of the sheet can be measured at any point, $T(x, y)$. Calculating this temperature is a typical computational physics problem. A quantity, such as temperature, pressure, or light intensity, that is defined by a single number (scalar) at every point in space is known as a *scalar field*. Other types of fields are possible. Consider the force on a small test charge in the vicinity of other charges, $\mathbf{F}(x, y)$. Because the force is a vector, this situation defines a *vector field*. We now describe a framework to display two-dimensional scalar and vector fields.

One way to store and display information in two-dimensional simulations is to divide space into a discrete grid and to store data at every grid point. A scalar field sampled at every grid point can be stored in a 2D array. This array is passed to an object in the 2D display package to produce the visualization.

Figure 8.1 shows the two-dimensional diffraction pattern from a rectangular aperture. This pattern has an analytic solution:

$$f(x, y) = I_0 \left(\frac{\sin x}{x} \right)^2 \left(\frac{\sin y}{y} \right)^2 , \qquad (8.1)$$

where we have chosen dimensionless units for x and y. Listing 8.1 plots this solution using a Scalar2DFrame. All that is required is that an array of values and a world coordinate scale be passed to the Scalar2DFrame using the setAll method.

Listing 8.1 DiffractionApp plots the analytic solution to the diffraction pattern from a 2D rectangular aperture.

```
package org.opensourcephysics.manual.ch08;
import org.opensourcephysics.frames.*;

public class DiffractionApp {
  public static void main(String[] args) {
    Scalar2DFrame frame = new Scalar2DFrame("x", "y", "Raster Frame");
    int nx = 512, ny = 512;
    double[][] data = new double[nx][ny];
```

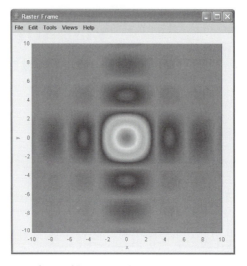

Figure 8.1 Diffraction pattern from a 2D rectangular aperture illuminated by a monochromatic light source. The color changes from blue to red as the energy density increases.

```
frame.setAll(data, -10, 10, -10, 10); // sets data and scale
for(int i = 0;i<nx;i++) {
  double x = frame.indexToX(i);    // gets the x-value for the index
  double xAmp = (x==0) ? 1 : Math.sin(x)/x;
  for(int j = 0;j<ny;j++) {
    double y = frame.indexToY(j); // gets the y-value for the index
    double yAmp = (y==0) ? 1 : Math.sin(y)/y;
    double amp = xAmp*yAmp;
    // sqrt of intensity gives better visibility
    data[i][j] = Math.sqrt(amp*amp);
  }
}
frame.setAll(data); // sends new data to the visualization
frame.setVisible(true);
frame.setDefaultCloseOperation(javax.swing.JFrame.EXIT_ON_CLOSE);
  }
}
```

A scalar field has a scale given by the locations of its corners in world units. This scale is set when data are passed to the frame. If the scale is not specified, the previously set scale is used.

```
frame.setAll(data,-10, 10, -10, 10);  // sets data and scale
frame.setAll(data);                    // sets only data
```

Because it is convenient to be able to convert between array indices and the corresponding world coordinates, OSP scalar and vector field visualizations define array index to world coordinate conversion methods. For example, the following code fragment computes the x-value of the tenth column and the array index closest to $x = 0.5$.

```
double x = frame.indexToX(10); // world coordinates for index 10
int i = frame.xToIndex(0.5);   // index closest to x=0.5
// similar methods for y
```

Table 8.1   Two-dimensional visualization components in the
2D display package.

---

org.opensourcephyysics.display2d

BinaryLattice	An array of ones and zeros shown as a two-color grid.
CellLattice	An array of 256 possible values shown as multicolor grid rectangles.
ByteRaster	An array of 256 possible values shown using one pixel per value.
GridPlot	An array of numbers shown as multicolored rectangles.
ComplexGridPlot	An array of complex numbers shown as a multicolor grid. Color and intensity are used to represent phase and magnitude, respectively.
ComplexInterpolatedPlot	A variant of a complex grid plot whose complex numbers have been interpolated at every pixel. This interpolation smooths the image.
ComplexSurfacePlot	A 3D surface whose height is proportional to the magnitude of a complex number and whose color represents phase.
ContourPlot	A contour plot.
GrayscalePlot	An array of numbers shown as a gray-scale grid.
InterpolatedPlot	A variant of a grid plot whose numbers have been interpolated at every pixel. This interpolation smooths the image.
SiteLattice	An array of 256 possible values shown as circles at the intersections of grid lines.
SurfacePlot	A 3D surface whose height is proportional to a scalar.
VectorPlot	A vector field plot.

Because these conversions use the current scale, it is important that the scale be set before these methods are invoked. This is why the setAll method is invoked using the uninitialized array with minimum and maximum values before indexToX and indexToY methods are invoked. The setAll method is invoked a second time without a scale after the diffraction pattern has been computed.

The Scalar2DFrame class plots arrays of floating point numbers. The frame's Views menu allows us to display this data as a grid plot, a contour plot, an interpolated plot, or a surface plot. The 2D display package (org.opensourcephysics.display2d) provides visualizations for other types of fields as shown in Table 8.1. For example, a 2D vector field or a complex 2D scalar field require at least two numbers (components).

Table 8.2 shows the rendering time in milliseconds for some OSP 2D visualizations. The times shown were obtained using the Sun Microsystems Java 1.4 VM running on Windows 2000 and Windows XP using a 400 MHz Pentium II and a dual 2.6 GHz Pentium 4 processor, respectively. These times are for rendering only. The overhead needed to generate the data is not included.

Using the timing results in Table 8.2 as a guide, we see that it may not be possible to produce a ten frame per second animation on a low-end machine using some visualizations

Table 8.2 Rendering times in milliseconds (± 10 ms) for two-dimensional visualization components. The first number is the time recorded on a 400 MHz Pentium II. The second number is the time recorded on a 2.6 GHz dual-processor Pentium 4. The drawing panel size is 300×300 pixels.

Component	16 x 16	32 x 32	64 x 64
BinaryLattice	20|10	30|10	40|10
CellLattice	20|0	30|0	30|0
ByteRaster	10|0	10|0	10|0
GridPlot	20|0	20|10	30|10
ComplexGridPlot	20|0	20|10	30|10
ComplexInterpolatedPlot	120|50	120|50	120|50
ComplexSurfacePlot	100|30	200|70	600|200
ContourPlot	140|70	480|200	2000|600
InterpolatedPlot	120|50	120|50	120|50
SurfacePlot	100|30	200|70	600|200
VectorPlot	50|15	110|60	410|200

if the grid density is high or if multiple views are being drawn. Notice that the rendering times for lattice and raster components are insensitive to grid size because data are copied quickly and directly from an array into an image.

There are several ways to improve performance after a program is running correctly. The easiest technique is to void unnecessary screen repainting by performing several computation steps before the panel's `redraw` or `render` methods are invoked. Methods, such as `indexToX`, that are invoked frequently but perform straightforward and routine computations can be optimized for the given application and the optimized code can be in-lined to remove the method call. Java 2D components such as the `WritableRaster` can be used directly. If a significant portion of a visualization is static, you should consider creating a background image as described in Section 8.5. Special components , such as `DataRaster` described in Section 7.6, have been programmed so that very little computation is required when data are added. And finally, a program may render a low-resolution visualization when the program is running and high-resolution visualization when the animation stops.

8.2 ■ IMAGES AND RASTERS

An image in which pixels are color coded can be used to generate a visually appealing image of a scalar field. The computation is carried out using an $n \times m$ array in which every array element corresponds to an image pixel. The Java `java.awt.image` package contains a `WritableRaster` class that enables a program to access image pixels.[1] In fact, a Java raster is used in the `DataRaster` class described in Section 7.6. A number of the visualizations in the 2D display package convert their data into an offscreen image that is later copied to an output device. `BinaryLattice`, `GridPlot`, and `GrayscalePlot` use this technique. After the image pixels are colored, the image is copied to a computer display or

---

[1] The term raster was originally applied to the pattern of horizontal lines that appeared on a television or computer screen.

printer. This rendering time for images is fast because image-based operations are usually performed on a video card and not by the main CPU.

The ByteRaster class in the 2D display package simplifies the process of converting an array of bytes onto an image. A byte variable is a signed integer that takes on 256 values ranging from Byte.MIN_VALUE=-128 to Byte.MAX_VALUE=127 so that it can be stored in an eight-bit memory location.[2] Incrementing the maximum value wraps around to produce the minimum value. Because casting an int to a byte folds the value into the range $[-128, 127]$, we sometimes perform a computation using integers and cast the results to bytes to produce a color palette that repeats every 256 values.

Listing 8.2 shows the ByteRaster class being used to display the ever popular Mandelbrot set. The Mandelbrot set consists of points c in the complex plane that obey the following rule:

1. Start with the complex number $z = 0 + i0$.
2. Generate a new complex number z' by multiplying z by itself and adding the result to c:

$$z' = z^2 + c. \tag{8.2}$$

3. Repeat steps (1) and (2). If the complex number z goes toward infinity, then c is not a member of the Mandelbrot set. All numbers that remain bounded are members of the set.

It can be shown that if the magnitude of z is greater than 2, then z will approach infinity. The MandelBrotApp code assumes that the number c is in the Mandelbrot set if $|z| < 2$ after 256 iterations. In order to show how rapidly a number fails the test, we color the pixel corresponding to the number of iterations. Note how the byte raster is assigned a scale using the setAll method. This method passes the data and sets the dimensions of the byte raster in world units. The plotting panel adjusts its gutters and scale to insure that its minimum and maximum match the byte raster's dimension as shown in Figure 8.2.

The RasterFrame class in the frames package is a convenient wrapper for an image. The DiffractionRasterApp program shows this frame being used to display the diffraction of light from a two-dimensional aperture. Because computer displays have a limited intensity range and because the human eye does not respond linearly to intensity, the program plots the square root of the light intensity rather than the intensity.

Listing 8.2 The Mandelbrot program.

```
package org.opensourcephysics.manual.ch08;
import org.opensourcephysics.display.*;
import org.opensourcephysics.display2d.ByteRaster;

public class MandelbrotApp implements Runnable {
  static final int SIZE = 300; // the image size in pixels
  PlottingPanel panel =
      new PlottingPanel("re", "im", "Mandelbrot Set");
  DrawingFrame frame = new DrawingFrame(panel);
  ByteRaster byteRaster = new ByteRaster(SIZE, SIZE);
  byte[][] data = new byte[SIZE][SIZE];
```

[2]A byte is always a signed number in Java. Java does not support an unsigned 8-bit data type.

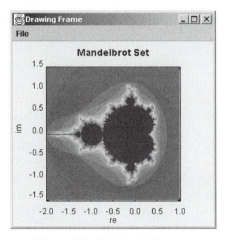

Figure 8.2   The Mandelbrot set.

```
double reMin = -2, reMax = 1, imMin = -1.5, imMax = 1.5;

MandelbrotApp() {
  frame.setDefaultCloseOperation(javax.swing.JFrame.EXIT_ON_CLOSE);
  panel.addDrawable(byteRaster);
  frame.setVisible(true);
  new Thread(this).start(); // creates and starts the thread
}

int iteration(double re, double im) {
  // algorithm: take z, square it, add the number (re,im)
  int count = 0;
  double a = 0, b = 0, temp = 0;
  while(count<255) {
    temp = a;
    a = a*a-b*b+re;
    b = 2*temp*b+im;
    if(a*a+b*b>4.0) {
      break;
    }
    count++;
  }
  return(10*count); // expand the scale by ten
}

public void run() {
  for(int iy = 0;iy<SIZE;iy++) {
    double im = imMin+(imMax-imMin)*iy/(double) SIZE;
    for(int ix = 0;ix<SIZE;ix++) {
      double re = reMin+(reMax-reMin)*ix/(double) SIZE;
      data[ix][iy] = (byte) (-128+iteration(re, im));
    }
    try { // yield after every row to give other threads a chance
      Thread.yield();
    } catch(Exception e) {}
  }
```

```
      byteRaster.setAll(data, reMin, reMax, imMin, imMax);
      panel.repaint();
   }

   public static void main(String[] args) {
      new MandelbrotApp();
   }
}
```

8.3 ■ LATTICES

Arrays of ordinal data such as integers or bytes occur frequently in computer simulations, such as the Ising model, random walks, percolation, and cellular autonoma such as the *Game of Life*. Open Source Physics lattice components, such as `BinaryLattice`, `CellLattice`, and `SiteLattice`, are designed to display this type of data. The `LatticeFrame` class in the frames package is a composite object that can be switched between site and cell lattice visualizations using the View menu. `LatticeFrameApp` is available on the CD and demonstrates how this frame is used.

Unlike the raster class introduced in Section 8.2, the image generated by a lattice can be resized. In other words, a 32×32 lattice that is added to a 320×320 drawing panel draws every element using a 32×32 square of pixels. Lattice and raster APIs are otherwise very similar as shown in Table 8.3. The program shown in Listing 8.3 displays the 32×32 lattice shown in Figure 8.3.

Listing 8.3   A binary lattice of random ones and zeros.

```
package org.opensourcephysics.manual.ch08;
import org.opensourcephysics.display.*;
import org.opensourcephysics.display2d.CellLattice;

public class CellLatticeApp {
   static final int SIZE = 32;

   public static void main(String[] args) {
      PlottingPanel panel = new PlottingPanel("x", "y", "Byte Lattice");
      DrawingFrame frame = new DrawingFrame(panel);
      CellLattice lattice = new CellLattice(SIZE, SIZE);
      byte[][] data = new byte[SIZE][SIZE];
      for(int iy = 0;iy<SIZE;iy++) {
         for(int ix = 0;ix<SIZE;ix++) {
            data[ix][iy] = (byte) (256*Math.random());
         }
      }
      lattice.setAll(data, -1, 1, 1, -1);
      // sets a block of cells to new values
      lattice.setBlock(4, 8, new byte[12][9]);
      panel.addDrawable(lattice);
      frame.setDefaultCloseOperation(javax.swing.JFrame.EXIT_ON_CLOSE);
      frame.setVisible(true);
      lattice.showLegend();
   }
}
```

Table 8.3   The `ByteLattice` interface defines methods common to the raster and lattice classes.

org.opensourcephysics.display2d.ByteLattice	
getValue	Gets the value of a cell or a site.
setBlock	Sets a block of cells or sites.
setValue	Sets a cell or site value.
setCol	Sets values within a single column.
setColorPalette	Sets the color palette using an array of colors.
setGridLineColor	Sets the grid color if the show grid lines option has been enabled.
setIndexedColor	Sets the color in a palette associated with a single value.
setMinMax	Sets world coordinates for the upper left and lower right corners.
setRow	Sets values within a single row.
setShowGridLines	Shows a grid if the cell size is larger than two pixels.
showLegend	Shows a legend of colors and values.

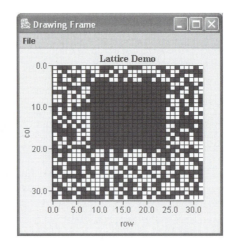

Figure 8.3   The 32 × 32 binary lattice visualization generated by Listing 8.3.

We use *byte* arrays because they do not require much memory and provide a convenient range of 256 colors. The `ByteLattice` interface in the 2D display package defines a common interface that is implemented in a number of classes: `CellLattice`, `SiteLattice`, and `ByteRaster`. Classes that display floating point data do not implement this interface but implement similar methods (see Section 8.4).

The default byte lattice color palette shades values from blue toward red as the data value increases from −128 to 127. Zero is green. In order to customize the default palette, lattice classes implement the `setIndexedColor` method to set the color associated with a single value. For example, we can set the color associated with zero to black as follows:

```
lattice.setIndexedColor(0,Color.black);
```

The entire color palette can be set by passing an array of colors to the setColorPalette and the setGridLineColor method.

```
lattice.setColorPalette(new Color[]
     {Color.blue,Color.green,Color.red,Color.white});
lattice.setGridLineColor(Color.black);
```

Values with undefined palette colors are rendered as black and extra palette colors are ignored. A BinaryLattice stores two palette values. Zero corresponds to the first palette color and one corresponds to the second palette color.

Lattice and raster classes use the first index as the x-index and the second index as the y-index. The [0][0] data value is drawn at the lattice x-minimum and y-minimum location in a drawing panel. Because the default x and y scales increase to the right and up, respectively, the [0][0] data value usually corresponds to the lower left-hand corner of the visualization. The position of the [0][0] element can be changed by setting the lattice minimum and maximum values using either the setAll or the setMinMax methods . Because lattice classes implement the measurable interface, setting $y_{min} > y_{max}$ not only sets the lattice minimum and maximum values but also flips the image if the axis is autoscaled.

```
method!sample code}
lattice.setAll(data,-1,1,1,-1);   // sets data and scale
lattice.setMinMax(-1,1,1,-1);     // sets scale
```

The BinaryLattice class is implemented differently from a ByteLattice. This class minimizes storage by creating a buffer using a Java *packed raster*. Because every *bit* of this raster corresponds to a single cell, every *byte* stores eight cells. Although the code uses shift operators to change cell values, users need not decipher these details because lattice cells are accessed using high-level abstractions for setting and getting values.

```
int ix=3, iy=4;
int val=lattice.getValue(ix, iy);          // gets a cell value
lattice.setValue(ix, iy,1);                // sets a cell value to one
lattice.setBlock(ix, iy, new int[8][4]); // zeros a block of cells
```

BinaryLattice and ByteLattice classes can superimpose grid lines to highlight cell boundaries. However, grid lines are not drawn if the cell size is less than four pixels and rasters cannot draw grid lines because their cell size is always one pixel. The visibility and color of grid lines are set using the setShowGrid and setGridColor methods.

```
lattice.setShowGridLines(true);
lattice.setGridLineColor(Color.black);
```

Test programs for the lattice classes are BinaryLatticeApp, CellLatticeApp, and SiteLatticeApp. These programs are not listed but are available on the CD.

8.4 ■ PLOT2D INTERFACE

Various classes in the 2D display package accept arrays of floating-point numbers to create visualizations such as grid plots, contour plots, and 3D surface plots. These classes implement the Plot2D interface, as diagrammed in Figure 8.4, and therefore have a similar API. To provide an overview of the API, we now show a simple example of a concrete implementation. The GridPlot class defines a visualization that shows the structure of a scalar field by

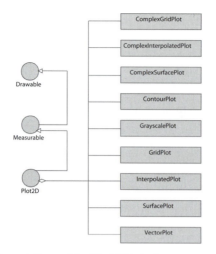

Figure 8.4   Concrete implementations of the `Plot2D` interface.

drawing multicolored rectangles with color representing the value of the field. Listing 8.4 uses this component to plot a gaussian scalar field:

$$U(x, y) = e^{-(x^2+y^2)} . \tag{8.3}$$

Listing 8.4   A scalar field plotting program.

```
package org.opensourcephysics.manual.ch08;
import org.opensourcephysics.display.*;
import org.opensourcephysics.display2d.*;

public class GridPlotApp {
  static final int SIZE = 32;

  public static void main(String[] args) {
    DrawingPanel plottingPanel =
        new PlottingPanel("x", "y", "Grid Plot");
    DrawingFrame frame = new DrawingFrame(plottingPanel);
    double[][] data = new double[SIZE][SIZE];
    GridPlot plot = new GridPlot();
    plot.setAll(data, -1.5, 1.5, -1.5, 1.5); // sets data and scale
    for(int i = 0;i<SIZE;i++) {           // calculate field
      double x = plot.indexToX(i);        // x-coordinate at this index
      for(int j = 0;j<SIZE;j++) {
        double y = plot.indexToY(j);      // y-coordinate at this index
        data[i][j] = Math.exp(-(x*x+y*y)); // field value
      }
    }
    plot.setAll(data); // sets the data
    plottingPanel.addDrawable(plot);
    frame.setDefaultCloseOperation(javax.swing.JFrame.EXIT_ON_CLOSE);
    frame.setVisible(true);
  }
}
```

Table 8.4   The `Plot2D` interface defines methods common to 2D visualizations that use the `GridData` storage model.

---

org.opensourcephysics.display2d.Plot2D

`indexToX`	Gets the x world coordinate for the given index. Similar method for y.
`xToIndex`	Gets closest index from the given x world coordinate. Similar method for y.
`setAll`	Sets the data and scales the data model. The grid is resized as needed.
`setAutoscaleZ`	Sets autoscaling parameters for the amplitude.
`setColorPalette`	Sets the color palette using an array of colors.
`setFloorCeilColor`	Sets the floor and ceiling colors.
`setGridData`	Sets the data storage object.
`setGridLineColor`	Sets the gridline or contour line color.
`setIndexes`	Specifies the ordering of data, such as amplitude and phase, when using multi-component data.
`setPaletteType`	Sets a predefined color palette.
`setShowGridLines`	Draws the gridlines or contour lines when true.
`setVisible`	Sets the visibility of the plot.
`showLegend`	Shows a legend of colors and values.
`update`	Updates the object's state. The `update` method should be invoked after the grid's data or scale is changed.

Because OSP visualizations implement the `drawable` interface, the example begins by instantiating a drawing panel and adding a plot to the panel. The data's world coordinates are stored when the `setAll` method is invoked with a data array followed by minima and maxima. This allows us to later invoke the `indexToX` and `indexToY` methods.

Visualizations that implement the `plot2d` interface are designed to display arrays of floating-point numbers using the methods shown in Table 8.4. Each floating-point array is stored in an object referred to as a *data model* that contains scale and other information. This data model can be created in a number of ways. (Section 8.8 describes data models.) If a data model has already been instantiated, it can be passed to the visualization in the constructor. If a visualization is instantiated without a data model, then the tool will automatically create an appropriate model when the `setAll` method is invoked. This automatic creation of the data model requires that the program allocate the data array. The data array can be used over and over in the computation because the visualization copies the array's values into the data model. Figure 8.5 shows screen shots of four scalar field visualizations that implement the `Plot2D` interface.

Many visualizations contain a *color mapper* to convert data to colors. A gray-scale color mapper, for example, interprets a number as a gray level, while a dual-shade color mapper interprets data by shading colors from blue to red. The smallest and largest data values that can be translated to color are the mapper's *floor* and *ceiling* values. This range is referred to as the *z-range*. Values below the floor or above the ceiling are mapped to special colors. Setting the autoscale property to true will adjust the z-range to match the data. If this property is false, the user must supply a valid range. These display options are set using

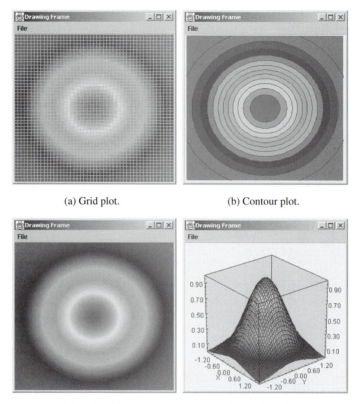

(a) Grid plot.   (b) Contour plot.

(c) Interpolated plot.   (d) Surface plot.

Figure 8.5   Visualizations of the scalar field $U(x, y) = e^{-(x^2+y^2)}$.

the `setAutoscaleZ` and `setFloorCeilColor` methods as follows:

```
double floor=-1, ceil=1;
Color floorColor=Color.black, ceilColor=Color.white;
// plot is a visualization that implements Plot2D
plot.setPaletteType (ColorMapper.DUALSHADE);
plot.setAutoscaleZ(false, floor, ceil);
plot.setFloorCeilColor(floorColor, ceilColor);
plot.showLegend();
```

8.5 ■ SCALAR FIELDS

Grid Plot

The `GridPlot` shown in Figure 8.5(a) draws a rectangle centered at the measurement point using a color that is determined by the value of the field. Listing 8.4 shows how a grid plot is created and used. This plot is similar to a lattice except that it uses floating-point numbers rather than bytes to store data. When the `setAll` method is invoked, the visualization copies the array into a data model that contains additional information such as a scale.

The grid plot's color mapper supports various color palettes. The DUALSHADE palette, for example, shades pixels from blue toward red as the values change from floor to ceiling. A GRAYSCALE palette shades pixels from black to white.

```
plot.setPaletteType(ColorMapper.DUALSHADE);
```

Consult the ColorMapper class for other color palettes.

Custom palettes are created by passing an array of colors to the setColorPalette method. For example, the following code fragment creates a three-color palette that divides the floor to ceiling range into three equal intervals. Note that five colors may appear onscreen because data values below the floor or above the ceiling are colored using special floor and ceiling colors.

```
plot.setColorPalette(new color[]={Color.RED,Color.GREEN,Color.BLUE});
```

It is easy to use a grid plot in place of a lattice. However, because a grid plot is designed to show a field of floating-point numbers, it uses more memory than a lattice and it is difficult to check for equality of values due to roundoff. In general, a lattice is preferred when values are countable (ordinal).

An advantage of GridPlot is that it is very fast as shown in Table 8.2. We sometimes use it as an alternative to more complicated visualizations, such as contour plots, during an animation. Because a grid plot displays the structure of the underlying data, it is also useful for debugging. The grid plot clearly shows if an algorithm fails at a point because values are not interpolated or smoothed.

Contour Plot

The ContourPlot class draws a contour plot of a two-dimensional scalar field.[3] In fact, the only change required to Listing 8.4 in order to produce Figure 8.5(b) is to change the type of component that is being constructed.

```
plot = new ContourPlot();
```

It is, of course, possible to customize a plot and these customizations will depend on the type of plot. The default number of contour lines is twelve, but this number can be changed. Note that because the shape of a contour line is sensitive to a contour's z-level, it is often desirable not to autoscale the z-axis. This is especially true for scalar fields that contain singularities, such as charged particles in electrostatics, because even a small change in the location of a grid point can produce large changes in the scalar field's values at the grid points. The following code fragment specifies five fixed contour levels at $-2, -1, 0, 1$, and 2.

```
// plot must be of type ContourPlot to set number of levels
plot = setNumberOfLevels(5);
plot.setAutoscaleZ(false, -2, 2);   //disable autoscaling
```

Complex visualizations, such as contour plots, are computationally intensive and may not be well suited for real-time animations. We can, however, use a DrawableBuffer introduced in Section 7.5. The buffer contains the contour plot that provides the back-

[3]The ContourPlot and SurfacePlot components are based on the Surface Plotter program by Yanto Suryono.

ground image for the panel. The entire panel is repainted when the interactive circle is being dragged, but this is fast because repainting merely copies the background image and paints a circle. If the contour data were to change, we require that the program invoke the buffer's `invalidateImage` method to generate a new background. `BufferedContourApp` demonstrates this buffering technique and is available on the CD.

Interpolated Plot

An interpolated plot is similar to a grid plot except that color values are interpolated at every pixel. This interpolation blurs the grid's cell boundaries. The only change required to Listing 8.4 to produce the interpolated plot shown in Figure 8.5(c) is to change the component being constructed.

```
plot = new InterpolatedPlot();
```

Because interpolation from the data grid is performed at every pixel, the time required to render an interpolated plot is insensitive to the number of grid points. Rendering time is, however, roughly proportional to the number of pixels in the drawing panel because an interpolation must be performed at every pixel.

Surface Plot

A `SurfacePlot` shows a scalar field using a three-dimensional color-coded mesh. The height (z-value) is proportional to the field magnitude. Again, the only change required to Listing 8.4 to produce the surface plot shown in Figure 8.5(d) is to change the component being constructed. Viewing angles and other display options can be set after the surface plot is instantiated.

```
surfacePlot = new SurfacePlot();
surfacePlot.setRotationAngle(125);   // default is 125 degrees
surfacePlot.setElevationAngle(10);   // default is 10 degrees
surfacePlot.setDistance(200.0);      // default is 200 units
surfacePlot.set2DScaling(8.0);       // default is 8 pixels per unit
```

Because it is desirable to be able to adjust a surface plot's viewing angle using a mouse, we have defined a `SurfacePlotMouseController` that implements the Java `MouseListener` and `MouseMotionListener` interfaces. This class handles mouse events that rotate, zoom, and translate the surface plot using drag, shift-drag, and control-drag mouse actions, respectively. Keyboard events are also supported. A complete surface plot test program is shown in Listing 8.5. Note that the `SurfacePlotMouseController` is added to the panel using standard Java `addMouseListener` and `addMouseMotionListener` methods.

Listing 8.5   Surface plot test program.

```
package org.opensourcephysics.manual.ch08;
import org.opensourcephysics.display.*;
import org.opensourcephysics.display2d.*;

public class SurfacePlotApp {
  static final int SIZE = 32;

  public static void main(String[] args) {
    DrawingPanel drawingPanel = new DrawingPanel();
    DrawingFrame frame = new DrawingFrame(drawingPanel);
    double[][] data = new double[SIZE][SIZE];
```

```
SurfacePlot plot = new SurfacePlot();
plot.setAll(data, -1.5, 1.5, -1.5, 1.5); // sets the data and scale
for(int i = 0;i<SIZE;i++) {                // calculate field
  double x = plot.indexToX(i);
  for(int j = 0;j<SIZE;j++) {
    double y = plot.indexToY(j);
    data[i][j] = Math.exp(-(x*x+y*y));    // magnitude
  }
}
plot.setAll(data); // sets the data
drawingPanel.addDrawable(plot);
SurfacePlotMouseController mouseController =
    new SurfacePlotMouseController(drawingPanel, plot);
drawingPanel.addMouseListener(mouseController);
drawingPanel.addMouseMotionListener(mouseController);
frame.setDefaultCloseOperation(javax.swing.JFrame.EXIT_ON_CLOSE);
frame.setVisible(true);
  }
}
```

8.6 ■ VECTOR FIELDS

The VectorPlot class shows a vector field. To use this class we instantiate a multidimensional array to store components of the vector. The first array index indicates the component, the second index indicates the column or x position, and the third index indicates the row or y position. The vectors in the visualization are set by passing the array to the frame using the setAll method.

Because a Java multidimensional array is an array of arrays, it is easy to reference subarrays which contain data for a single component.

```
double[][][] data = new double[2][32][32];
double[][]   xdata = data[0];
double[][]   ydata = data[1];
```

Note that the xdata and ydata subarrays are not allocated. These identifiers are merely references (pointers) to memory that is allocated in the first statement.

The VectorPlotApp program displays a circulating vector field whose magnitude increases in direct proportion to the distance from the origin:

$$\mathbf{A} = -r \sin\theta\hat{\mathbf{x}} + r\cos\theta\hat{\mathbf{y}} = -y\hat{\mathbf{x}} + x\hat{\mathbf{y}}. \tag{8.4}$$

Because we have found that using color rather than length to represent field strength produces a more effective representation of magnitude over a wider range of values, the VectorPlot class uses color-coded arrows. Figure 8.6 shows a black and white reproduction of the plot produced by Listing 8.6. Arrows in this plot have a fixed length that is chosen to fill the plot's viewing area. Section 8.8 describes how to program a data model to obtain the traditional association between arrow length and field magnitude.

Listing 8.6 Vector field test program.

```
package org.opensourcephysics.manual.ch08;
import org.opensourcephysics.display.*;
import org.opensourcephysics.display2d.*;
```

Figure 8.6   A vector plot with arrows of equal length. Running the `VectorPlotApp` program shows how color is used to show magnitude.

```
public class VectorPlotApp {
  static final int SIZE = 24;

  public static void main(String[] args) {
    DrawingPanel drawingPanel = new DrawingPanel();
    DrawingFrame frame = new DrawingFrame(drawingPanel);
    drawingPanel.setSquareAspect(true);
    double[][][] data = new double[2][SIZE][SIZE];
    VectorPlot plot = new VectorPlot();
    plot.setAll(data, -1, 1, -1, 1);
    for(int i = 0;i<SIZE;i++) {
      double x = plot.indexToX(i);    // the x location
      for(int j = 0;j<SIZE;j++) {
        double y = plot.indexToY(j); // the y location
        data[0][i][j] = -y;
        data[1][i][j] = x;
      }
    }
    plot.setAll(data);
    drawingPanel.addDrawable(plot);
    frame.setDefaultCloseOperation(javax.swing.JFrame.EXIT_ON_CLOSE);
    frame.setVisible(true);
    plot.showLegend();
  }
}
```

8.7 ■ COMPLEX FIELDS

Fields of complex numbers, as shown in Figure 8.7, occur frequently in electromagnetism and in quantum mechanics. Because real and imaginary numbers are similar to vector components, complex scalar fields are stored similarly to vector fields. We again use an array with three indices to store a complex field.

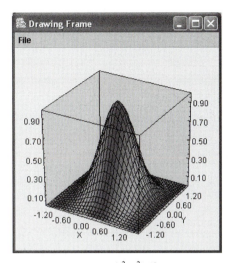

Figure 8.7 A complex scalar field $\psi(x, y) = e^{-(x^2+y^2)}e^{i5x}$. Phase is shown as color on the computer screen.

`ComplexGridPlot` and `ComplexInterpolatedPlot` use brightness to show a field's magnitude while `ComplexSurfacePlot` uses height. All three complex field visualizations use color to show phase starting with blue for positive real, red for positive imaginary, yellow for negative real, and green for negative imaginary. Listing 8.7 shows how the complex surface plot is used to display a Gaussian quantum wave function with a phase modulation (momentum boost) in the x direction:

$$\psi(x, y) = e^{-(x^2+y^2)}e^{i5x}. \tag{8.5}$$

Listing 8.7 Complex scalar field test program.

```
package org.opensourcephysics.manual.ch08;
import org.opensourcephysics.display.*;
import org.opensourcephysics.display2d.*;

public class ComplexSurfacePlotApp {
  final static int SIZE = 32;

  public static void main(String[] args) {
    DrawingPanel drawingPanel = new DrawingPanel();
    drawingPanel.setShowCoordinates(false);
    DrawingFrame frame = new DrawingFrame(drawingPanel);
    double[][][] data = new double[2][32][32];
    double[][] xdata = data[0];
    double[][] ydata = data[1];
    ComplexSurfacePlot plot = new ComplexSurfacePlot();
    plot.setAll(data, -1.5, 1.5, -1.5, 1.5);
    for(int i = 0;i<SIZE;i++) {
      double x = plot.indexToX(i);           // the x location
      for(int j = 0;j<SIZE;j++) {
```

```
        double y = plot.indexToY(j);         // the y location
        double amp = Math.exp(-2*(x*x+y*y)); // magnitude
        xdata[i][j] = amp*Math.cos(5*x);     // real component
        ydata[i][j] = amp*Math.sin(5*x);     // imaginary component
      }
    }
    plot.setAutoscaleZ(false, 0, 1);
    plot.setAll(data);
    drawingPanel.addDrawable(plot);
    drawingPanel.repaint();
    SurfacePlotMouseController mouseController =
        new SurfacePlotMouseController(drawingPanel, plot);
    drawingPanel.addMouseListener(mouseController);
    drawingPanel.addMouseMotionListener(mouseController);
    frame.setDefaultCloseOperation(javax.swing.JFrame.EXIT_ON_CLOSE);
    frame.setVisible(true);
  }
}
```

8.8 ■ DATA MODELS

Storing data in an array of numbers is simple but lacks the flexibly of a data model that stores an array along with additional information such as a scale and color. The `GridData` interface (see Table 8.5) in the 2D display package overcomes this limitation by modeling an arbitrary number of measurements (components) on a two-dimensional grid using a variety of different array configurations. Using a data model allows us to separate the data from visualizations such as `GridPlot`, `ContourPlot`, `SurfacePlot`, and `VectorPlot`.

There are many ways to encapsulate and store field data using a grid. We can allocate memory to store both the field and the coordinates of each grid point or we can store only the field and compute coordinates using the grid's stored minimum and maximum values. Likewise, we can store just the vector components or we can store components and computed values such as magnitude or phase, etc. The `GridData` interface in the 2D display package is designed to support these types of options. The `ArrayData`, `PointData`, and `FlatData` classes are concrete implementations of the `GridData` interface.

The `VectorPlot` class described in Section 8.6 uses three values from the data model to draw an arrow at a grid point: a color parameter, the x-component of the arrow, and the y-component of the arrow. The `ArrayData` class stores these values using three $n \times m$ subarrays. These subarrays are created within `ArrayData` by allocating a multidimensional array with three indices.

```
// values for n by m vector field using ArrayData
double[][][] data = new double[3][n][m];
```

Each subarray can be accessed by dereferencing the first index.

```
double[][] arrow_color = new double[0];
double[][] arrow_x = new double[1];
double[][] arrow_y = new double[2];
```

Values at a grid point are obtained by accessing each subarray.

```
// values at i,j grid point
double x:                          // x-coordinate must be computed
```

Table 8.5   Methods in the `GridData` interface.

org.opensourcephysics.display2d.GridData	
getBottom	Gets the world units of the last row of data.
getData	Gets the multidimensional array containing the raw data.
getDx	Gets the change in x between data rows.
getDy	Gets the change in y between data columns.
getLeft	Gets the world units of the first column of data.
getNx	Gets the number of entries along the x-ordinate.
getNy	Gets the number of entries along the y-ordinate.
getRight	Gets the world units of the last column of data.
getTop	Gets the world units of the first row of data.
getZRange	Gets the maximum and minimum values for a sample or component.
indexToX	Gets the x world coordinate for the given index. Similar method for y.
interpolate	Estimates a sample at an untabulated point (x, y) using bilinear interpolation of grid point data.
setScale	Sets world coordinates for the grid's rows and columns assuming that the data values are for the edge of a cell.
setCellScale	Sets world coordinates for the grid's rows and columns by assuming that the data values are for the center of a cell.
xToIndex	Gets closest index from the given x world coordinate. Similar method for y.

```
double y;                            // y-coordinate must be computed
double color = arrow_color[i][j]; // same as data[0][i][j]
double x_length = arrow_x[i][j];  // same as data[1][i][j]
double y_length = arrow_y[i][j];  // same as data[2][i][j]
```

The `PointData` class stores a vector field differently. This class models a grid point as a one-dimensional subarray that includes the coordinates of the grid point as the first two elements of this array. Values at a grid point are obtained by dereferencing the data array's first two indices.

```
// values for n by m vector field using PointData
double[][][] data = new double[n][m][5];
// values at i,j grid point
double[] grid_point=data[i][j];
double x = grid_point[0];         // same as data[i][j][0]
double y = grid_point[1];         // same as data[i][j][1]
double color = grid_point[2];     // same as data[i][j][2]
double x_length = grid_point[3];  // same as data[i][j][3]
double y_length = grid_point[4];  // same as data[i][j][4]
```

A point's x- and y-values are computed using methods in the `PointData` class. These coordinate values are usually not changed outside the class but may be used by a computation that has access to the data array.

The `FlatData` class does away with multidimensional arrays and stores all values in a single one-dimensional array. These values are stored in *row major* order.

```
// values for n by m vector field using FlatData
double[] data = new double[n*m*3];
// values at i,j grid point
double x;  // x-coordinate must be computed
double y;  // y-coordinate must be computed
double color=data[i*n+m];
double x_length=data[i*n+m+1];
double y_length=data[i*n+m+2];
```

The data model is usually passed to a visualization in its constructor, but it can later be changed using the setGridData method. Note that the same GridData object can be used by more than one visualization.

Consider again the vector field that was used in Listing 8.7. This field is sampled on a 32×32 grid spanning a region of physical space from $x_{min} < x < x_{max}$ and $y_{min} < y < y_{max}$. A GridPointData model for this field is instantiated and used as follows:

```
// 3 samples at every grid point
GridData griddata = GridPointData(32,32,3);
// compute coordinates for array
griddata.setScale(xmin,xmax,ymin,ymax);
VectorPlot plot= new VectorPlot(griddata);
```

We break encapsulation and retrieve a reference to the storage object's data array using the getData method for computational speed.

```
double[][][] data = griddata.getData();
```

We must, of course, know the storage model that is being used to properly access the array. Because we know that we are using a GridPointData object, we can compute the vector field using the stored x- and y-values.

```
for(int i = 0, nx = data.length; i<nx; i++) {
   for(int j = 0, ny = data[0].length; j<ny; j++) {
      double x = data[i][j][0]; // grid point x location
      double y = data[i][j][1]; // grid point y location
      double r = Math.sqrt(x*x+y*y);
      data[i][j][2] = r;                    // magnitude determines color
      data[i][j][3] = (r==0)? 0 : -y/r;  // horizontal
      data[i][j][4] = (r==0)? 0 : x/r;   // vertical
   }
}
VectorPlot plot = new VectorPlot(griddata);
drawingPanel.addDrawable(plot);
```

The color parameter is set equal to the field's magnitude and the arrow's x- and y-components are set equal to the direction cosines so that the VectorPlot draws arrows of equal length whose color is indicative of field strength. The following code fragment shows how to produce a visualization that uses field magnitude to determine the arrow length.

```
data[i][j][2] = 1;    // constant color
data[i][j][3] = -y;   // horizontal arrow component
data[i][j][4] = x;    // vertical arrow component
```

Because a complex field is similar to a vector field, complex field visualizations also use three values from the data model: visual magnitude (intensity), a real component, and an imaginary component. Although these values are clearly redundant from a physics point of view, a third parameter allows us to use a nonlinear intensity scale in the visualization in order to see detail in regions where the complex field is weak.

   Scalar fields require only a single value from a data model. If we wish to store a scalar field using a 32×32 grid that bounds a region two world units on a side, we can instantiate either a `GridPointData` object,

```
GridData griddata=new GridPointData(32,32,1);
griddata.setScale(-1,1,-1,1); //world units
```

or an `ArrayData` object,

```
GridData griddata=new ArrayData(32,32,1);
griddata.setScale(-1,1,-1,1); // world units
```

or a `FlatData` object,

```
GridData griddata=new FlatData(32,32,1);
griddata.setScale(-1,1,-1,1); // world units
```

The number of data components is the last argument in the above constructors. We then create a visualization by instantiating a visualization and passing it the grid data.

```
ContourPlot contour = new ContourPlot(griddata);
DrawingPanel panel = new DrawingPanel();
panel.addDrawable(contour);
```

Because any data model can be used with a 2D visualization, we can choose a data model that fits the numerical algorithm and the physical model.

8.9 ■ OVERLAYS

Because drawable components are rendered in the order that they are added to a drawing panel, visualizations can be overlaid. The gradient of a scalar field $A(x, y)$ defines a vector field. Figure 8.8 shows an interpolated plot of the scalar field $U(x, y) = y * x$ overlaid with its gradient.

   The gradient of a scalar field is a vector field that points in the direction that the scalar field is increasing most rapidly. In a two-dimensional Cartesian coordinate system, the components of the gradient are equal to the derivative of the scalar field along the x- and y-axes, respectively. (See Chapter 10 for a discussion of derivatives.) Listing 8.8 again shows the usefulness of encapsulation and abstraction by hiding the details of how this calculation is done. We merely send a scalar field to the `gradient` method in the `Util2D` class in the 2D display package. This method returns a vector field whose grid points match the scalar field. Note that we have changed the vector field's color map to `BLACK` so as not to blend with the interpolated plot's colors.

> **Listing 8.8**   A vector field overlayed on an interpolated plot
> of the scalar field.

```
package org.opensourcephysics.manual.ch08;
import org.opensourcephysics.display.*;
import org.opensourcephysics.display2d.*;
import javax.swing.JFrame;

public class GradientOverlayApp {
  public static void main(String[] args) {
    PlottingPanel drawingPanel =
        new PlottingPanel("x", "y", "Gradient");
    drawingPanel.setSquareAspect(false);
```

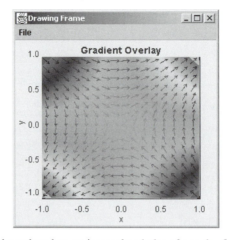

Figure 8.8   A vector field overlayed on an interpolated plot of a scalar field $U(x, y) = yx$.

```
DrawingFrame frame = new DrawingFrame(drawingPanel);
GridPointData pointdata =
    new GridPointData(16, 16, 1); // scalar field
pointdata.setScale(-1, 1, -1, 1);
double[][][] data = pointdata.getData();
for(int i = 0, row = data.length;i<row;i++) {
  for(int j = 0, col = data[0].length;j<col;j++) {
    double x = data[i][j][0]; // the x location
    double y = data[i][j][1]; // the y location
    data[i][j][2] = y*x;       // magnitude
  }
}
InterpolatedPlot plot = new InterpolatedPlot(pointdata);
GridPointData vecData2d = Util2D.gradient(pointdata, 1);
VectorPlot vectorplot = new VectorPlot(vecData2d);
vectorplot.setPaletteType(VectorColorMapper.BLACK);
drawingPanel.addDrawable(plot);          // add scalar field
drawingPanel.addDrawable(vectorplot); // overlay vector field
frame.setDefaultCloseOperation(JFrame.EXIT_ON_CLOSE);
frame.setVisible(true);
  }
}
```

The ComplexContourPlot class in the 2D display package uses inheritance and composition to define an object that overlays contour lines to a complex interpolated plot. The interpolated plot shows the phase of the complex scalar field and the contour plot overlays lines showing the field's magnitude.

8.10 ■ PROGRAMS

The following examples are in the org.opensourcephysics.manual.ch08 package. Not all programs are described in the narrative.

BinaryLatticeAnimationApp

`BinaryLatticeAnimationApp` tests a binary lattice by rapidly assigning random values to each cell using an animation thread.

BinaryLatticeApp

`BinaryLatticeApp` creates a 32×32 lattice of random values and then sets a 16×16 block of cells to zero. See Section 8.3.

BufferedContourApp

`BufferedContourApp` superimposes a dragable circle on a static background image of a contour plot. Buffering is described in Section 7.2.

ByteRasterApp

`ByteRasterApp` tests the `ByteRaster` class by creating random data in a 256×256 raster.

CalcGridPlotApp

`CalcGridPlotApp` tests a grid plot by allowing the user to enter an arbitrary function $f(x, y)$ using a simple user interface. Other programs that plot scalar fields using arbitrary functions are `CalcContourPlotApp` and `CalcSurfacePlotApp`.

CellLatticeApp

`CellLatticeApp` tests the cell lattice by displaying the default color palette. See Section 8.3.

ComplexSurfacePlotApp

`ComplexSurfacePlotApp` displays a complex scalar field using a complex surface plot. This program is described in Section 8.7. Other programs that display complex scalar fields are: `ComplexGridPlotApp`, `ComplexContourApp`, and `ComplexInterpolatedApp`.

ContourPlotApp

`ContourPlotApp` creates a contour plot of a scalar field $U(x, y) = x*y$.

DiffractionRasterApp

`DiffractionRasterApp` demonstrates how to use `RasterFrame` by computing a single-slit diffraction pattern.

DLAApp

`DLAApp` models diffusion limited aggregation using a byte raster.

GameOfLifeApp

`GameOfLifeApp` uses a binary lattice to model the game of life. Right-click on the lattice to toggle life on and off.

GaussianSurfacePlotApp

`GaussianSurfacePlotApp` displays a Gaussian scalar field using a surface plot. This program is described in Section 8.5. Other programs that display examples of scalar fields are: `GaussianGridPlotApp`, `GaussianContourApp`, and `GaussianInterpolatedApp`.

GrayscaleApp

`GrayscaleApp` displays a scalar field using a gray-scale plot. This plot looks similar to a grid plot with a gray-scale color palette. However, it uses a different rendering model.

GridPlotApp

`GridPlotApp` displays a Gaussian scalar field using a grid plot. This program is described in Section 8.8.

MandelbrotApp

`MandelbrotApp` displays the Mandelbrot set using a byte raster. This program is described in Section 8.2.

RandomWalkApp

`RandomWalkApp` uses a binary lattice to simulate a random walk.

SiteLatticeApp

`SiteLatticeApp` tests the `SiteLattice` class by displaying the default color palette.

VectorPlotApp

`VectorPlotApp` displays a vector field. This program is described in Section 8.6. The `CalcVectorPlotApp` program allows the user to enter an arbitrary vector field **A** using a simple user interface.

CHAPTER

9

Differential Equations and Dynamics

We describe how to solve dynamical problems using systems of first-order differential equations. Ordinary differential equation (ODE) solvers are defined in the *numerics* package (org.opensourcephysics.numerics) along with other numerical analysis tools.

9.1 ■ OVERVIEW

Many physical models, such as Newton's second law, describe the relationship between an unknown function and its derivatives as shown in Figure 9.1. For example, we can compute the velocity $v(t)$ of a particle falling near Earth as a function of time as follows:

$$\frac{dv}{dt} = -g - bv, \tag{9.1}$$

where $g = 9.8 \, \text{m/s}^2$ is the constant acceleration of gravity. This equation is referred to as a *rate equation* because it expresses the rate of change of the variable as a function of variables and parameters. The quantity being differentiated (velocity) is called the *dependent variable* and the quantity with respect to which it is differentiated (time) is called the *independent variable*. Equation 9.1 is a *first-order* equation because the highest derivative is a first derivative.

Equation 9.1 has the following general form:

$$\dot{x} \equiv \frac{dx}{dt} = f(x, t). \tag{9.2}$$

We would like to solve Equation 9.2 to find $x(t)$ for $0 \leq t \leq t_f$. The aim of this chapter is to approximate this solution by finding the value of x^{i+1} at time $t_{i+1} = t_i + \Delta t$ given the value of x^i at time t_i. We start at $x^0 = x(0)$ and obtain $x^1 = x(\Delta t)$. Once x^1 is obtained, the process is repeated to find x^2 and so on until the final time t_f. Note that we use a superscript to specify the value of x at a particular time step and so that we can later use a subscript to specify variables, such as a position or velocity component, using the generic x_i.

In general, Newton's second law for particle motion in one dimension is a second-order differential equation that takes the form

$$\frac{d^2x}{dt^2} = F(x, v, t)/m. \tag{9.3}$$

This equation can be converted to two first-order equations by considering both position x

145

Figure 9.1   Differential equations can be solved using the *ODE* interface to define the physics and the *ODESolver* interface to define the numerical algorithm.

and velocity v to be unknown functions of time:

$$\dot{x} = v \qquad\qquad (9.4a)$$

$$\dot{v} = F(x, v, t)/m . \qquad\qquad (9.4b)$$

The initial position x^0 and initial velocity v^0 are advanced by Δt to obtain x^1 and v^1, respectively. The process of defining extra variables such as v to reduce the order of a differential equation (but increase the number of equations) can be extended to higher-order derivatives as needed. Because higher-order differential equations can be converted into a system of first-order equations, we need only consider general methods of solving systems of first-order equations.

Additional particles or additional spacial dimensions add additional equations. In general, a system of particles results in a system of first-order ordinary differential equations with an independent time t variable and with N dependent variables that can be written as

$$\dot{x}_0 = f_0(x_0, x_1, \ldots, x_{N-1}, t) \qquad\qquad (9.5a)$$

$$\dot{x}_1 = f_1(x_0, x_1, \ldots, x_{N-1}, t) \qquad\qquad (9.5b)$$

$$\vdots$$

$$\dot{x}_{N-1} = f_N(x_0, x_1, \ldots, x_{N-1}, t). \qquad\qquad (9.5c)$$

This system is said to be autonomous if the functions f_i do not explicitly depend on the independent variable. If we think of the variables x_i as components of a state vector $\mathbf{x}$, this system can be written compactly as

$$\dot{\mathbf{x}} = \mathbf{f}(\mathbf{x}). \qquad\qquad (9.6)$$

The variable that is used to take the derivative can be any independent parameter but is often the time t. In order to treat real-world problems where the vector function $\mathbf{f}(\mathbf{x})$ is often an explicit function of time, we treat the time as an additional variable and add the trivial rate equation $dt/dt = 1$. Our convention is to list the independent variable last in the state array. We often do this even for autonomous equations because there is little additional computation and because it allows us to automatically track the independent variable in the state array when using adaptive step-size algorithms as described in Section 9.5.

In classical (Newtonian) physics, the position and velocity (momentum) space that we have defined is called *phase space*. The trajectory of a particle or a system of particles through phase space is completely determined if we know the initial state and an expression for the rate (force). Consider the equation of motion in two dimensions for a particle falling under the influence of gravity with a constant acceleration $g = 9.8 \, \text{m/s}^2$. This equation can be written as

$$\dot{x} = v_x \tag{9.7a}$$

$$\dot{v}_x = 0 \tag{9.7b}$$

$$\dot{y} = v_y \tag{9.7c}$$

$$\dot{v}_y = -9.8 \tag{9.7d}$$

$$\dot{t} = 1. \tag{9.7e}$$

Another common example is the equation of motion for a driven one-dimensional harmonic oscillator with mass m and spring constant k. This equation can be written as

$$\dot{x} = v \tag{9.8a}$$

$$\dot{v} = -\frac{k}{m}x + A \sin \omega_0 t \tag{9.8b}$$

$$\dot{t} = 1. \tag{9.8c}$$

And finally, the equation of motion for a particle in orbit about a gravitational center of attraction $\mathbf{F_g}(\mathbf{r}) = -(1/r^2)\hat{\mathbf{r}}$ can be written in dimensionless units as

$$\dot{x} = v_x \tag{9.9a}$$

$$\dot{v}_x = -\frac{x}{r^3} \tag{9.9b}$$

$$\dot{y} = v_y \tag{9.9c}$$

$$\dot{v}_y = -\frac{y}{r^3} \tag{9.9d}$$

$$\dot{t} = 1, \tag{9.9e}$$

where we have used $\cos\theta = x/r$ and $\sin\theta = y/r$ to resolve the unit vector $\hat{\mathbf{r}}$ into its xy-components.

We will use the examples above to test numerical algorithms in the following sections.

9.2 ■ ODE INTERFACE

The ODE interface defined in the numerics package enables us to encapsulate a system of first-order ordinary differential equations (9.6) in a Java class. This interface contains two methods, getState and getRate, as shown in Listing 9.1. The getState method returns a state array $(x_0, x_1, \ldots, x_{n-1})$ where the variables represent either position or velocity. The getRate method evaluates the derivatives using the given state array and stores the result in the given rate array $(\dot{x}_0, \dot{x}_1, \ldots, \dot{x}_{n-1})$. Because most numerical algorithms evaluate the rate multiple times as they advance the system by Δt, the given state is usually not the current state of object.

> **Listing 9.1**   The ODE interface is used to solve systems of first-order differential equations.

```
package org.opensourcephysics.numerics;
public interface ODE {
```

```
   public double[] getState();
   public void getRate(double[] state, double[] rate);
}
```

A Java class that implements the ODE interface for the driven harmonic oscillator system (Equation 9.8) is shown in Listing 9.2. The Falling class uses the ODE interface to model a two-dimensional falling particle. This class is available on the CD but is not shown here because it is similar to Listing 9.2.

Listing 9.2   An implementation of the ODE interface that models the driven simple harmonic oscillator.

```
package org.opensourcephysics.manual.ch09;
import org.opensourcephysics.numerics.ODE;

public class SHO implements ODE {
   // state array contains =[x, vx, t]
   double[] state = new double[] {5.0, 0.0, 0.0};
   double omega = 1, amp = 1.0;
   double k = 1, m = 1.0;

   public double[] getState() { return state; }

   public void getRate(double[] state, double[] rate) {
      rate[0] = state[1];
      rate[1] = -k/m*state[0]+amp*Math.sin(omega*state[2]);
      rate[2] = 1;
   }
}
```

A convention that we find useful is to place the velocity rate immediately after the position rate. For example, the dynamic variables for two particles should be stored in an array that is ordered as follows $[x_1, v_{x1}, y_1, v_{y1}, x_2, v_{x2}, y_2, v_{y2}, t]$. This ordering enables us to efficiently code certain numerical algorithms (such as the Verlet method) because the differential equation solver can assume that a velocity rate follows every position rate in the state array. Consult the documentation or the solver's source code to determine if a particular variable ordering is necessary.

9.3 ■ ODE SOLVER INTERFACE

There are many possible algorithms to advance a system of first-order differential equations from an initial state to a final state. The ODESolver interface shown in Listing 9.3 defines a set of methods that enables us to define algorithms to solve differential equations in objects that implement the ODE interface.

Listing 9.3   The ODESolver interface defines numerical algorithms to solve differential equations.

```
public interface ODESolver {
   public void initialize(double stepSize);
   public double step(); //returns step size
   public void setStepSize(double stepSize);
   public double getStepSize();
}
```

An ODE solver's `initialize` method sets the initial step size and allocates arrays to store temporary values. This method should be called after the ODE object has been created or if the number of equations within the ODE object changes. Because adaptive algorithms are free to change the step size, the `step` method returns this parameter. We also provide the `setStepSize` and `getStepSize` methods to set and read the step-size parameter. The `setStepSize` method does not allocate or initialize arrays within the solver.

Differential equations can be solved by creating a solver and repeatedly invoking the solver's `step` method. Consider again the equation of motion for the harmonic oscillator that is coded in Listing 9.2. These equations are solved in the `SHOApp` program shown in Listing 9.4. The important point to realize is that we simply need to instantiate a different solver to switch numerical methods. Run this example with various `ODESolvers` and note the different results. Note in particular that amplitude of oscillation increases when using the `Euler` solver. The explicit Euler method is often numerically unstable even with small values of the step size Δt. It is useful as a pedagogic tool, for debugging, and to perform the first step in numerical methods that are not self-starting.

Listing 9.4 Driven harmonic oscillator program.

```
package org.opensourcephysics.manual.ch09;
import org.opensourcephysics.numerics.*;

public class SHOApp {
  public static void main(String[] args) {
    double time = 10; // solution range
    double dt = 0.1;  // ode step size
    ODE ode = new SHO();
    ODESolver ode_solver = new RK45(ode);
    ode_solver.initialize(dt);
    double[] state = ode.getState();
    while(time>0) {
      String xStr = "x = "+state[0];
      String vStr = "  v = "+state[1];
      String tStr = "  t = "+state[2];
      System.out.print(xStr+vStr+tStr+"\n");
      time -= ode_solver.step();
    }
  }
}
```

In summary, an ODE solver's `step` method advances the state of the oscillator by the time step Δt using a numerical method. Listing 9.5 in the next section shows the details of how the Euler ODE solver is written. It is, of course, up to the programmer to determine if the chosen ODE solver is accurate and stable when applied to the given problem. Section 9.4 describes various ODE solvers that are implemented in the numerics package.

9.4 ■ ALGORITHMS

The numerics package defines implementations of the `ODESolver` interface as shown in Figure 9.2 and Table 9.1. The simplest of these is the `Euler` solver which approximates Equation 9.10 as

$$\mathbf{x}^{i+1} = \mathbf{x}^i + \mathbf{f}(\mathbf{x}^i)\Delta t. \tag{9.10}$$

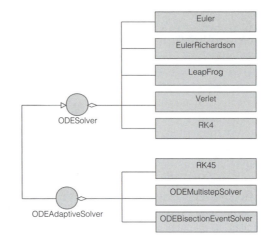

Figure 9.2   ODE algorithms in the numerics package implement the `ODESolver` interface and the `ODEAdaptiveSolver` interface.

Table 9.1   Ordinary differential equation solvers are implemented in the numerics package.

org.opensourcephysics.numerics	
Euler	Euler algorithm.
Verlet	Verlet algorithm.
EulerRichardson	Euler–Richardson second-order algorithm with fixed step size.
RK4	Runge–Kutta fourth-order algorithm with fixed step size.
RK45	An adaptive step size algorithm based on fourth- and fifth-order Runge–Kutta–Fehlberg methods run in tandem.
ODEBisectionEventSolver	A solver that deals with StateEvents using a bisection root finder.
ODEMultistepSolver	A solver that takes a succession of adaptive steps to produce a fixed step size.

This algorithm is coded in Listing 9.5. It starts with the current value of the nth variable at the ith step x_n^i and assumes that the rate of change is constant over the interval Δt. The key to understanding this method is the Taylor-series expansion of $\mathbf{x}(t)$:

$$\mathbf{x}(t + \Delta t) = \mathbf{x}(t) + \Delta t \dot{\mathbf{x}}(t) + \frac{\Delta t^2}{2!} \ddot{\mathbf{x}}(t) + \frac{\Delta t^3}{3!} \dddot{\mathbf{x}}(t) + \cdots . \qquad (9.11)$$

We get the Euler method by discarding (truncating) all but the first two terms of the series. In other words, the Euler method only agrees with the exact solution if the first derivative is constant and higher-order derivatives are zero.

Listing 9.5 Euler's method for solving differential equations. Only the step method is shown.

```
// ODESolver methods are also defined in the abstract superclass
package org.opensourcephysics.numerics;

public class Euler extends AbstractODE {
  public Euler(ODE ode) {
    super(ode);
  }

  public double step() {
    double[] state = ode.getState();
    ode.getRate(state, rate);
    for(int i = 0;i<numEqn;i++) {
      state[i] = state[i]+stepSize*rate[i];
    }
    return stepSize;
  }
}
```

The Euler method has an error per step that is proportional to the discarded term which is proportional to Δt^2. Halving the step size decreases this *local error* by a factor of four. Unfortunately, we also have to take twice as many steps so the *global error* over a fixed interval is proportional to Δt. The exponent that determines how the error depends on step size is known as the *order* of the algorithm. It is written as $O(\Delta t^m)$. Because error accumulates as the integration proceeds from the initial time t_0 to the final time t_f, the global error of the Euler method is first order, $O(\Delta t^1)$. Although it is tempting to choose a very small value of Δt to reduce this error, a great many time steps will be required. Because of the long computation time, the accumulation of arithmetic error due to a computer's finite precision, and the instability noted when solving the simple harmonic oscillator model, the Euler method is not recommended.

There are three general approaches to improving the accuracy of the Euler method:

- **Taylor** series methods, such as Euler–Richardson, can achieve higher accuracy through higher derivatives of f_i.

- **multistep** methods, such as Verlet, can achieve higher accuracy by using information about the rate at multiple time steps $\{\ldots, t_{n-2}, t_{n-1}, t_n, t_{n+1}, \ldots\}$.

- **Runge–Kutta** methods can achieve higher accuracy by evaluating the rate equation at intermediate steps between t_n and t_{n+1}.

Taylor-Series Methods

Taylor-series methods include higher-order terms of the Taylor expansion of the solution. Truncating the series expansion at second order is particularly intuitive for Newton's second law problems because the second derivative of the position is the force per unit mass, $a = F/m$. Physicists simplify this expansion further by dropping the third term in the velocity rate thereby assuming constant acceleration throughout the time step Δt. Using physics notation, the Taylor series expansion for position and velocity is written as

$$x_n^{i+1} = x_n^i + v_n^i \Delta t + \frac{1}{2} a_n \Delta t^2 \qquad (9.12a)$$

$$v_n^{i+1} = v_n^i + a_n \Delta t . \qquad (9.12b)$$

The truncation error is now third order for position but only second order for velocity. The resulting approximation, known as the *midpoint* method, is very appealing because it can be shown to be equivalent to using the average velocity throughout the interval:

$$x_n^{i+1} = x_n^i + \frac{v_n^{i+1} + v_n^i}{2} \Delta t. \qquad (9.13)$$

The midpoint algorithm is exact for projectile problems, but it suffers from the same instability as Euler's method for other types of problems. A better Taylor-series method is the Euler–Richardson algorithm which uses the state at the beginning of the interval to estimate the rate at the midpoint:

$$x_{\text{mid}} = x_n^i + \frac{1}{2} v_n^i \Delta t \qquad (9.14a)$$

$$v_{\text{mid}} = v_n^i + \frac{1}{2} a(x^i, v^i) \Delta t \qquad (9.14b)$$

$$x_n^{i+1} = x_n^i + v_{\text{mid}} \Delta t + O(\Delta t^3) \qquad (9.14c)$$

$$v_n^{i+1} = v_n^i + a(x_{\text{mid}}, v_{\text{mid}}) \Delta t + O(\Delta t^3). \qquad (9.14d)$$

The numerics package contains an implementation of this algorithm.

Multistep Methods

Multistep algorithms use information from multiple time steps to calculate a new state. If the algorithm only requires information from previous states to calculate, then the algorithm is said to be *explicit*. If the algorithm uses the new state, that is, if x^{i+1} appears within the rate equation, then the algorithm is said to be *implicit*. Although implicit algorithms require that a system of equations (9.6) be solved for x^{i+1} at every time step, they are usually very stable. This section presents a simple algorithm known as the *Verlet* method. Even though the Verlet method is only third order in position and second order in velocity, it is widely used in molecular dynamics simulations.

The Verlet method is easy to justify, although we will not do so here. It requires only a single evaluation of the rate and may be written as

$$x_n^{i+1} = 2x_n^i - x_n^{i-1} + \frac{1}{2} a_n \Delta t^2 + O(\Delta t^4) \qquad (9.15a)$$

$$v_n^{i+1} = \frac{x_n^{i+1} - x_n^{i-1}}{2\Delta t} + O(\Delta t^2). \qquad (9.15b)$$

The Verlet algorithm is implicit because the new value x_n^{i+1} appears on the right-hand side of the velocity equation. The system of linear equations is, however, trivial. The new position is computed first, and this value is used to compute the velocity.

The Verlet method's position error is fourth order, and although the velocity is only second order, it does a good job of conserving energy. Implementing algorithm 9.15 does have a small disadvantage in that it is not self-starting because the initial conditions are only known at $t = 0$ and not at $t = -\Delta t$. We must use another method to compute the first step.

A mathematically equivalent version of the original Verlet algorithm is given by

$$x^{i+1} = x^i + v^i \Delta t + \frac{1}{2} a^i \Delta t^2 \tag{9.16a}$$

and

$$v^{i+1} = v^i + \frac{1}{2}(a^{i+1} + a^i)\Delta t. \tag{9.16b}$$

We see that (9.16), known as the *velocity* form of the Verlet algorithm, is self-starting and minimizes roundoff errors. (See *An Introduction to Computer Simulation Methods, Third Edition* by Harvey Gould, Jan Tobochnik, and Wolfgang Christian for a discussion of finite difference methods for the solution of Newton's equation of motion.) Because this form of the Verlet algorithm is the most commonly used, we have implemented it in the `Verlet` ODE solver in the numerics package. Because the Verlet method is implicit, the rate equation for position and velocity are different and the `Verlet` ODE solver *requires* that the state and rate arrays alternate position and velocity.

Runge–Kutta Methods

One of the most popular methods for solving ODEs is Runge–Kutta. It achieves the accuracy of Taylor-series methods without the calculation of higher derivatives. All variables are treated the same. Although there are many variations of the Runge–Kutta approach, we will only examine the fourth-order Runge–Kutta since it balances simplicity and power.

Runge–Kutta fourth-order is the "Volkswagen" of ODE solvers; it may not have extra horsepower, but it often gets the job done. In the RK4 algorithm, the derivative is computed at the beginning of the time interval, twice in the middle, and again at the end of the interval.

The algorithm starts by calculating the change in state using Δt as the step size:

$$\mathbf{d}_1 = \mathbf{f}(\mathbf{x}^i)\Delta t. \tag{9.17}$$

Half this change in state is added to the initial state to produce an intermediate state $\mathbf{x}^i + \frac{\mathbf{d}_1}{2}$. This intermediate state is then used to calculate another change in state:

$$\mathbf{d}_2 = \mathbf{f}(\mathbf{x}^i + \mathbf{d}_1/2)\Delta t. \tag{9.18}$$

This process of estimating intermediate state and rate is repeated,

$$\mathbf{d}_3 = \mathbf{f}(\mathbf{x}^i + \mathbf{d}_2/2)\Delta t \tag{9.19}$$

$$\mathbf{d}_4 = \mathbf{f}(\mathbf{x}^i + \mathbf{d}_3)\Delta t, \tag{9.20}$$

before the final state is calculated using a weighted average of all four results:

$$\mathbf{x}^{n+1} = \mathbf{x}^n + \frac{\mathbf{d}_1 + 2\mathbf{d}_2 + 2\mathbf{d}_3 + \mathbf{d}_4}{6} + O(\Delta t^5). \tag{9.21}$$

Mathematicians tell us—and we believe them—that this weighted average is fifth-order accurate in the step size for a single step. (The algorithm is globally accurate to fourth order.) The method that implements this algorithm is shown in Listing 9.6.

Listing 9.6   A fourth-order Runge–Kutta step.

```
public double step(){
  double state[]=ode.getState();
  ode.getRates(state,rates1);
  for(int i = 0;i<numEqn; i++) {
    k1[i] = state[i]+stepSize*rates1[i]/2;
  }
  ode.getRates(k1,rates2);
  for(int i = 0;i<numEqn; i++) {
    k2[i] = state[i]+stepSize*rates2[i]/2;
  }
  ode.getRates(k2,rates3);
  for(int i = 0;i<numEqn; i++) {
    k3[i] = state[i]+stepSize*rates3[i];
  }
  ode.getRates(k3,rates4);

  for(int i = 0;i<numEqn; i++) {
    state[i] = state[i]+stepSize*(rates1[i]+2*rates2[i]+
        2*rates3[i]+rates4[i])/6.0;
  }
  return stepSize;
}
```

9.5 ■ ADAPTIVE STEP SIZE

A major problem in solving ODEs lies in determining an appropriate step size. It is not efficient to use the same small time step throughout a simulation if the rate of change of the system is unknown or if the rate varies widely during the system's evolution. Instead, an *adaptive* (variable) time-step algorithm is suggested, as shown in Figure 9.3. Adaptive step-size algorithms estimate the local error and increase or reduce the step size if the error is greater or less than a predetermined tolerance, respectively. The AdaptiveSolver interface shown in Listing 9.7 extends ODE solvers to express this additional capability.

Listing 9.7   The adaptive ODE solver interface.

```
public interface ODEAdaptiveSolver extends ODESolver {
  public void setTolerance(double tol);

  public double getTolerance();
}
```

One possible adaptive technique is to advance the state using different step sizes and compare results. The first computation is performed using the full time step Δt. The second computation is done using two half steps $\Delta t/2$. If the two answers agree to within the specified tolerance, then we assume that the solution has the desired accuracy and accept the result. Otherwise, the time step is reduced and the process is repeated. Although this scheme usually works, it is very inefficient.

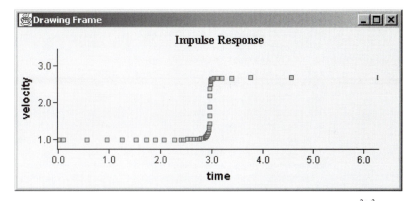

Figure 9.3 The velocity of a particle acted on by an impulsive force $f(t) = e^{-t^2/\epsilon^2}$. The time step is adjusted automatically to maintain a specified tolerance.

In the 1960s, E. Fehlberg discovered that it was possible to evaluate five intermediate rates to obtain a Runge–Kutta solution that has the interesting property that the same intermediate rates can be used to evaluate a fourth-order and a fifth-order algorithm. It is, therefore, possible to run a fourth- and a fifth-order numerical algorithm in tandem. Fehlberg recognized that the difference between these two algorithms provided a good estimate of the fourth-order method's error. Suppose that we have an estimated error of 10^{-3} but that we require an error of no more than 10^{-8}. Because the local error is $O(\Delta t^5)$ we can decrease the step size by a factor of 10:

$$\Delta t' = \left(\frac{10^{-8}}{10^{-3}}\right)^{1/5} \Delta t = \Delta t/10. \tag{9.22}$$

Conversely, we would increase the step size if the error is too small. The RK45 differential equation solver implements this technique to adjust the step size until the predetermined tolerance is reached. (See the source code for additional details.)

Figure 9.3 shows the velocity of a particle influenced by a force that varies rapidly, $F(x) = \epsilon/(\epsilon^2 + x^2)$. The particle has an initial velocity $v = 1$. Note the nonuniform increments along the time axis. At first there is little variation in the force and the step size increases. As the particle approaches $x = 0$, the force changes rapidly and the adaptive solver increases and decreases the step size as needed. Finally, the particle resumes its uniform force-free motion and the step size becomes very large. Listing 9.8 shows the code for this example.

Listing 9.8 Simulation of a particle acted on by an impulsive force.

```java
package org.opensourcephysics.manual.ch09;
import org.opensourcephysics.display.*;
import org.opensourcephysics.numerics.*;
import java.awt.Color;

public class AdaptiveStepApp {
  public static void main(String[] args) {
    PlottingPanel panel =
        new PlottingPanel("time", "velocity", "Impulse Response");
```

```
     DrawingFrame frame = new DrawingFrame(panel);
     panel.setSquareAspect(false);
     panel.setPreferredMinMaxY(0.8, 3.5);
     Dataset dataset = new Dataset();
     dataset.setMarkerShape(Dataset.SQUARE);
     dataset.setMarkerColor(new Color(255, 128, 128, 128), Color.red);
     panel.addDrawable(dataset);
     ODE ode = new Impulse();
     ODEAdaptiveSolver ode_solver = new RK45(ode);
     ode_solver.initialize(0.1); // the initial step size
     ode_solver.setTolerance(1.0e-4);
     while(ode.getState()[0]<12) {
       dataset.append(ode.getState()[2], ode.getState()[1]);
       ode_solver.step();
     }
     frame.setDefaultCloseOperation(javax.swing.JFrame.EXIT_ON_CLOSE);
     frame.setVisible(true);
     panel.repaint();
   }
}

class Impulse implements ODE {
  double epsilon = 0.01;
  double[] state = new double[] {-3.0, 1.0, 0.0}; // x,v,t

  public double[] getState() { return state;}

  public void getRate(double[] state, double[] rate) {
    rate[0] = state[1];
    rate[1] = epsilon/(epsilon*epsilon+state[0]*state[0]);
    rate[2] = 1;
  }
}
```

9.6 ■ MULTISTEPPING

Adaptive step-size algorithms are a poor choice whenever we wish to display a system's
state at predetermined times such as when we wish to show an integral table or an animation.
We have defined two ODE solvers that perform multiple adaptive steps in order to advance
the independent variable by a fixed step size. Because the step size is chosen so as to achieve
the desired accuracy, the last step will almost always overshoot. The ODEMultistepSolver
automatically reduces the step size if the remaining time is less than the optimum step size.
The ODEInterpolationSolver always takes the optimum step size and then interpolates
the final state. This interpolation does not affect the accuracy of subsequent steps because
the solver maintains the exact state internally. These two ODE solvers are too long to list
here, but they are defined in the numerics package.

In order to test the accuracy of our multistep ODE solvers, we recast a one-dimensional
integral as a differential equation. Consider the following indefinite integral:

$$F(t) = \int_{a}^{t} f(x)\,dx \quad \text{where } F(a) = 0. \tag{9.23}$$

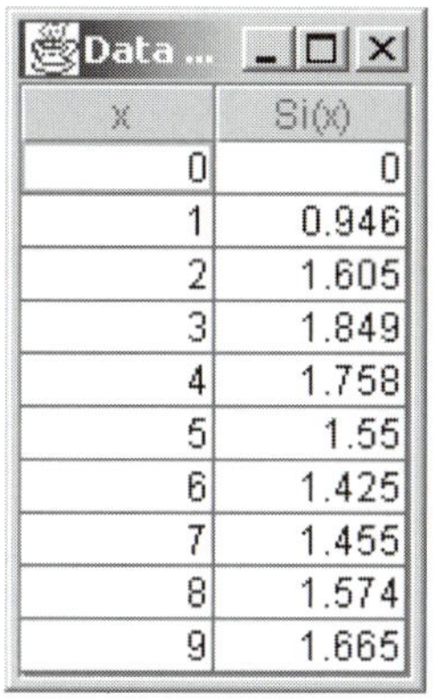

x	Si(x)
0	0
1	0.946
2	1.605
3	1.849
4	1.758
5	1.55
6	1.425
7	1.455
8	1.574
9	1.665

Figure 9.4 The Si function evaluated at ten points.

Differentiating with respect to t produces the following first-order differential equation:

$$\frac{dF(t)}{dt} = f(t). \tag{9.24}$$

We now consider the *sine integral* because it cannot be evaluated using standard functions but is tabulated in reference books, such as the classic *Handbook of Mathematical Functions* (1964) by M. Abromowitz and I. A. Stegun:

$$Si(z) = \int_0^z \frac{\sin x}{x}\, dx. \tag{9.25}$$

A typical value of this integral is $Si(10) = 1.6583475942$. Listing 9.9 solves this integral and tabulates the values. The result is shown in Figure 9.4 using a data table.

Listing 9.9 `SineIntegralApp` displays a table of integral values.

```
package org.opensourcephysics.manual.ch09;
import org.opensourcephysics.frames.*;
import org.opensourcephysics.numerics.*;

public class SineIntegralApp implements ODE {
  double[] state = new double[] {0.0, 0.0};

  public double[] getState() {
    return state;
  }

  public void getRate(double[] state, double[] rates) {
    rates[0] = 1;                     // independent variable
    rates[1] = integrand(state[0]); // function to integrate
  }

  public double integrand(double x) {
    if(x==0) {
      return 1;
    } else {
```

```
            return Math.sin(state[0])/state[0];
        }
    }

    public static void main(String[] args) {
        TableFrame frame = new TableFrame("Sine Integral: Si(x)");
        frame.setColumnNames(0, "x");
        frame.setColumnNames(1, "Si(x)");
        ODE ode = new SineIntegralApp();
        // RK45MultiStep hides the adaptive step size from the user
        ODEAdaptiveSolver ode_method = new RK45MultiStep(ode);
        ode_method.setTolerance(1.0e-5);
        ode_method.setStepSize(1.0);
        for(int i = 0;i<10;i++) {
            frame.appendRow(new double[] {ode.getState()[0],
                    ode.getState()[1]});
            ode_method.step();
        }
        frame.setDefaultCloseOperation(javax.swing.JFrame.EXIT_ON_CLOSE);
        frame.setVisible(true);
        frame.refreshTable();
    }
}
```

9.7 ■ HIGH-ORDER ODE SOLVERS

Andrew Gusev and Yuri B. Senichenkov at Saint Petersburg Polytechnic University, Russia have contributed a collection of high-order differential equation solvers to the OSP library based on the Runge–Kutta method. These solvers are shown in Table 9.2 and are defined in the org.opensourcephysics.ode package. The code for this optional package and for the examples in this section are available on the CD in the ospode.zip archive.

Dopri853 is an explicit Runge–Kutta adaptive step-size method that integrates a system of ordinary differential equations using a Runge–Kutta approach developed by Dorman and Prince. (See P. J. Prince and J. R. Dorman, "High order embedded Runge–Kutta formulae," J. Comp. Appl. Math. **7**, 67–75 (1981).) This solver is an eighth-order accurate integrator and requires 13 function evaluations per integration step. Additional information about the Dorman and Prince formulas can be found in Hairer, Norsett, and Wanner (1993).

The Radau5 solver is designed to solve stiff systems of first-order ordinary differential equations of the form $M\dot{\mathbf{x}} = f(\mathbf{x}, t)$ with a possibly singular matrix M using an implicit Runge–Kutta method of order five. The method is described in Hairer and Wanner (1996).

Although the Radau5 is not recommended for nonstiff problems, both Dopri853 and Radau5 solvers can be used and the solutions carefully compared if the user does not know the stiffness of the equations. Because Radau5 is an implicit ODE solver, it takes many more calculations to compute a model solution using this solver than with Dopri853. The user should balance the time required to calculate solutions against the need for the increased solution accuracy.

The Dopri853 and Radau5 solvers implement the ODEAdaptiveSolver interface and can be used directly if a fixed step size is not required.

Table 9.2 High-order ODE solvers are implemented in an optional OSP package.

org.opensourcephysics.ode	
Dopri5	Dorman-Prince fifth-order ODE solver with variable step size.
Dopri853	Dorman-Prince 8/5/3 (12 stage) ODE solver with adaptive step size.
ODEInterpolationSolver	Adjusts its internal step size in order to obtain the desired accuracy but returns an interpolated result at a fixed step size.
Radau5	Implicit Runge–Kutta method of order five for stiff equations.

```
// ode implements the ODE interface
ODEAdaptiveSolver ode_solver = new Dopri5( ode );
ODEAdaptiveSolver ode_solver = new Dopri853( ode );
ODEAdaptiveSolver ode_solver = new Radau5( ode );
```

These algorithms can also be multistepped to produce a fixed step size using static methods in the MultistepSolvers class.

```
// ode implements the ODE interface
ODEAdaptiveSolver ode_solver = MultistepSolvers.MultistepDopri5(ode);
ODEAdaptiveSolver ode_solver = MultistepSolvers.MultistepDopri853(ode);
ODEAdaptiveSolver ode_solver = MultistepSolvers.MultistepRadau5(ode);
```

Another solver developed by the St. Petersburg group combines the adaptive and fixed step-size approaches. The ODEInterpolationSolver class adjusts its internal step size in order to obtain the desired accuracy and interpolates the final state in order to maintain a fixed step size.

```
// ode implements the ODE interface
ODEAdaptiveSolver ode_solver =
    ODEInterpolationSolver.InterpolationDopri5(ode);
ODEAdaptiveSolver ode_solver =
    ODEInterpolationSolver.InterpolationDopri853(ode);
ODEAdaptiveSolver ode_solver =
    ODEInterpolationSolver.InterpolationRadau5(ode);
```

The ODEInterpolationSolver class makes a copy of the initial state of the system and advances this state array. If the program requests the state at some time t_f the internal state may, in fact, be at time greater than t_f. The return values are interpolated from previously computed states that are stored in the solver. This interpolation is accurate and fast and allows the solver to always integrate the differential equations at the optimal step size. Interpolation can produce a significant increase in performance. Run the short example shown in Listing 9.10. The test program evaluates the rate equation 218 times when using the interpolation solver and 613 times when using the multistep solver.

Listing 9.10 ODEInterpolationSolverApp test program.

```
package org.opensourcephysics.manual.ch09;
import org.opensourcephysics.ode.*;
import org.opensourcephysics.numerics.ODEAdaptiveSolver;
```

```
public class ODEInterpolationSolverApp {
  public static void main(String[] args) {
    ODETest ode = new ODETest();
    ODEAdaptiveSolver ode_solver =
        ODEInterpolationSolver.Dopri853(ode);
    ode_solver.setStepSize(1.0);
    ode_solver.setTolerance(1e-6);
    double time = 0;
    double[] state = ode.getState();
    while(time<25) {
      System.out.println("x1 = "+state[0]+" \t t="+time
            +" \t  step size="+ode_solver.getStepSize());
      time += ode_solver.step();
    }
    System.out.println("rate evaluated #: "+ode.n);
  }
}
```

9.8 ■ CONSERVATION LAWS

Although it may appear that we should always use the highest-order differential equation solver available, this is not the case. Consider a system of N particles with pairwise interactions. A three-dimensional model with N particles has $6N$ first-order differential equations, and computing the force for any given state requires evaluation of $\sim N^2/2$ pairwise interactions. The force computation will, of course, be performed many times in a single step if a high-order method is used.

Because the inverse square law in classical mechanics is the prototype for pairwise interactions, we study it to learn about the stability and accuracy of differential equation solvers. The InverseSquare class implements the ODE interface to solve the central force problem (9.9). This class is available on the CD but is not shown here because it is similar to other ODE models such as the simple harmonic oscillator. The InverseSquareApp program shown in Listing 9.11 uses the InverseSquare class to show a simulation of a particle acted on by a $1/r^2$ force pulling toward the center.

InverseSquareApp allows the user to select one of five numerical methods: Euler, Verlet, Runge–Kutta fourth-order, Runge–Kutta–Fehlberg, or multistepping. Default initial conditions are set so as to produce a circular orbit with radius $r = 1$. Running the program shows both the particle motion and the variation in total energy ΔE as a function of time, where the total energy is given by the value of the Hamiltonian H in dimensionless units:

$$H = \frac{1}{2}v^2 + \frac{1}{r} . \tag{9.26}$$

Figure 9.5 compares the energy variation during ten orbits for the Verlet and RK45 solvers using a time step of $\Delta t = 0.01$ and an ODE tolerance of 10^{-3}. Because the RK45 algorithm uses an adaptive step size, we read the tolerance parameter from the control when instantiating this solver. The tolerance parameter has no effect on fixed step-size algorithms such as Verlet. Running the program shows that the adaptive step-size algorithm produces a faster animation because Δt is increased to achieve the desired tolerance. The total energy is, however, not constant but slowly drifts to lower values. Decreasing the tolerance lowers the rate of energy loss but does not change the overall behavior.

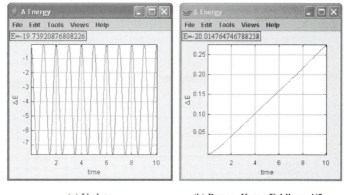

(a) Verlet.   (b) Runge–Kutta–Fehlberg 4/5.

Figure 9.5   The error in total energy depends on the ODE solver.

Listing 9.11   `InverseSquareApp` creates a simulation of a particle acted on by a $1/r^2$ force pulling toward the center.

```
package org.opensourcephysics.manual.ch09;
import org.opensourcephysics.controls.*;
import org.opensourcephysics.display.*;
import org.opensourcephysics.frames.*;
import org.opensourcephysics.numerics.*;

public class InverseSquareApp extends AbstractSimulation {
  DisplayFrame orbitFrame =
      new DisplayFrame("x (AU)", "y (AU)", "Particle Orbit");
  PlotFrame energyFrame =
      new PlotFrame("time", "$\\Delta$ E", "$\\Delta$ Energy");
  InverseSquare particle = new InverseSquare();
  double initialEnergy;

  public InverseSquareApp() {
    orbitFrame.addDrawable(particle);
    orbitFrame.setPreferredMinMax(-2.5, 2.5, -2.5, 2.5);
    orbitFrame.setSquareAspect(true);
    energyFrame.setConnected(true);
    energyFrame.setMarkerShape(0, Dataset.NO_MARKER);
  }

  public void doStep() {
    particle.doStep(); // advances time
    double energy = particle.getEnergy();
    energyFrame.append(0, particle.getTime(), initialEnergy-energy);
    orbitFrame.setMessage("t = "
        +decimalFormat.format(particle.state[4]));
    energyFrame.setMessage("E="+energy, DrawingPanel.TOP_LEFT);
  }

  public void initialize() {
    String solver =
        control.getString("ODE Solver").toLowerCase().trim();
```

```
      if(solver.equals("euler")) {
        particle.odeSolver = new Euler(particle);
      } else if(solver.equals("verlet")) {
        particle.odeSolver = new Verlet(particle);
      } else if(solver.equals("rk4")) {
        particle.odeSolver = new RK45(particle);
      } else if(solver.equals("rk45")) {
        ODEAdaptiveSolver adaptiveSolver = new RK45(particle);
        adaptiveSolver.setTolerance(
            control.getDouble("adaptive stepsize solver tolerance"));
        particle.odeSolver = adaptiveSolver;
      } else if(solver.equals("rk45multistep")) {
        ODEAdaptiveSolver adaptiveSolver = new RK45MultiStep(particle);
        adaptiveSolver.setTolerance(
            control.getDouble("adaptive stepsize solver tolerance"));
        particle.odeSolver = adaptiveSolver;
      } else {
        control.print("Solver not found. Valid solvers are:
            Euler, Verlet, RK4, RK45, RK45MultiStep");
      }
      particle.odeSolver.setStepSize(control.getDouble("dt"));
      double x = control.getDouble("x");
      double vx = control.getDouble("vx");
      double y = control.getDouble("y");
      double vy = control.getDouble("vy");
      particle.initialize(new double[] {x, vx, y, vy, 0});
      orbitFrame.setMessage("t = 0");
      initialEnergy = particle.getEnergy();
      energyFrame.setMessage("E="+initialEnergy, DrawingPanel.TOP_LEFT);
  }

  public void reset() {
    control.setValue("ODE Solver", "Verlet");
    control.setValue("x", 1);
    control.setValue("vx", 0);
    control.setValue("y", 0);
    control.setValue("vy", "2*pi");
    control.setValue("dt", 0.01);
    control.setValue("adaptive stepsize solver tolerance", 1e-3);
    enableStepsPerDisplay(true);
    initialize();
  }

  public static void main(String[] args) {
    SimulationControl.createApp(new InverseSquareApp(), args);
  }
}
```

Although the Verlet algorithm also does not keep the total energy constant, it is better than the higher-order algorithm because the energy oscillates thereby conserving the average energy. Algorithms that preserve a geometric property on phase space, such as energy, are known a symplectic algorithms. Because speed of computation and the conservation of energy are crucial, the Verlet algorithm is often used in molecular dynamics simulations with large numbers of particles (molecules). Statistical averages are of primary interest while the detailed trajectories of individual particles are unimportant in these types of simulations.

Figure 1.1 shows the trajectories of two electrons in a classical helium model. The electrons repel each other and are attracted toward the central core. The program that produced this figure is named `HeliumApp` and is available on the CD. The rate equation is defined in the `Helium` class and is hard-wired for the helium model as follows:

```
public void getRate(double[] state, double[] rate ){
  // state[]: x1, vx1, y1, vy1, x2, vx2, y2, vy2, t
  double deltaX=(state[4]-state[0]);        // x12 separation
  double deltaY=(state[6]-state[2]);        // y12 separation
  double dr_2=(deltaX*deltaX+deltaY*deltaY); // r12 squared
  double dr_3=Math.sqrt(dr_2)*dr_2;         // r12 cubed

  rate[0]=state[1];                         // x1 rate
  rate[2]=state[3];                         // y1 rate

  double r_2=state[0]*state[0]+state[2]*state[2]; // r1 squared
  double r_3=r_2*Math.sqrt(r_2);            // r1 cubed
  rate[1]=-2*state[0]/r_3-deltaX/dr_3;      //vx1 rate
  rate[3]=-2*state[2]/r_3-deltaY/dr_3;      //vy1 rate

  rate[4]=state[5];                         // x2 rate
  rate[6]=state[7];                         // y2 rate
  r_2=state[4]*state[4]+state[6]*state[6];  // r2 squared
  r_3=r_2*Math.sqrt(r_2);                   // r2 cubed
  rate[5]=-2*state[4]/r_3+deltaX/dr_3;      // vx2 rate
  rate[7]=-2*state[6]/r_3+deltaY/dr_3;      // vy2 rate
  rate[8]=1;                                // time rate
}
```

The complexity of the helium rate makes it apparent that a more systematic approach is needed to model the N-body problem. The `PlanarNBodyApp` class on the CD models a number of N-body problems with periodic orbits. The `getRate` method in the `PlanarNBody` class is much simpler because this class implements the `computeForce` method to compute the net force on each particle. Each particle has x, v_x, y, and v_y dynamical variables, and these variables are accessed one particle at a time in a loop.

```
public void getRate(double[] state, double[] rate) {
  computeForce(state); // force array alternates fx and fy
  for(int i=0; i<n; i++){
    int i4=4*i;
    rate[i4]   = state[i4+1];    // x rate is vx
    rate[i4+1] = force[2*i];     // vx rate is fx
    rate[i4+2] = state[i4+3];    // y rate is vy
    rate[i4+3] = force[2*i+1];   // vy rate is fy
  }
  rate[state.length-1]=1;     // time rate is last
}
```

The `NBodyApp` program models an arbitrary number of gravitationally interacting particles but includes a number of additional features such as the ability to save and restore systems of particles using an extensible markup language (xml) data file as discussed in Chapter 12. `NBodyApp` is used as a curriculum development example in Chapter 15.

9.9 ■ COLLISIONS AND STATE EVENTS

Solving differential equations in the presence of collisions is difficult. One way to model a collision is to add a force component perpendicular to the surface of contact. Consider a particle colliding with the floor. We can add a spring force to the rate equation with a spring constant k so that the particle will push away from the floor when it falls below the surface:

$$F_{spring}(z) = -kz \quad (z < 0). \tag{9.27}$$

We can avoid interpenetration of the particle into the floor by choosing a large value for k, but we will then have to take many time steps during the collision resulting in what is known as a *stiff* ODE. A stiff ODE is characterized by more than one time scale. In our example, the time scale for projectile motion is of the order of seconds, whereas the time scale for the collision is of the order of milliseconds. Even sophisticated ODE solvers, such as Runge–Kutta, may experience difficulty in solving this type of dynamics because they assume that values in the state array vary smoothly. Nonpenetrating collisions violate this assumption.

A better way to model a nonpenetrating collision is to detect the collision and then respond to it using the impulse approximation to instantaneously change the particle's velocity. This is a typical example of what we call a *state event* as defined in Listing 9.12. The simplest collision (state event) to consider is an *elastic collision* without friction. The parallel component of the particle's velocity is unaffected and the normal component of the velocity is negated. In an *inelastic collision* the normal component is multiplied by the negative of a number r between zero and one called the *coefficient of restitution*. If $r = 1$, the collision is elastic and if $r = 0$, the collision is said to be totally inelastic.

Listing 9.12   A *StateEvent* defines an illegal condition and a corrective action for an ODE solver.

```
package org.opensourcephysics.numerics;
public interface StateEvent {
  public double getTolerance();

  // evaluate returns negative for illegal state
  public double evaluate( double[] state);

  public boolean action(); // action when state is illegal
}
```

If a collision occurs at time t_c, we must stop the ODE solver and compute new velocities based on the physics of the collision. However, before we can solve the physics, we must deal with the geometric issue of detecting the time of contact between the bodies. Suppose that t_c lies between t_i and $t_{i+1} = t_i + \Delta t$. A simple way of determining t_c is to use the method of *bisection*. If we detect interpenetration at $t_i + \Delta t$, we reset the ODE solver back to t_i and compute the state at $t_i + \Delta t/2$. If the interpenetration has not yet occurred, we know the collision lies between $t_i + \Delta t/2$ and $t_i + \Delta t$ and we advance the simulation from $t_i + \Delta t/2$ to $t_i + 3\Delta t/4$. Otherwise, it lies between t_i and $t_i + \Delta t/2$ and we advance the simulation from t_i to $t_i + \Delta t/4$. This process is repeated until the collision time is computed to within some suitable numerical tolerance. The ODEBisectionEventSolver class in the numerics package implements this algorithm.

The ODEBisectionEventSolver is a differential equation solver that responds to an illegal state by determining the time when the state becomes illegal and taking an appropriate corrective action. The illegal condition and the action are encapsulated in an object that implements the StateEvent interface. The key to this approach is to define an event function $f(t)$ that returns $f(t) \geq 0$ when a state is legal and $f(t) < 0$ when a state is illegal.

The evaluate method returns a value of an event function $f(t)$. Finding the time when a state becomes illegal consists of finding the root of this function.

In order to properly deal with numerical approximations, an event is assumed to have taken place when $f(t) \leq -\epsilon$ and a given state will be considered legal if $f(t) \geq +\epsilon$, where the tolerance ϵ is the value returned by the getTolerance method. The event solver will attempt to reduce δt using the method of bisection to find a time t_c that brings the system to a state where $|f(t_c)| < \epsilon$. The event solver then invokes the action method.

The action method must bring the system to a state such that either $f(t_c) \geq \epsilon$ or that this condition will become true after a finite number of iterations.

Several possible state events may be defined for the same ODE. In this case, if more than one event takes place in the same interval $[t, t + \delta t]$, the solver must search for the one that takes place first and apply the corresponding action. If two of the events take place at the very same instant, they will be dealt with by the solver in an arbitrary order. Care must be taken not to produce infinite loops by events that trigger a second event which in turn triggers again the first event. This will result in the computer hanging at this point. A second caveat consists in the so-called *Zeno* problem described below.

Elastic Collision Event

The ElasticCollision class shown in Listing 9.13 models a floor-ball collision using the height of the ball above the floor minus the ball's radius as the event function. Notice that we only trigger an event if the ball is falling.

Listing 9.13 An elastic collision state event.

```
class ElasticCollision implements StateEvent {

  public double getTolerance() {return TOL;}

  public double evaluate(double[] state) {
     return(state[1]<0)        // v must be negative
            ? state[0]-radius  // y should be >= radius
            : TOL;             // return a legal state
  }

  public boolean action() {
     state[1] = -state[1];     // make vy positive
     stopAtCollision = true;   // false continues integration
     return stopAtCollision;   // stop the integration
  }
} // end of inner class ElasticCollision
```

The state event action is triggered when the event function is less than zero. This reverses the velocity, sets the stopAtCollision boolean, and returns. The returned value tells the solver whether it should stop the computation at the exact moment of the event or continue solving the ODE for the rest of the prescribed step Δt.

The BallFloorCollision class shown in Listing 9.14 uses the ElasticCollision class as an inner class to trigger events in an ODEBisectionEventSolver. The solver is instantiated using the Runge–Kutta fourth-order algorithm and the elastic collision event is created and added to the solver in the constructor. The `BallFloorCollisionApp` is available on the CD but is not shown here because it is similar to other applications.

Listing 9.14   A floor-ball collision can be modeled using a `StateEvent`.

```
package org.opensourcephysics.manual.ch09;
import org.opensourcephysics.display.*;
import org.opensourcephysics.numerics.*;
import java.awt.*;

public class BallFloorCollision implements ODE, Drawable {
  static final double TOL = 0.001;
  static final double g = 9.8;                    // acceleration of gravity
  double[] state = new double[] {10, 0, 0}; // y,v,t
  double radius = 1, dt = 0.1;
  boolean stopAtCollision = false;
  DrawableShape box = DrawableShape.createRectangle(0, -0.5, 10, 1);
  DrawableShape ball = DrawableShape.createCircle(0, 0, 2*radius);
  // a solver that supports state events

  ODEBisectionEventSolver solver =
      new ODEBisectionEventSolver(this, RK4.class);

  public BallFloorCollision() {
    // choose one of the following two StateEvents
    // solver.addEvent(new ElasticCollision());
    solver.addEvent(new InelasticCollision());
    solver.initialize(dt);
    ball.setMarkerColor(new Color(128, 128, 255), Color.BLUE);
  }

  void doStep() {
    solver.step();
  }

  public double[] getState() {
    return state;
  }

  public void getRate(double[] state, double[] rate) {
    rate[0] = state[1]; // dy/dt = v
    rate[1] = -g; // fails with inelastic collisions due to Zeno effect
    rate[2] = 1.0;      // dt/dt = 1
  }

  public void draw(DrawingPanel panel, Graphics g) {
    ball.setXY(0, state[0]);
    ball.draw(panel, g);
    box.draw(panel, g);
  }

  private class ElasticCollision implements StateEvent {
    public double getTolerance() {
      return TOL;
    }
```

```
    public double evaluate(double[] state) {
      return(state[1]<0)        // v must be negative
            ? state[0]-radius // y should be >= radius
            : TOL;                // return a legal state
    }

    public boolean action() {
      state[1] = -state[1];    // make vy positive
      stopAtCollision = true; // change this to false to conintinue
      return stopAtCollision; // collision stops the integration step
    }
  } // end of inner class ElasticCollision

  private class InelasticCollision implements StateEvent {
    double r = 0.8; // coefficient of restitution

    public double getTolerance() {
      return TOL;
    }

    public double evaluate(double[] state) {
      return(state[1]<0)        // v must be negative
            ? state[0]-radius // y should be >= radius
            : TOL;                // return a legal state
    }

    public boolean action() {
      state[1] = -r*state[1];    // make vy positive and reduce its value
      stopAtCollision = false; // change this to true to stop
      return stopAtCollision; // collision stops the integration step
    }
  } // end of inner class InelasticCollision
}
```

Bisection

Although the bisection method is one with a rather slow convergence rate for the general problem, it presents several advantages for this particular situation. First, it only assumes that f is continuous when computing the root; actually, in many real problems, our `evaluate` method returns nondifferentiable values.[1] Second, the lack of speed is compensated by the smaller number of evaluations of the function $f(t)$. When evaluating the function is expensive (in terms of computations), the bisection method compares better to faster convergent methods (which require several evaluations of $f(t)$ for each iteration step). Third, the method allows simple separation of events when more than one takes place in the same interval. Finally, because the step size Δt for most ODEs is already small, convergence is usually quickly attained.

The `BallFloorCollision` class contains a second state event `InelasticCollision`. This event's action method reduces the velocity of the falling ball at every bounce in order to model more realistic physics. The model will, however, fail if the elastic event solver is used due to what is known as a *Zeno* problem. In the real world, the ball's velocity is zero as the ball reaches its equilibrium state.

---

[1]The state function $f(t)$ may even be discontinuous if it does not trigger a state event near the discontinuity. The inelastic collision example is discontinuous when the velocity changes sign.

Events take place at decreasingly smaller time intervals as the ball approaches equilibrium. In other words, when the vertical velocity is very small, rebounding takes place in a sequence of t_n such that $t_{n+1} - t_n \to 0$. The solution to this problem is to change the physics embodied in the getRate method when the ball is in resting contact with the floor. Otherwise we obtain infinite force during interpenetration or we must model the deformation of the ball and the floor. The net force (acceleration) of the ball is zero when the ball is in resting contact with the floor and the computation of rate[1] should reflect this condition. Inserting the following code fragment into the BallFloorCollision class solves the Zeno problem because the ball no longer falls below the floor and therefore does not trigger state events when in resting contact.

```
rate[1] = (state[0]-radius>0) // resting on floor changes physics
          ? -g    // dv/dt = -g when falling
          : 0;    // dv/dt = 0 when resting
```

There is a vast body of literature dealing with collision detection and resting contact. Needless to say, we have only scratched the surface. If the motion occurs in two dimensions, then the constraints act normal to the surface of contact. Simple examples of two-dimensional simulations are the BallBoxCollisionApp and the BallsInBoxApp programs available on the CD.

9.10 ■ PROGRAMS

The following examples are in the org.opensourcephysics.manual.ch09 package.

AdaptiveStepApp

AdaptiveStepApp demonstrates the advantage of an adaptive step-size ODESolver by integrating a particle's response to short-duration force as described in Section 9.5.

BallBoxCollisionApp

BallBoxCollisionApp simulates a bouncing ball in a box using StateEvent objects to constrain the position of the ball (see Section 9.9).

BallFloorCollisionApp

BallFloorCollisionApp models a particle colliding with the floor using a StateEvent. See Section 9.9.

BallsInBoxApp

BallsInBoxApp simulates interacting (colliding) hard disks constrained within a box and falling under the influence of gravity. See Section 9.9.

HeliumApp

HeliumApp models electron orbits in a classical helium atom as described in Section 9.8.

InverseSquareApp

InverseSquareApp models a particle orbiting under the influence of an inverse square $1/r^2$ force. (See Section 9.8.)

NBodyApp

NBodyApp models an arbitrary number of particles interacting through a $1/r^2$ force (see Section 9.8).

PlanarNBodyApp

PlanarNBodyApp models special cases of the gravitational few-body problem with closed periodic orbits (see Section 9.8).

SHOApp

SHOApp tests the ODESolver interface using a simple harmonic oscillator as described in Section 9.3.

SineIntegralApp

SineIntegralApp creates a table of values by integrating the function $\sin x/x$ as described in Section 9.6.

The following examples require the org.opensourcephysics.ode package from Saint Petersburg Polytechnic University, Russia. This optional package is described in Section 9.7 and is available on the CD.

ComparisonApp

ComparisonApp displays the difference between an exact solution and a numerical solution.

DopriApp

DopriApp tests the Dorman–Prince differential equation solver implementations.

ODEInterpolationSolverApp

ODEInterpolationSolverApp tests the ODE interpolation solver implementations.

ODEMultistepSolverApp

ODEMultistepSolverApp tests the ODE multistep solver implementations.

CHAPTER
10

Numerics

The *numerics* package (`org.opensourcephysics.numerics`) defines numerical analysis tools. Because we do not seek to implement a large scientific subroutine library, this package emphasizes numerical algorithms and APIs that we have found to be important in the teaching of computational physics.

10.1 ■ FUNCTIONS

One of the most fundamental concepts in mathematics is that of a function. A *function* is a rule f that gives a well-defined output y corresponding to some well-defined input x, specifically

$$y = f(x) \quad \text{for some domain of } x. \tag{10.1}$$

The variable y is called the dependent variable and the variable x is called the independent variable. The set of permissible values of x is called the *domain* and the set of all possible outputs is called the *range*. Figure 10.1 shows a simple function plotter that is able to evaluate keyboard input.

The numerics package defines the `Function` interface shown in Listing 10.1 that allows us to program functions with domain and range restricted to values allowed by Java data type `double`.

Listing 10.1   The `Function` interface is defined in the numerics package.

```
package org.opensourcephysics.numerics;
public interface Function {
  public double evaluate(double x);
}
```

We can use the `Function` interface to implement real-valued functions such as the amplitude of a single slit diffraction pattern ($f(x) = \sin x / x$).

```
public class SlitPattern implements Function {
  public double evaluate (double x) {
    return (x==0)? 1 :Math.sin (x)/x;
  }
}
```

The numerics package also contains interfaces for functions of more than one variable. `MultiVarFunction` defines an interface for an object that accepts an array of numbers and returns a number, while `VectorFunction` defines an interface for an object that accepts

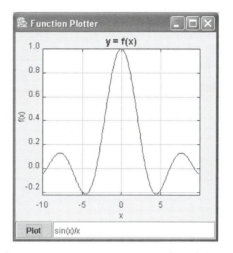

Figure 10.1 A function plotter uses a parser to convert a string of characters into a `Function`.

an array of numbers and returns an array of numbers. The `RectangularPattern` class models a multivariable function for the diffraction pattern from a rectangular aperture $f(x, y) = (\sin x/x)(\sin y/y)$ with a height to width ratio of 0.4.

```
public class RectangularPattern implements MultiVarFunction {
    double ratio = 0.4;
    public double evaluate (double[] x) {
        SlitPattern sp = new SlitPattern();
        return sp.evaluate(x[0])*sp.evaluate(ratio*x[1]);
    }
}
```

To display a function in a drawing panel, we must evaluate and plot a series of points $(x, f(x))$ for a given interval. Although we can use a loop to add points to a dataset, we can also use the `FunctionDrawer` class in the display package. The `FunctionDrawer` evaluates the given function by computing values at the specified number of points within the domain.

```
// drawingPanel is a DrawingPanel
boolean filled = false;    // area under function not filled
int n = 200;               // number of data points
double xmin=-10, xmax=10;  // plotting domain
Function f = new SlitPattern();
drawingPanel.addDrawable(
    new FunctionDrawer(f,xmin, xmax, n, filled));
```

If the `FunctionDrawer` is instantiated with only a single parameter $f(x)$, then the domain is undefined. The `FunctionDrawer` draws the function by evaluating points within the drawing panel's x-minimum and maximum values.

The `functionFill` utility method in the numerics `Util` class saves us the trouble of writing a loop to evaluate a function and storing values in an array. This utility method takes the following input parameters: (1) a function, (2) minimum and maximum values for the independent variable, and (3) an integer specifying the number of data points. It returns a two-column array containing x- and y-values which can then be passed to a dataset or a data table for display.

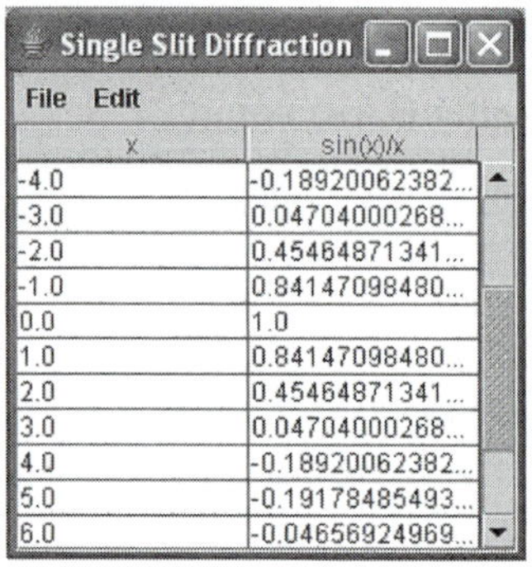

x	sin(x)/x
-4.0	-0.18920062382...
-3.0	0.04704000268...
-2.0	0.45464871341...
-1.0	0.84147098480...
0.0	1.0
1.0	0.84147098480...
2.0	0.45464871341...
3.0	0.04704000268...
4.0	-0.18920062382...
5.0	-0.19178485493...
6.0	-0.04656924969...

Figure 10.2 A table of single-slit diffraction values.

```
// f is an object that implements Function
double[][] array=Util.functionFill(f,-10,10,new double[2][21]);
```

Creating the table of the single-slit diffraction values shown in Figure 10.2 takes just a few lines of code as shown in Listing 10.2. The code uses a `TableFrame` (see Section 3.8) and an anonymous class to define the `Function`.

Listing 10.2 `SlitTableApp` creates a data table using a `Function` to fill an array.

```
package org.opensourcephysics.manual.ch10;
import org.opensourcephysics.frames.*;
import org.opensourcephysics.numerics.*;

public class SlitTableApp {
  public static void main(String[] args) {
    TableFrame frame = new TableFrame("Single Slit Diffraction");
    Function f = new Function() { // create anonymous class
      public double evaluate(double x) {
        return(x==0) ? 1 : Math.sin(x)/x;
      }
    }; // end of Function
    double[][] array =
         Util.functionFill(f, -10, 10, new double[2][21]);
    frame.setRowNumberVisible(false);
    frame.appendArray(array);
    frame.setColumnNames(0, "x");
    frame.setColumnNames(1, "sin(x)/x");
    frame.setVisible(true);
    frame.setDefaultCloseOperation(javax.swing.JFrame.EXIT_ON_CLOSE);
  }
}
```

Many classes in the numerics package define methods that instantiate `Function` objects or implement `Function` themselves. For example, the `Util` class contains convenience methods for creating a constant function $f(x) = c$, a linear function $f(x) = mx + b$, and a Gaussian $f(x) = \frac{1}{\sigma\sqrt{2\pi}} e^{-(x-x_0)^2/2\sigma^2}$. The `Polynomial` and `CubicSpline` classes implement `Function`. We can take the derivative, integrate, or Fourier analyze these objects because the algorithms that perform these mathematical operations are written independent of the type of function that they will receive.

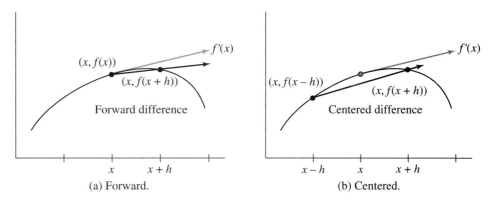

Figure 10.3 The true derivative of a function $f'(x)$, shown as a light arrow, can be approximated by evaluating a function at two or more points. The (a) forward and (b) centered approximations to the first derivative are shown as dark arrows.

10.2 ■ DERIVATIVES

If a function $f(x)$ is evaluated at two points x and $x + h$, we can compute the slope of the line that passes through those points. The value of the slope as $h \to 0$, shown as a light arrow in Figure 10.3, is the first derivative of the function:

$$f'(x) = \frac{df(x)}{dx} \equiv \lim_{h \to 0} \frac{f(x+h) - f(x)}{h} \quad \text{(first derivative)}. \qquad (10.2)$$

We can approximate this definition by using a small value of h, subtracting the terms in the numerator, and performing the division. This approximation is known as a first-order forward finite difference. Higher-order approximations can be made. The `Derivative` class in the numerics package defines second-order approximations to the first derivative by evaluating the function $f(x)$ at various points in the vicinity of x where h is the interval over which the approximation is made.

The following first-derivative approximations are second order in h but use the intervals to the right of x, centered on x, and to the left of x. The centered approximation is the most efficient, but the other methods are useful if the function $f(x)$ is only defined to one side of x:

$$\frac{df(x)}{dx} \approx \frac{-f(x+2h) + 4f(x+h) - 3f(x)}{2h} \qquad (10.3a)$$

$$\frac{df(x)}{dx} \approx \frac{f(x+h) - f(x-h)}{2h} \quad \text{(centered)} \qquad (10.3b)$$

$$\frac{df(x)}{dx} \approx \frac{f(x-2h) - 4f(x-h) + 3f(x)}{2h}. \qquad (10.3c)$$

These formulas are coded in the `Derivative` class.

```
static public double forward(Function f, double x, double h){
  return (-f.evaluate(x+2*h) +
          4*f.evaluate(x+h) - 3*f.evaluate(x))/h/2.0;
}
```

```
static public double centered(Function f,double x, double h){
   return (f.evaluate (x + h) - f.evaluate(x - h)) / h/2.0;
}

static public double backward(Function f, double x, double h){
   return (f.evaluate(x-2*h) -
          4*f.evaluate(x-h) + 3*f.evaluate(x))/h/2.0;
}
```

The Derivative class also implements a first-derivative algorithm based on Romberg's implementation of Richardson's deferred approach to the limit. This algorithm computes the derivative multiple times using ever smaller values of h. The sequence of results can be extrapolated to the limit $h \to 0$, and this extrapolation can also be used to estimate an error. If the estimated error is greater than the given tolerance, then the calculation is repeated and another term is added to the sequence. This algorithm is implemented in the romberg method. Computing multiple approximations and calculating the tolerance does, of course, require extra computer cycles.

Finite difference approximations can be derived for higher-order derivatives and for partial derivatives of multivariable functions. The most useful is the centered finite difference approximation for the second derivative:

$$\frac{d^2 f(x)}{dx^2} \approx \frac{f(x+h) - 2f(x) + f(x-h)}{2h}. \tag{10.4}$$

This formula is implemented in the Derivative class.

```
static public double second(Function f, double x, double h) {
   return(f.evaluate(x+h)-2*f.evaluate(x)+f.evaluate(x-h))/h/h;
}
```

Static methods in the Derivative class are useful for evaluating derivatives at points, but it is sometimes convenient to create a derivative Function. The getFirst method returns a Function object that does just that. (See Section 10.6 for how to compute the derivative of a polynomial.)

```
static public Function getFirst(final Function f, final double h) {
   return new Function() {
      public double evaluate(double x) { // in-line the code for speed
         return(f.evaluate(x+h)-f.evaluate(x-h))/h/2.0;
      }
   }; // end of Function
}
```

The getSecond method is defined in a similar manner and returns a second-derivative Function.

Gaussian functions (normal distributions) occur frequently in physics and the Util class in the numerics package defines the gaussian method that instantiates a Gaussian Function $f(x) = \frac{1}{\sigma\sqrt{2\pi}} e^{-(x-x_0)^2/2\sigma^2}$ with a specified center x_0 and width σ. The full width at half maximum is $2\sqrt{2\ln 2}\sigma \approx 2.3548\sigma$. We plot the derivative of this function as shown in Figure 10.4 using numerical derivatives and a FunctionDrawer as shown in Listing 10.3. Note the high level of abstraction that has been achieved by using the Function interface.

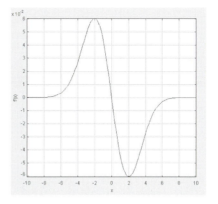

Figure 10.4 The numerical derivative of a Gaussian Function.

Listing 10.3 GaussianDerivativeApp plots the numerical derivative
of a Gaussian Function.

```
package org.opensourcephysics.manual.ch10;
import org.opensourcephysics.display.*;
import org.opensourcephysics.frames.*;
import org.opensourcephysics.numerics.*;

public class GaussianDerivativeApp {
  public static void main(String[] args) {
    PlotFrame frame =
        new PlotFrame("x", "f'(x)", "Derivative of Gaussian");
    Function f = Util.gaussian(0, 2.0); // center=0; sigma=2
    Function df = Derivative.getFirst(f, 0.001);
    frame.addDrawable(new FunctionDrawer(df, -10, 10, 200, false));
    frame.setVisible(true);
    frame.setDefaultCloseOperation(javax.swing.JFrame.EXIT_ON_CLOSE);
  }
}
```

10.3 ■ INTEGRALS

Evaluating a definite integral for an arbitrary function $\int_a^b f(x)\,dx$ is more complicated than evaluating the derivative. The integral of a function can be visualized as the area under the curve from a to b as shown in Figure 10.5. Unfortunately, even simple functions, such as $\sin x/x$, cannot be integrated analytically and we must resort to numerical methods. The Integral class implements a number of algorithms for the approximation of integrals including the trapezoidal method, Simpson's method, Romberg's method, and the ODE approach described in Section 9.5.

The rectangular approximation is the simplest integral approximation. It computes the area assuming the value of the function $f(x)$ is constant between x and $x + \Delta x$:

$$\int_a^b f(x) \approx \sum_{i=0}^{n-1} f(x_i)\,\Delta x. \tag{10.5}$$

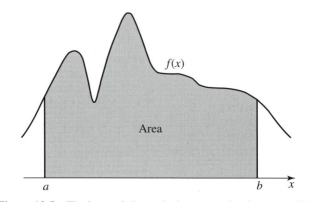

Figure 10.5   The integral F equals the area under the curve $f(x)$.

Because the error shown in Figure 10.6 is first order in Δx, the rectangular approximation is used only for illustration. The trapezoidal approximation assumes a linear change between $f(x)$ and $f(x + \Delta x)$, and Simpson's approximation assumes a polynomial fit between sampled points.

IntegralApp shown in Listing 10.4 tests the accuracy of algorithms in the Integral class and estimates their efficiency by printing the number of function evaluations needed to obtain the desired precision. Run the test program and notice that Romberg's method is the most efficient for high-precision results. Simpson's methods are more robust because they merely evaluate a sum, and these algorithms are usually sufficient.

Listing 10.4   IntegralApp tests the integration algorithms.

```
package org.opensourcephysics.manual.ch10;
import org.opensourcephysics.numerics.*;

public class IntegralApp {
  static final double LN2 = Math.log(2);

  public static void main(String[] Args) {
    Function f = new IntegralTestFunction();
    double a = 1, b = 2;
    double tol = 1.0e-10; // double has 16 significant digits
    double area = Integral.ode(f, a, b, tol);
    System.out.println("ODE area="+area+"  err="+(area-LN2));
    System.out.println("counter="+IntegralTestFunction.c);
    IntegralTestFunction.c = 0;
    area = Integral.trapezoidal(f, a, b, 2, tol);
    System.out.println("Trapezoidal area="+area+"  err="+(area-LN2));
    System.out.println("counter="+IntegralTestFunction.c);
    IntegralTestFunction.c = 0;
    area = Integral.simpson(f, a, b, 2, tol);
    System.out.println("Simpson area="+area+"  err="+(area-LN2));
    System.out.println("counter="+IntegralTestFunction.c);
    IntegralTestFunction.c = 0;
    area = Integral.romberg(f, a, b, 2, tol);
    System.out.println("Romberg area="+area+"  err="+(area-LN2));
```

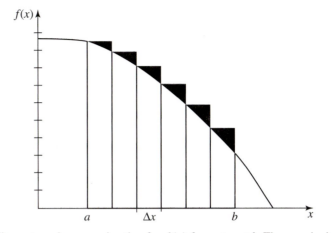

Figure 10.6 The rectangular approximation for $f(x)$ for $a \leq x \leq b$. The error is shaded.

```
      System.out.println("counter="+IntegralTestFunction.c);
      IntegralTestFunction.c = 0;
   }
}

class IntegralTestFunction implements Function {
   static long c = 0;

   public double evaluate(double x) {
      c++;
      return 1.0/x;
   }
}
```

10.4 ■ PARSERS

An object that evaluates a string of characters and produces a numerical result is known as a mathematical expression *parser*. Parsers are very useful since they allow a user to change a mathematical function without recompiling the program. Two expression parsers are included on the Open Source Physics CD. One parser was written by Yanto Suryono and another by Nathan Frank. The first parser is very fast but is restricted to real numbers, while the second supports both real and complex data types. Table 10.1 lists the functions recognized by Suryono's parser.

To hide implementation details and to have a consistent—albeit minimal— API to parse functions of real variables, we defined an abstract class named MathExpParser. A MathExpParser class implements both the Function and MultiVarFunction interfaces and declares abstract methods to set the function and the independent variable strings. The SuryonoParser class is a concrete implementation of a MathExpParser, and this class is used whenever we need a real-valued parser. The ParsedFunction class that is used in Section 2.3 to process user input contains a SuryonoParser object.

Table 10.1   The OSP expression parser accepts common one- and two-parameter functions.

abs(a)	acos(a)	acosh(a)	asin(a)	asinh(a)
atan(a)	atanh(a)	ceil(a)	cos(a)	cosh(a)
exp(a)	frac(a)	floor(a)	int(a)	log(a)
random(a)	round(a)	sign(a)	sin(a)	sinh(a)
sqr(a)	sqrt(a)	step(a)	tan(a)	tanh(a)
atan2(a,b)	max(a,b)	min(a,b)	mod(a,b)	

```
String fx="sin(x)/x";  // usually read from a keyboard
try {
  function = new ParsedFunction(fx, "x");
} catch(ParserException ex) { // user input errors are common
  // set f(x)=0 if there is an error
  function = Util.constantFunction(0);
}
```

Using a parser is straightforward. Define a character string and pass it to the parser either in the constructor or using the setFunction method as shown in Listing 10.5. Because users often make typing mistakes when entering mathematical expressions, the setFunction method throws an exception if the string does not represent a valid expression. Catching the exception allows the program to inform the user of the mistake and take corrective action.

Listing 10.5   ParseRealApp parses a string to produce a function.

```
package org.opensourcephysics.manual.ch10;
import org.opensourcephysics.numerics.*;

public class ParseRealApp {
  public static void main(String[] args) {
    MathExpParser parser = MathExpParser.createParser();
    try {
      parser.setFunction("sin(2*pi*x)", new String[] {"x"});
    } catch(ParserException ex) {
      System.out.println(ex.getMessage());
    }
    for(double x = 0, dx = 0.1;x<1;x += dx) {
      System.out.println("x="+Util.f2(x)+" \t y="+parser.evaluate(x));
    }
  }
}
```

The complex expression parser is distributed in the jep.zip code archive. You must download and import this package into your Java development environment before running the ParseComplexApp example shown in Listing 10.6.

Listing 10.6   ParseComplexApp parses a string to produce a function.

```
package org.opensourcephysics.manual.ch10;
import org.opensourcephysics.numerics.*;
import org.nfunk.*;
import org.nfunk.jep.type.Complex;
```

```
public class ParseComplexApp {
  public static void main(String[] args) {
    JEParser parser = null;
    try {
      parser = new JEParser("e^(i*x)", "x", JEParser.MAKE_COMPLEX);
    } catch(ParserException ex) {
      System.out.println(ex.getMessage());
    }
    for(double x = 0, dx = 0.1;x<1;x += dx) {
      Complex z = parser.evaluateComplex(x);
      System.out.println("x="+Util.f2(x)+" \t re="+Util.f3(z.re())
          +" \t im="+Util.f3(z.im())));
    }
  }
}
```

The `RealWaveApp` and `ComplexWaveApp` programs are more elaborate examples that use real and complex parsers to animate time-dependent wave functions $f(x, t)$. They are not shown here but are available on the CD.

10.5 ■ FINDING THE ROOTS OF A FUNCTION

The roots of a function $f(x)$ are the values of the variable x for which the function $f(x)$ is zero. Even an apparently simple problem, such as finding the energy roots of the quantum finite well,

$$f(x) = \cot x - x = 0, \tag{10.6}$$

cannot be solved analytically for x.

Whatever the function and whatever the approach to root finding, the first step should be to learn as much as possible about the function or functions involved. Analyzing the corresponding physics and mathematics is the best approach. One technique is to plot a function to help guess the approximate locations of the roots as shown in Figure 10.7.

Newton's (or the Newton–Raphson) method for finding a root is based on replacing the function by the first two terms of the Taylor expansion of $f(x)$ about the root x. If our initial guess for the root is x_0, we can write $f(x) \approx f(x_0) + (x - x_0)f'(x_0)$. If we set $f(x)$ equal to zero and solve for x, we find $x = x_0 - f(x_0)/f'(x_0)$. If we made a good choice for x_0, the resultant value of x should be closer to the root. The general procedure is to calculate successive approximations as follows:

$$x_{n+1} = x_n - \frac{f(x_n)}{f'(x_n)}. \tag{10.7}$$

If this series converges, it converges very quickly. However, if the initial guess is poor or if the function has closely spaced multiple roots, the series may not converge.

Newton's algorithm is implemented in the `Root` class in the numerics package with two different signatures as shown in Listing 10.7. The more efficient method requires a derivative function. The other method takes a numerical derivative using the Romberg's method. Both methods return not-a-number (`Double.NaN`) if the method fails to converge.

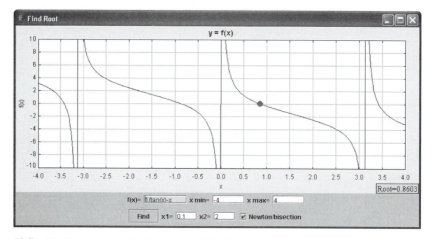

Figure 10.7   Finding the roots of a function is difficult because a function such as $f(x) = \cot x - x = 0$ can have many roots.

Listing 10.7   The `newton` method is defined in the `Root` class
in the `numerics` package.

```
public static double newton( Function f, double x, double tol) {
  int count = 0;
  while (count < MAX_ITERATIONS) {
    double xold = x; // save the old value to test for convergence
    double df = 0;
    try {
      df = Derivative.romberg(
          f, x, Math.max(0.001, 0.001 * Math.abs(x)), tol / 10);
    }
    catch (NumericMethodException ex) {
      return Double.NaN; // did not converge
    }
    x -= f.evaluate(x) / df;
    if (Util.relativePrecision(Math.abs(x - xold), x) < tol)return x;
    count++;
  }
  return Double.NaN; // did not converge in max iterations
}

public static double newton(final Function f, final Function df,
                      double x, final double tol) {
  int count = 0;
  while (count < MAX_ITERATIONS) {
    double xold = x; // save the old value to test for convergence
    // approximate the derivative using centered difference
    x -= f.evaluate(x) / df.evaluate(x);
    if (Util.relativePrecision(Math.abs(x - xold), x) < tol)return x;
    count++;
  }
  return Double.NaN; // did not converge in max iterations
}
```

The simplest root-finding algorithm is the *bisection* method shown in Listing 10.8. The algorithm works as follows:

1. Choose values x_{left} and x_{right}, with $x_{\text{left}} < x_{\text{right}}$ such that $g^{(p)}(x_{\text{left}})g^{(p)}(x_{\text{right}}) < 0$. There must be a value of x such that $g^{(p)}(x) = 0$ in the interval $[x_{\text{left}}, x_{\text{right}}]$.

2. Choose the midpoint $x_{\text{mid}} = x_{\text{left}} + \frac{1}{2}(x_{\text{right}} - x_{\text{left}}) = \frac{1}{2}(x_{\text{left}} + x_{\text{right}})$ as the guess for x^*.

3. If $g^{(p)}(x_{\text{mid}})$ has the same sign as $g^{(p)}(x_{\text{left}})$, then replace x_{left} by x_{mid}; otherwise, replace x_{right} by x_{mid}. The interval for the location of the root is now reduced.

4. Repeat steps 2 and 3 until the desired level of precision is achieved.

The bisection algorithm is also implemented in the `Root` class in the numerics package. It can be used with any function, but it requires that the root be bracketed by two values having opposite signs.

Listing 10.8   The `bisection` method is defined in the `Root` class in the numerics package.

```
public static double bisection(final Function f, double x1,
                               double x2, final double tol) {
   int count = 0;
   int maxCount = (int)(Math.log(Math.abs(x2 - x1)/tol)/Math.log(2));
   maxCount = Math.max(MAX_ITERATIONS, maxCount) + 2;
   double y1 = f.evaluate(x1), y2 = f.evaluate(x2);
   if (y1 * y2 > 0) { // y1 and y2 must have opposite signs
      return Double.NaN; // interval does not contain a root
   }
   while (count < maxCount) {
      double x = (x1 + x2) / 2;
      double y = f.evaluate(x);
      if(Util.relativePrecision(Math.abs(x1-x2), x)<tol) return x;
      if (y * y1 > 0) { // replace endpoint that has the same sign
         x1 = x;
         y1 = y;
      }
      else {
         x2 = x;
         y2 = y;
      }
      count++;
   }
   return Double.NaN; // did not converge in max iterations
}
```

There are many root-finding techniques, but all can be made to fail with appropriately chosen functions. The bisection algorithm is guaranteed to converge if we can find an interval where the function changes signs. However, it may be slow. Newton's algorithm is fast but may not converge. A hybrid algorithm that keeps track of the interval between the most negative and the most positive values while iterating is a good approach. The algorithm stars with Newton's method. If the solution jumps outside the given interval, Newton's method is interrupted and the bisection algorithm is used for one step. But any root algorithm can fail, so be sure to test the root at the end of the iteration process to check that the algorithm actually converged.

The `RootApp` program shown in Listing 10.9 compares Newton's algorithm and the bisection method for finding the root(s) of $f(x) = x^2 - 2$. Both algorithms succeed in finding the positive root $\sqrt{2}$ with the given parameters. However, if we set x1=0, Newton's algorithm fails, and if we set x1=-10 and x2=10, the bisection algorithm fails for obvious reasons. The `RootFinderApp` program provides a graphical user interface for exploring the properties of these two algorithms. This program is not shown here but is available on the CD.

Listing 10.9   `RootApp` compares Newton's algorithm and the bisection root-finding methods.

```
package org.opensourcephysics.manual.ch10;
import org.opensourcephysics.numerics.*;

public class RootApp {
  public static void main(String[] Args) {
    Function f = new Function() {
      public double evaluate(double x) {
        double f = x*x-2;
        return f;
      }
    };
    double x1 = 1, x2 = 2;
    double tol = 1.0e-3;
    double root = Root.bisection(f, x1, x2, tol);
    System.out.println("bisection method root="+root);
    root = Root.newton(f, x1, tol);
    System.out.println("Newton's method root="+root);
  }
}
```

10.6 ■ POLYNOMIALS

A polynomial is a function that is expressed as

$$p(x) = \sum_{i=0}^{n} a_i x^i, \tag{10.8}$$

where n is the *degree* of the polynomial and the n constants a_i are the *coefficients*. The evaluation of (10.8) as written is very inefficient because x is repeatedly multiplied by itself and the entire sum requires $\mathcal{O}(N^2)$ multiplications. A more efficient algorithm was published in 1819 by W. G. Horner.[1] It uses a factored polynomial and requires only n multiplications and n additions and is now known as *Horner's rule*. It is written as follows:

$$p(x) = a_0 + x\Big[a_1 + x\big[a_2 + x[a_3 + \cdots]\big]\Big]. \tag{10.9}$$

Using the correct algorithm for this simple task can dramatically reduce processor time if large polynomials are repeatedly evaluated.

[1]This method of evaluating polynomials by factoring was already known to Newton.

Table 10.2 Methods for manipulating polynomial coefficients are defined in the `Polynomial` class.

org.opensourcephysics.numerics.Polynomial	
`add(double a)`	Adds a scalar to this polynomial and returns a new polynomial.
`add(Polynomial p)`	Adds a polynomial to this polynomial and returns a new polynomial.
`deflate(double r)`	Reduces the degree of this polynomial by removing the given root r.
`derivative()`	Returns the derivative of this polynomial.
`integral(double a)`	Returns the integral of this polynomial having the value a at $x = 0$.
`subtract(double a)`	Subtracts a scalar from this polynomial and returns a new polynomial.
`subtract(Polynomial p)`	Subtracts a polynomial from this polynomial and returns a new polynomial.

Polynomials are important computationally because most analytical functions can be approximated as a polynomial using a Taylor-series expansion. Polynomials can be added, multiplied, integrated, and differentiated analytically and the result is still a polynomial. This property makes them ideally suited for object-oriented programming. The `Polynomial` class in the numerics package implements many of these algebraic operations (see Table 10.2). Listing 10.10 shows a test program that demonstrates how to calculate and display a polynomial's roots.

Listing 10.10 The `PolynomialApp` class tests the `Polynomial` class.

```java
package org.opensourcephysics.manual.ch10;
import org.opensourcephysics.controls.*;
import org.opensourcephysics.display.*;
import org.opensourcephysics.numerics.*;

public class PolynomialApp extends AbstractCalculation {
  PlottingPanel drawingPanel =
      new PlottingPanel("x", "f(x)", "Polynomial Visualization");
  DrawingFrame drawingFrame = new DrawingFrame(drawingPanel);
  double xmin, xmax;
  Polynomial p;

  public void resetCalculation() {
    control.setValue("coefficients", "-2,0,1");
    control.setValue("xmin", -10);
    control.setValue("xmax", 10);
  }

  public void calculate() {
    xmin = control.getDouble("xmin");
    xmax = control.getDouble("xmax");
    String[] coef = control.getString("coefficients").split(",");
    p = new Polynomial(coef);
    plotAndCalculateRoots();
  }
}
```

```
void plotAndCalculateRoots() {
  drawingPanel.clear();
  drawingPanel.addDrawable(new FunctionDrawer(p));
  double[] range = Util.getRange(p, xmin, xmax, 100);
  drawingPanel.setPreferredMinMax(xmin, xmax, range[0], range[1]);
  drawingPanel.repaint();
  double[] roots = p.roots(0.001);
  control.clearMessages();
  control.println("polynomial="+p);
  for(int i = 0, n = roots.length;i<n;i++) {
    control.println("root="+roots[i]);
  }
}

public void derivative() {
  p = p.derivative();
  plotAndCalculateRoots();
}

public static void main(String[] args) {
  Calculation app = new PolynomialApp();
  CalculationControl c = new CalculationControl(app);
  c.addButton("derivative", "Derivative",
              "The derivative of the polynomial.");
  app.setControl(c);
}
}
```

It is always possible to construct a polynomial that passes through a set of n data points (x_i, y_i) by creating a *Lagrange interpolating polynomial* as follows:

$$p(x) = \sum_{i=0}^{n} \frac{\prod_{i \neq j}(x - x_j)}{\prod_{i \neq j}(x_i - x_j)} y_i . \tag{10.10}$$

For example, three data points generate the second-degree polynomial:

$$p(x) = \frac{(x - x_1)(x - x_2)}{(x_0 - x_1)(x_0 - x_2)} y_0 + \frac{(x - x_0)(x - x_2)}{(x_1 - x_0)(x_1 - x_2)} y_1 + \frac{(x - x_0)(x - x_1)}{(x_2 - x_0)(x_2 - x_1)} y_3 . \tag{10.11}$$

Note that terms multiplying the y values will be zero at the sample points except for the term multiplying the sample point's abscissa y_i. Various computational tricks can be used to speed the evaluation of (10.10), but these will not be discussed here (see Besset or Press et al). We have implemented polynomial interpolation using a generalized Horner expansion in the LagrangeInterpolator class in the numerics package. Listing 10.11 tests this implementation by sampling the $\sin x$ at five data points and fitting these points to a polynomial. The output is shown in Figure 10.8.

> **Listing 10.11**   The LagrangeInterpolatorApp program tests the LagrangeInterpolator class by sampling an arbitrary function and fitting the samples to a polynomial.

```
package org.opensourcephysics.manual.ch10;
import org.opensourcephysics.controls.*;
import org.opensourcephysics.display.*;
```

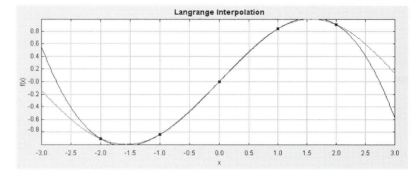

Figure 10.8 A polynomial approximation to the sine function. The polynomial's coefficients were computed by constructing a Lagrange interpolating polynomial passing through the points indicated by black markers. The computed coefficients are approximately: $c[0]=0.0$, $c[1]=0.97$, $c[2]=0$, $c[3]=-0.13$, $c[4]=0$.

```
import org.opensourcephysics.numerics.*;

public class LagrangeInterpolatorApp extends AbstractCalculation {
  PlottingPanel drawingPanel =
      new PlottingPanel("x", "f(x)", "Langrange Interpolation");
  DrawingFrame drawingFrame = new DrawingFrame(drawingPanel);
  Dataset dataset = new Dataset();

  public LagrangeInterpolatorApp() {
    dataset.setConnected(false);
  }

  public void resetCalculation() {
    control.setValue("f(x)", "sin(x)");
    control.setValue("sample start", -2);
    control.setValue("sample stop", 2);
    control.setValue("n", 5);
    control.setValue("random y-error", 0);
    calculate();
  }

  public void calculate() {
    String fstring = control.getString("f(x)");
    double a = control.getDouble("sample start");
    double b = control.getDouble("sample stop");
    double err = control.getDouble("random y-error");
    int n = control.getInt("n"); // number of intervals
    double dx = (n>1) ? (b-a)/(n-1) : 0;
    Function f;
    try {
      f = new ParsedFunction(fstring);
    } catch(ParserException ex) {
      control.println(ex.getMessage());
      return;
    }
    drawingPanel.clear();
    dataset.clear();
```

```
double[] range = Util.getRange(f, a, b, 100);
drawingPanel.setPreferredMinMax(a-(b-a)/4, b+(b-a)/4,
                                range[0], range[1]);
drawingPanel.addDrawable(dataset);
FunctionDrawer func = new FunctionDrawer(f);
func.color = java.awt.Color.RED;
drawingPanel.addDrawable(func);
double x = a;
for(int i = 0;i<n;i++) {
  dataset.append(x, f.evaluate(x)*(1+err*(-0.5+Math.random())));
  x += dx;
}
LagrangeInterpolator interpolator =
    new LagrangeInterpolator(dataset.getXPoints(),
                             dataset.getYPoints());
drawingPanel.addDrawable(new FunctionDrawer(interpolator));
drawingPanel.repaint();
drawingFrame.setVisible(true);
double[] coef = interpolator.getCoefficients();
for(int i = 0;i<coef.length;i++) {
  control.println("c["+i+"]="+coef[i]);
}
}

public static void main(String[] args) {
  Calculation app = new LagrangeInterpolatorApp();
  Control c = new CalculationControl(app);
  app.setControl(c);
}
}
```

Lagrange interpolation should be used cautiously. If the degree of the polynomial is high, if the distance between points is large, or if the points are subject to experimental error, the resulting polynomial can oscillate wildly. Press et al. recommend that the degree of an interpolating polynomial be small. If the tabulated data is accurate but large, we often use multiple polynomials constructed from a small number of nearest neighbor points. *Cubic spline* interpolation uses polynomials in this way.

A cubic spline is a third-order polynomial that is required to have a continuous second derivative with neighboring splines at its endpoints. Because it would be inefficient to store a large number of Polynomial objects, the CubicSpline class in the numerics package stores the coefficients for the multiple polynomials needed to fit a data set in a single array. The CubicSplineApp program tests this class, but it is not shown here because it too is similar to LagrangeInterpolatorApp.

If the sample data are inaccurate, we often compute the coefficients for a polynomial of lower degree that passes as close as possible to the sample points. This fitting procedure is often used to construct an *ad hoc* function that describes experimental data. The PolynomialLeastSquareFit class in the numerics package implements such a fitting algorithm (see Besset) and the PolynomialFitApp program tests this class. It is not shown because it is similar to LagrangeInterpolatorApp.

Suppose we are given a table of $y_i = f(x_i)$ and are asked to determine the value of x that corresponds to a given y. In other words, how do we find the inverse function $x = f^{-1}(y)$? An interpolation routine that does not require evenly spaced ordinates, such as the `CubicSpline` class, provides an easy and effective solution. The following code uses this technique to define an arcsine function.

Listing 10.12   The `Arcsin` class demonstrates how to use interpolation to define an inverse function.

```
package org.opensourcephysics.manual.ch10;
import org.opensourcephysics.numerics.*;

public class Arcsin {
  static Function arcsin;

  private Arcsin() {} // probibit instantiation; all methods are static

  static public double evaluate(double x) {
    if(x<-1||x>1) {
      return Double.NaN;
    } else {
      return arcsin.evaluate(x);
    }
  }

  static { // static block is executed when class is first loaded
    int n = 10;
    double[] xd = new double[n];
    double[] yd = new double[n];
    double x = -Math.PI/2, dx = Math.PI/(n-1);
    for(int i = 0;i<n;i++) {
      xd[i] = x;
      yd[i] = Math.sin(x);
      x += dx;
    }
    arcsin = new CubicSpline(yd, xd);
  }
}
```

10.7 ■ FAST FOURIER TRANSFORM (FFT)

The discrete Fourier transform is used throughout science and mathematics to convert a grid of spacial or temporal data into the frequency domain. If we sample a complex function (take data) at evenly spaced intervals Δ, we generate a sequence of N complex numbers

$$\{g_n\} = \{g_0, g_1, \ldots, g_{N-2}, g_{N-1}\}, \tag{10.12}$$

where every value g_n has a real and an imaginary part. Set the imaginary part equal to zero if the data are real valued. (We describe a better way to handle real-valued data later.) We compute Fourier coefficients h_k by evaluating the following sum:

$$h_k = \sum_{n=0}^{N-1} g_n e^{-i2\pi nk/N}, \tag{10.13}$$

thereby producing a new sequence of complex numbers

$$\{h_k\} = \{h_{-N/2}, h_{-N/2+1}, \ldots, h_0, \ldots, h_{N/2-1}, h_{N/2}\}. \tag{10.14}$$

This new sequence can be converted back into the original sequence using the inverse Fourier transform:

$$g_n = \frac{1}{N} \sum_{k=-N/2}^{N/2} h_n e^{i2\pi nk/N}. \tag{10.15}$$

The sequence of values (10.14) correspond to the amplitudes of complex exponentials $e^{i2\pi g_n x}$ that make up the original data (10.12). We use complex exponentials because these functions are the most general and because sine and cosine transformations can be performed by reordering the data. Sine, cosine, and exponential functions are, of course, related through Euler's (De Moivre's) formula:

$$e^{ix} = \cos x + i \sin x. \tag{10.16}$$

If Δ represents the spacial interval in 10.12 measured in meters, then the interval between values in 10.14 is in cycles per meter. If Δ represents a temporal interval measured in seconds, then the interval in 10.14 is in cycles per second.

We make use of the fast Fourier transform (FFT) algorithm to implement the discrete Fourier transformation. A detailed discussion of this transformation my be found in most numerical methods textbooks including *Numerical Recipes* by Press et al. We have adapted an open source implementation of this algorithm for use in the numerics package.[2] The FFTApp program shown in Listing 10.13 tests the FFT class using a single frequency to generate a sequence of Fourier coefficients (10.14).

Listing 10.13   The FFTApp program tests the FFT class by transforming a sinusoidal function into the frequency domain.

```
package org.opensourcephysics.manual.ch10;
import org.opensourcephysics.numerics.*;

public class FFTApp {
  public static void main(String[] args) {
    int numpts = 10;
    double[] data = new double[2*numpts];
    FFT fft = new FFT(numpts);
    double x = 0, dx = 1.0/numpts;
    double cycles = 1;
    for(int i = 0;i<numpts;i++) {
      data[2*i] = Math.cos(cycles*2*Math.PI*x);
      data[2*i+1] = Math.sin(cycles*2*Math.PI*x);
      x += dx;
    }
    System.out.println("Data before FFT.");
    printData(data);
    fft.transform(data);
    System.out.println("Data after FFT.");
```

[2]The implementation is a based on code contributed by Brian Gough to the GNU Scientific Library (GSL). It was converted to Java by Bruce Miller at NIST.

```
    printData(data);
    fft.inverse(data);
    System.out.println("Data after FFT inverse.");
    printData(data);
  }

  static void printData(double[] data) {
    for(int i = 0, n = data.length/2;i<n;i++) {
      String re = Util.f4(data[2*i]);
      String im = Util.f4(data[2*i+1]);
      System.out.println("i="+i+"\t re="+re+"\t im="+im);
    }
    System.out.println(); // blank line separator
  }
}
```

Run the FFTApp program with various input frequencies (number of cycles) and observe the following:

- Complex data are represented by two double values in the input array. Thus, N data points are represented by a double array with dimension $2N$.

- N complex data points produce N complex frequency coefficients.

- The transform of the given harmonic function produces a single nonzero coefficient if the data contains an integer number of cycles. Otherwise, there will be leakage into adjacent coefficients. Although the input amplitude equals one, the output amplitude is greater than one and is equal to the number of data points.

- The ordering of the output does not correspond to the frequency. There is a discontinuity in frequency half way through the transformed data.

- The inverse transform reproduces the original data.

The ordering and scaling conventions of FFT algorithms vary across disciplines and this often causes unnecessary confusion. We use a common numerical convention known as the engineering convention that scales the FFT output by the number of data points N. The output frequencies are arranged in what is known as *wrap-around* order (see Table 10.3).

Wrap-around ordering may appear strange, but it is computationally efficient. We can easily filter an audio signal by transforming the data, removing the unwanted frequencies, and applying the inverse transformation if we know where a given frequency occurs in the output array. The getWrappedDf convenience method generates an array of frequencies in wrap-around order.

Wrap-around order is, however, awkward when visualizing the frequency spectrum, and we have therefore implemented the toNaturalOrder method to rearrange and normalize the data. The getNaturalDf generates the corresponding array of frequencies. The FFTPlotApp program shown in Listing 10.14 plots the Fourier coefficients of the step function in natural order.

The Fourier transformation has a number of symmetries that can easily be checked using the FFTApp program. For example:

- If the data are real valued and even, then transform is real and the coefficient of any positive frequency equals the coefficient of the corresponding negative frequency.

Table 10.3   Wrap-around ordering of data in the space (time) domain and the spacial (temporal) frequency domain.

Index	Space or Time	Frequency
0	$g[x = 0]$	$h[f = 0]$
1	$g[x = \Delta]$	$h[f = f_0]$
2	$g[x = 2\Delta]$	$h[f = 2 f_0]$
.		
$N/2$	$g[x = N * \Delta/2]$	$h[f = N * f_0/2 = -N * f_0/2]$
.		
$N - 2$	$g[x = (N - 2) * \Delta]$	$h[f = -2 * f_0]$
$N - 1$	$g[x = (N - 1) * \Delta]$	$h[f = -f_0]$

- If the data are imaginary valued and even, then transform is imaginary and the coefficient of any positive frequency equals the coefficient of the corresponding negative frequency.

- If the data are real valued and odd, then transform is imaginary and the coefficient of any positive frequency equals the negative of the coefficient of the corresponding negative frequency.

- If the data are imaginary valued and odd, then transform is real and the coefficient of any positive frequency equals the negative of the coefficient of the corresponding negative frequency.

These symmetries are evident if we transform a Heaviside step function as shown in Listing 10.14. The complementary colors (phase) in the output graph (Figure 10.9) illustrate the change in the sign of positive and negative frequency coefficients.

Listing 10.14   FFTPlotApp plots the Fourier spectrum of the step function using the fast Fourier transform algorithm.

```
package org.opensourcephysics.manual.ch10;
import org.opensourcephysics.display.*;
import org.opensourcephysics.numerics.FFT;

public class FFTPlotApp {
  public static void main(String[] args) {
    PlottingPanel panel = new PlottingPanel("f", "amp", null);
    DrawingFrame frame = new DrawingFrame("FFT", panel);
    ComplexDataset cdataset = new ComplexDataset();
    panel.addDrawable(cdataset);
    int numpts = 100;
    double[] data = new double[2*numpts];
    double xmin = -5, xmax = 5;
    double x = xmin, dx = (xmax-xmin)/numpts;
    for(int i = 0;i<numpts;i++) {
      data[2*i] = (x<Math.abs(2.5)) ? 1 : -1; // real
      data[2*i+1] = 0;                         // imaginary
      x += dx;
    }
    FFT fft = new FFT();
```

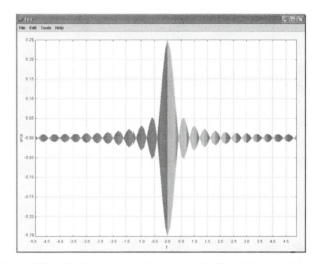

Figure 10.9 The Fourier transformation of a Heaviside step function.

```
fft.transform(data);
fft.toNaturalOrder(data);
cdataset.append(fft.getNaturalFreq(dx), data);
frame.setVisible(true);
frame.setDefaultCloseOperation(javax.swing.JFrame.EXIT_ON_CLOSE);
  }
}
```

If the data to be transformed are real valued, it is inefficient to construct and transform an array of complex numbers. Because doubling the size of the input array by adding complex components all of which are zero produces a symmetric frequency distribution, half of the FFT output is redundant. A more efficient algorithm transforms an input array of real numbers as if it was an array of complex numbers and then uses symmetries in the Fourier transformation to unscramble the resulting array thereby constructing an array containing only the positive frequency coefficients. Note that the number of real data points must be even. An alternative interpretation of the output is that the even numbered elements in the output array are the coefficients of the cosine terms and the odd numbered elements are the coefficients of the sine terms in a Fourier sine-cosine series expansion:

$$f_k = a_0 + a_1 \cos(Nkx/2) + \sum_{n=1}^{N/2-1} [a_{2n} \cos(nkx) + a_{2n+1} \sin(nkx)] \,. \tag{10.17}$$

The first element a_0 is the zero frequency (DC) component and the second element a_1 is the Nyquist frequency component. The Nyquist frequency is the highest frequency contained in the data due to the finite number of data points.

The FFTReal class in the numerics package computes the FFT of a real-valued function and returns a complex array containing the positive frequency coefficients. The FFTRealApp program shown in Listing 10.15 tests the FFTReal class by transforming a sinusoidal function. The FFTRealPlotApp program (available on the CD) transforms and plots the Heaviside step function.

Listing 10.15   The `FFTRealApp` computes the Fourier transformation of a sinusoidal function.

```
package org.opensourcephysics.manual.ch10;
import org.opensourcephysics.frames.*;
import org.opensourcephysics.numerics.*;

public class FFTRealApp {
  public static void main(String[] args) {
    TableFrame frame = new TableFrame("Real FFT");
    frame.setColumnNames(0, "frequency");
    frame.setColumnNames(1, "cos");
    frame.setColumnFormat(1, "#0.00");
    frame.setColumnNames(2, "sin");
    frame.setColumnFormat(2, "#0.00");
    int n = 16;
    double[] data = new double[n];
    double x = 0, dx = 1.0/n;
    for(int i = 0;i<n;i++) {
      data[i] = Math.sin(2*Math.PI*x);
      x += dx;
    }
    FFTReal realFFT = new FFTReal(n);
    realFFT.transform(data);
    double[] f = realFFT.getNaturalFreq(dx);
    for(int i = 0;i<n;i += 2) {
      frame.appendRow(new double[] {f[i/2], data[i], data[i+1]});
    }
    frame.setVisible(true);
    frame.setDefaultCloseOperation(javax.swing.JFrame.EXIT_ON_CLOSE);
  }
}
```

10.8 ■ PROGRAMS

The following examples are in the `org.opensourcephysics.manual.ch10` package.

ArcsinApp

`ArcsinApp` compares the interpolated inverse sine function to the exact result as described in Section 10.6.

CubicSplineApp

`CubicSplineApp` implements a visualization of cubic spline interpolation. Cubic splines are described in Section 10.6.

FFTApp

`FFTApp` tests the fast Fourier transform (FFT) as described in Section 10.7.

FFTPlotApp

FFTPlotApp computes and plots the fast Fourier transform (FFT) of the Heaviside step function as described in Section 10.7.

FFTRealApp

FFTRealApp computes the Fourier coefficients of a real-valued function as described in Section 10.7.

GaussianDerivativeApp

GaussianDerivativeApp plots the derivative of a Gaussian function centered at the origin as described in Section 10.7.

IntegralApp

IntegralApp tests numerical integration algorithms in the Integral class. See Section 10.3.

LagrangeInterpolatorApp

LagrangeInterpolatorApp implements a visualization of Lagrange interpolating polynomials as described in Section 10.6.

ParseComplexApp

ParseComplexApp parses a string of characters to produce a complex-valued function that can be evaluated. This example uses the optional JEParser package developed by Nathan Funk.

ParseRealApp

ParseRealApp parses a string of characters to produce a math function that can be evaluated as described in Section 10.4.

PolynomialApp

PolynomialApp tests the Polynomial class as described in Section 10.6.

RootApp

RootApp compares Newton's method and the bisection method for finding the root of a function as described in Section 10.5.

RootFinderApp

RootFinderApp is a program for exploring the properties of root-finding algorithms. See Section 10.5.

SlitTableApp

SlitTableApp creates a table of values for the function $f(x) = \sin x/x$.

11

Three-Dimensional Modeling

by Wolfgang Christian and Francisco Esquembre

The Open Source Physics library defines the OSP 3D API to create, organize, and manipulate 3D worlds for physics simulations. Although other high-level 3D packages, such as JOGL, can be incorporated into our framework, familiarity with theses packages is neither assumed nor required for the use of OSP 3D. The OSP 3D API is defined in the *core 3D* package (`org.opensourcephysics.display3d.core`). The *simple 3D* package (`org.opensourcephysics.display3d.simple3d`) implements this API using the standard Java Swing library.

11.1 ■ OVERVIEW

The Open Source Physics 3D API uses high-level constructs to create, organize, and manipulate 3D models. Because this API is defined in the core 3D package using Java interfaces, it can be implemented using almost any 3D library. OSP 3D simplifies 3D concepts to aid programmers in the fast, accurate creation of physics simulations. Familiarity with add-on libraries is neither assumed nor required for the use of OSP 3D.

Programming 3D physics simulations presents two challenges. The first challenge is understanding the physics. Interesting three-dimensional phenomena, such as the rigid-body dynamics of a spinning top shown in Figure 11.1, are mathematically challenging because they often require solving differential equations in reference frames that are attached to moving bodies. A physics-oriented 3D toolkit can help by providing facilities for transforming the dynamics to and from these noninertial frames.

The second challenge is the visualization itself. Advanced 3D libraries such as Java 3D or Java bindings for OpenGL (JOGL) are full-featured 3D programming languages. A programmer can use these advanced libraries to create textures, morphs, and animations that rival the latest Hollywood special effects. A basic scene, however, takes numerous lines of code to create. A *virtual universe* must be created, then a *view* into that universe, then a *canvas 3D* object for rendering onscreen. The list goes on, and we can't even see anything because we have no light source yet. All of these objects are created one at a time and each can be customized with surface textures and other display properties. While this structural concept makes the Java 3D or JOGL libraries extremely versatile, scene creation can become tedious for beginning programmers. In many physics simulations we aren't concerned with these effects.

As users become familiar with 3D programming, they can pick a particular implementation of OSP 3D API and use features in that implementation. We show you how to refine the

Figure 11.1   A top precessing due to an external torque.

appearance and put some special touches on a program using JOGL-specific features later in this chapter, but initially we just want to program interesting physics using the simple 3D package.

11.2 ■ SIMPLE 3D

The OSP 3D API is defined in the `org.opensourcephysics.display3d.core` package using Java interfaces and abstract classes. This API can be implemented using any library and we have implemented it in `org.opensourcephysics.display3d.simple3d` using only standard Java. We use the simple 3D package to study the OSP 3D API because programs written using this implementation will run on any Java enabled computer. To use other OSP 3D implementations, one must simply install the Java library that communicates with the graphics hardware and import the corresponding OSP 3D package.

The `Display3DFrame` class in the frames package makes it easy to create a simple 3D visualization as shown in Listing 11.1. This frame is composed of lower-level components that will be described in subsequent sections.

Listing 11.1   The `Box3DApp` class creates a box within a `Display3DFrame`.

```
package org.opensourcephysics.manual.ch11;
import org.opensourcephysics.display3d.simple3d.*;
import org.opensourcephysics.frames.*;
import java.awt.*;
import javax.swing.*;

public class Box3DApp {
  public static void main(String[] args) {
    // create a drawing frame and a drawing panel
    Display3DFrame frame = new Display3DFrame("3D Demo");
    frame.setPreferredMinMax(-10, 10, -10, 10, -10, 10);
    frame.setDecorationType(VisualizationHints.DECORATION_AXES);
    frame.setAllowQuickRedraw(false); // use shading when rotating
    Element block = new ElementBox();
    block.setXYZ(0, 0, 0);
    block.setSizeXYZ(6, 6, 3);
    block.getStyle().setFillColor(Color.RED);
    // divide into subblocks
    block.getStyle().setResolution(new Resolution(6, 6, 3));
    frame.addElement(block);
```

```
    frame.setVisible(true);
    frame.setDefaultCloseOperation(JFrame.EXIT_ON_CLOSE);
  }
}
```

It is not more difficult to create a 3D visualization than it is to create a 2D visualization because the `Display3DFrame` class behaves very much like its two-dimensional counterpart. The 2D `Drawable` interface has been replaced by the `Element` interface which is designed to mimic geometric shapes in the physical world. Elements have position and size properties and contain a `Style` object that controls an element's visual appearance such as color and resolution.

Some methods in the OSP 2D API, such as `setPreferredMinMax`, have been extended by adding parameters for the third dimension. New objects, such as `Camera`, enable the programmer to control the three-dimensional viewing perspective.

```
// camera position can be set using Cartesian or polar coordinates
frame.getCamera().setXYZ(0.0, 50.0, 0.0);
frame.getCamera().setAzimuthAndAltitude(0,Math.PI/4);
```

Elements, such as `ElementBox`, `ElementCylinder`, and `ElementEllipsoid`, are added to a container using the `addElement` method. Just as in the two-dimensional case, the high-level `Display3DFrame` is composed of lower-level objects, such as the `DrawingPanel3D` that fills the frame's viewing area. For changes to take effect, the program must draw the frame by invoking the `render` method. This method is invoked automatically by the `AbstractSimulation` class when the simulation is initialized and after every `doStep` method call.

Although OSP 3D is designed for three-dimensional visualizations, it can also show two-dimensional projections. These projections are available at runtime using the frame's menu. We can also project onto a coordinate plane by setting the projection mode programmatically.

```
frame.getCamera().setProjectionMode(Camera.MODE_PLANAR_YZ);
```

11.3 ■ DRAWING PANEL 3D

The `DrawingPanel3D` class defines a three-dimensional view with Cartesian world coordinates as seen in Figure 11.2. This panel can be added to any `JFrame`, but we usually place it into a `DrawingFrame3D` because this frame has been customized with a menu bar and other components that interact with a 3D panel.

```
// import a concrete implementation of OSP 3D
import org.opensourcephysics.display3d.simple3d.*;
// create a 3D world
DrawingPanel3D panel = new DrawingPanel3D();
DrawingFrame3D frame = new DrawingFrame3D();
frame.setDrawingPanel3D(panel);
// additional code places Elements into the 3D view
```

The `Display3DFrame` in the frames package that we introduced in Section 11.2 used the simple 3D implementation of OSP 3D, and we now use this same package in order to study the API in detail.

Because 3D scenes are almost always animated and because a 3D drawing may be complex, the `DrawingPanel3D` class always uses buffering as described in Chapter 7. The

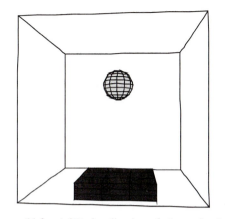

Figure 11.2 A 3D visualization of a bouncing ball.

3D scene is drawn into a hidden image that has the same dimensions as the panel, and this image is later copied to the screen by the operating system. Because the panel keeps track of changes to its properties and to its 3D objects by setting an internal boolean variable named `dirtyImage`, a program need only invoke the panel's `repaint` method and the hidden image will be updated automatically. This updating works in the following manner. The event dispatch thread invokes the `paintComponent` method, and this method quickly copies the background image to the screen. If the `dirtyImage` boolean is set, a timer event is initiated and after a short delay a new background image is rendered and copied. Multiple timer actions are combined (coalesced) into a single image update, and this rendering is aborted if the scene is repainted by another thread such as an animation thread.

Although `repaint` will work, programmers should invoke the panel's `render` method from within an animation thread for smoother animation and consistent frame rates. An easy way to do this is to subclass `AbstractSimulation` and let the simulation thread's `run` method automatically call the `render` method for animated frames. We do not need to explicitly set the animated property in the `DrawingFrame3D` because it is set in the frame's constructor. Listing 11.2 creates a bouncing ball simulation using the simulation architecture developed in Chapter 7. This animation may gain energy because we don't detect the exact moment of the collision as described in Section 9.9.

Listing 11.2 A bouncing ball animation using the OSP 3D API.

```
package org.opensourcephysics.manual.ch11;
import org.opensourcephysics.controls.AbstractSimulation;
import org.opensourcephysics.display3d.simple3d.*;

public class BounceApp extends AbstractSimulation {
  DrawingPanel3D panel = new DrawingPanel3D();
  DrawingFrame3D frame = new DrawingFrame3D(panel);
  Element ball = new ElementEllipsoid();
  Element floor = new ElementBox();
  double velocity = 0;

  public BounceApp() {
    panel.setPreferredMinMax(-5, 5, -5, 5, 0, 10);
```

```
      ball.setXYZ(0.0, 0.0, 9.0);
      ball.setSizeXYZ(2, 2, 2);
      ball.getStyle().setFillColor(java.awt.Color.YELLOW);
      // floor
      floor.setXYZ(0.0, 0.0, 0.0);
      floor.setSizeXYZ(5.0, 5.0, 1.0);
      // Add the objects to panel
      panel.addElement(ball);
      panel.addElement(floor);
      frame.setDefaultCloseOperation(javax.swing.JFrame.EXIT_ON_CLOSE);
      frame.setVisible(true);
   }

   protected void doStep() {
      double z = ball.getZ(), dt = 0.05;
      z += velocity*dt;   // moves the ball
      velocity -= 9.8*dt; // acceleration changes the velocity
      if(z<=1.0&&velocity<0) {
         velocity = -velocity;
      }
      ball.setZ(z);
   }

   public static void main(String[] args) {
      (new BounceApp()).startSimulation();
   }
}
```

`DrawingPanel3D` has a number of built in visual "hints" that help users orient their view of the 3D scene. There are, for example, axis decorations that show the Cartesian coordinate directions using arrows or using a rectangular bounding box. Red, green, and blue sides on the cube indicate the x, y, and z directions.

```
// decoration types are:
// DECORATION_NONE, DECORATION_AXES, DECORATION_CUBE
panel.getVisualizationHints().
      setDecorationType(VisualizationHints.DECORATION_CUBE);
```

The corners of the decoration cube show the locations of the world's minimum and maximum values. These values are set as follows:

```
panel.setPreferredMinMax( -5.0, 5.0, -5.0, 5.0, 0.0, 10.0);
```

Some visualization hints affect appearance and performance. If the scene is simple, it may not be necessary to switch to wire-mesh rendering when dragging with the mouse to position the camera. This quick-draw mode can be disabled if the scene is simple.

```
panel.getVisualizationHints().setAllowQuickRedraw(false);
```

11.4 ■ CAMERAS

We view a 3D world through a `Camera` that has a location and points in a direction known as the line of sight (LOS). The default LOS points from the camera toward the center of the drawing as determined by the average of the extrema along each coordinate. This makes

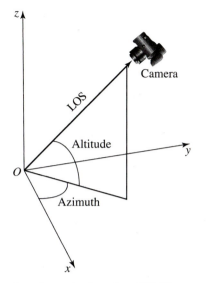

Figure 11.3 A `Camera` determines the projection of an OSP 3D scene onto a 2D view. The default line of sight (LOS) points from the camera toward the center of the drawing.

it easy to rotate the camera around a physical model that has been created near the center without loosing it. Click-dragging left to right rotates the azimuthal angle and click-dragging up to down rotates the altitude angle as shown in Figure 11.3. The default camera location is on the x-axis with a LOS along the axis toward the center. The location on the axis depends on the view's extremum values.

Although most users will adjust a 3D view using the mouse, it is straightforward to set the camera programmatically. Camera properties are set using accessor methods after obtaining a reference to the camera from the 3D panel.

```
Camera camera = (Camera) panel.getCamera();
camera.setPosition(x,y,z);
// an alternate to the above
camera.setAzimuthAndAltitude(theta,phi);
```

The `setPosition` method sets the camera's position in space using Cartesian coordinates. The `setAzimuthAndAltitude` method positions the camera using spherical polar coordinates. The azimuth is the rotation angle about the z-axis and the altitude is the angle above or below the xy-plane.

Just as a physical camera can have telephoto and wide-angle lenses that accentuate or diminish the three-dimensional perspective, our computer model can project the three-dimensional world onto the computer screen in a number of ways. Because a camera uses a mathematical transformation to map (project) a three-dimensional space onto a two-dimensional plane, these options are accessed using the `Camera` interface. (See Section 11.8 for a discussion of transformations.)

```
// projection modes are: MODE_PERSPECTIVE, MODE_NO_PERSPECTIVE,
// MODE_PLANAR_XY, MODE_PLANAR_XZ, and MODE_PLANAR_YZ
panel.getCamera().setProjectionMode(Camera.MODE_PLANAR_YZ);
```

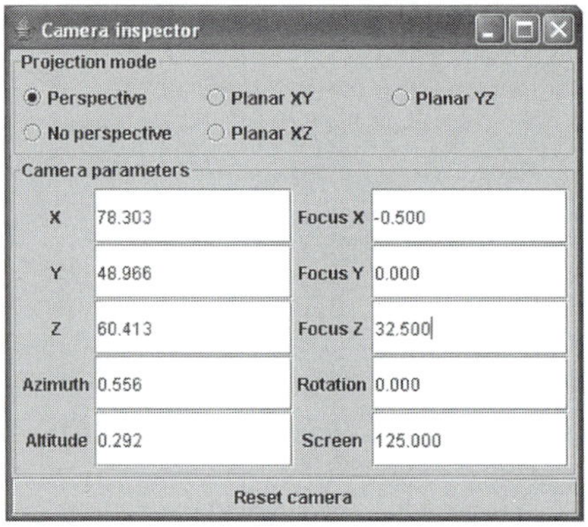

Figure 11.4   The camera inspector can be used to adjust the viewing position.

The camera's `getTransformation` method allows a program to project points from the world to screen coordinates should you need this low-level functionality.

Although adjusting the camera location is fairly intuitive using the mouse, you can use a `CameraInspector` available from the menu (see Figure 11.4) for precise positioning. Cartesian and angle coordinates are, of course, interrelated, so setting a value can affect the other values. The `CameraApp` test program available on the CD may also be helpful in understanding the camera's API.

11.5 ■ ELEMENTS AND GROUPS

A 3D simulation creates objects that implement the `Element` interface and adds these objects to a 3D container. The bouncing ball program shown in Listing 11.2 uses elements in a simulation. Objects are instantiated, added to a panel, and moved using a thread. Because `BounceApp` extends `AbstractSimulation` and the frame's animated property is set, it is not necessary to call the `render` method after objects have been moved. The `ElementApp` program is not shown here but is available on the CD. This program creates a scene using many of the 3D elements listed in Table 11.1.

Although specifying the orientation (rotation) of an element requires a modest amount of mathematics and will be described more fully in Section 11.8, sizing and positioning an element is easy using accessor methods such as `setXYZ` and `setSizeXYZ`.

```
Element box = new ElementBox();
box.setXYZ(10,10,0);         // position of center
box.setSizeXYZ(4, 4, 1);     // lengths of sides
Element arrow = new ElementArrow();
arrow.setZ(5);               // x and y remain at zero
arrow.setSizeXYZ(4, 4, 1);   // components along axes
```

The position of a symmetric object such as a box is its center of mass, and the position of an arrow is its tail.

An element contains a `Style` object that determines visual properties such as color. These properties are set as follows:

Table 11.1   Elements defined in the OSP 3D API.

org.opensourcephysics.display3d.core	
`ElementArrow`	An arrow with its tail located at its position. The `setSizeXYZ` method sets the arrow's components.
`ElementBox`	A box. The top and bottom can be open or closed.
`ElementCircle`	A circle that is drawn without any subdivisions or perspective.
`ElementCone`	A cone. The `setSizeXYZ` method sets the dimensions of the elliptical base and the height.
`ElementEllipsoid`	An ellipsoid that is drawn with subdivisions and perspective.
`ElementImage`	An image that is always drawn (projected) perpendicular to the line of sight.
`ElementPlane`	A plane.
`ElementPolygon`	A polygon that is created from an array of data points. A polygon can be closed (solid) or open (polyline).
`ElementSegment`	A straight line segment similar to arrow but without an arrowhead.
`ElementSphere`	An ellipsoid whose dimensions are of equal size. Changing any one dimension changes all other dimensions.
`ElementSpring`	A coil spring.
`ElementSurface`	A wire-mesh surface created from an array of data points.
`ElementText`	Text that can be justified left, right, and center with respect to the position.
`ElementTrail`	A trail of points. The maximum number of points that can be stored can be set.
`Group`	An element that is made up of other elements.

```
Style style = (Style) box.getStyle();
style.setFillColor(Color.RED);
style.setLineColor(Color.BLACK);
style.setLineWidth(3);
// divides box into subblocks
style.setResolution( new Resolution(5,5,2));
```

Most properties are self-explanatory, but the resolution property needs explanation because it is implementation dependent.

Three-dimensional objects are constructed using polygons, and it is sometimes difficult deciding how many polygons to use. The Resolution class stores information that can help an Element divide itself into smaller pieces that have a size no larger than some maximum unit. The precise meaning of the parameters passed into the Resolution constructor is left to each element and to each 3D implementation, but a parameter typically specifies the number of divisions in a coordinate direction. Arrows and lines use a Resolution object with a single parameter that specifies how many times they should divide themselves along their length.

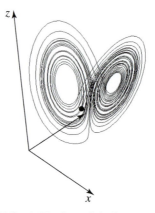

Figure 11.5   A 3D view of the Lorenz attractor.

Because the simple 3D package orders all polygons from far to near and then paints the polygons in that order, you may notice rendering artifacts when objects constructed with large polygons intersect. Increasing the resolution will reduce these artifacts, although too high of a resolution will increase rendering time. If truly high-resolution images are needed, it is best to use a 3D implementation such as JOGL or Java 3D that renders and smooths polygons using the graphics card. Almost all modern graphics cards support hardware-accelerated rendering using OpenGL graphics language drivers.

The `Lorenz` class shown in Listing 11.3 uses these ideas to show a three-dimensional graphical model of a two-dimensional layer of fluid that is heated from below as seen in Figure 11.5. Using a system of differential equations, the model has three state variables that are referred to as x, y, and z, although they have nothing to do with physical space. The x variable is a measure of the fluid flow velocity circulating around the cell, y is a measure of the temperature difference between the rising and falling fluid regions, and z is a measure of the difference in the temperature profile between the bottom and the top from the normal equilibrium temperature profile. The evolution of the fluid in this state space is defined by (11.1):

$$\frac{dx}{dt} = -\sigma x + \sigma y \tag{11.1a}$$

$$\frac{dy}{dt} = -xz + rx - y \tag{11.1b}$$

$$\frac{dz}{dt} = xy - bz. \tag{11.1c}$$

Typical input parameters for the Lorenz model are $\sigma = 10$, $b = 8/3$, and $r = 28$ with the initial condition $x_0 = 1$, $y_0 = 1$, $z_0 = 20$. The `getRate` and `getState` methods in the `Lorenz` class implement the `ODE` interface as described in Chapter 9.

The visualization shows the current (x, y, z) value as a small red ellipsoid and past positions as a trail. These objects are instantiated in the model's constructor and added to a `DrawingPanel3D`. The model solves the differential equations and uses the `trail.addPoint(x,y,z)` method to create the trail and the `ball.setXYZ(x,y,z)` method to update the ellipsoid.

> **Listing 11.3**   An implementation of the Lorenz model for convection in a two-dimensional fluid.

```
package org.opensourcephysics.manual.ch11;
import org.opensourcephysics.display3d.simple3d.*;
import org.opensourcephysics.numerics.*;

public class Lorenz implements ODE {
  DrawingPanel3D panel = new DrawingPanel3D();
  DrawingFrame3D frame = new DrawingFrame3D(panel);
  double[] state = new double[4];
  double a = 28.0, b = 2.667, c = 10.0;
  ODESolver ode_solver = new RK45MultiStep(this);
  Element ball = new ElementEllipsoid();
  ElementTrail trace = new ElementTrail();

  public Lorenz() {
    frame.setTitle("Lorenz Attractor");
    panel.setPreferredMinMax(-15.0, 15.0, -15.0, 15.0, 0.0, 50.0);
    panel.getVisualizationHints().setDecorationType(
        VisualizationHints.DECORATION_AXES);
    ball.setSizeXYZ(1, 1, 1);
    ball.getStyle().setFillColor(java.awt.Color.RED);
    panel.addElement(trace);
    panel.addElement(ball);
    ode_solver.setStepSize(0.01);
  }

  protected void doStep() {
    for(int i = 0;i<10;i++) {
      ode_solver.step();
      trace.addPoint(state[0], state[1], state[2]);
      ball.setXYZ(state[0], state[1], state[2]);
    }
    panel.setMessage("t="+Util.f3(state[3]));
  }

  public double[] getState() {
    return state;
  }

  public void initialize(double x, double y, double z) {
    state[0] = x;
    state[1] = y;
    state[2] = z;
    state[3] = 0; // time
    trace.clear();
    trace.addPoint(x, y, z);
    ball.setXYZ(x, y, z);
    frame.setVisible(true);
  }

  public void getRate(double[] state, double[] rate) {
    rate[0] = -(state[0]-state[1])*c;                      // x rate
```

```
    rate[1] = -state[1]-state[0]*state[2]+state[0]*a; // y rate
    rate[2] = (state[0]*state[1]-state[2])*b;         // z rate
    rate[3] = 1;                                       // time rate
  }
}
```

The `LorenzApp` class does the usual housekeeping by extending `AbstractSimulation` to create a simulation framework and instantiating a `Lorenz` model. It is available on the CD but is not shown here because it is similar to other simulations.

A `Group` is a special `Element` that allows us to combine multiple 3D elements to form a compound object. Because a `Group` is an `Element`, it acts like a single 3D object. Elements placed within a group are positioned and sized relative to their group's position and size. In addition, transforming a group (see Section 11.8) transforms every element within the group.

An easy way to create a compound object is to extend the 3D `Group` class and create the objects in the constructor. You should not, however, add the same `Element` to a group and a `DisplayPanel3D` or to two groups, but you can nest groups within groups. The `GroupApp` program on the CD demonstrates the use of a group by creating a large sphere and ten smaller spheres that rotate together.

11.6 ■ INTERACTIONS

Because users may wish to interact with a three-dimensional model in many different ways, we have defined a more sophisticated interaction API for 3D objects. This framework consists of the interfaces in the `org.opensourcephysics.display3d.core.interaction` package:

- `InteractionSource`. The interface for objects that can generate interaction events and send them to listeners.

- `InteractionListener`. The interface for objects that can listen to interaction events.

- `InteractionEvent`. The class for the event generated when the user interacts with a source.

- `InteractionTarget`. The interface for the particular target that has been hit in a source.

`BallInteractionApp` shown in Listing 11.4 demonstrates how this framework operates by enabling an interaction and providing a listener that restricts dragging to the xy-plane. The constructor registers the program as the listener and enables the ball's position interaction target. The `interactionPerformed` method implements the `InteractionListener` interface. Note that the listener sets the ball's z-coordinate to zero in order to disable dragging in this direction. If we wish to drag the particle in x, y, and z, we would still enable the position target, but we could remove the implementation of `InteractionListener` so as not to intervene in the 3D panel's positioning of the ball.

Listing 11.4   A program that allows users to drag a ball.

```
package org.opensourcephysics.manual.ch11;
import org.opensourcephysics.display3d.core.interaction.*;
import org.opensourcephysics.display3d.simple3d.*;

public class BallInteractionApp implements InteractionListener {
  DrawingPanel3D panel = new DrawingPanel3D();
  DrawingFrame3D frame = new DrawingFrame3D(panel);
  Element ball = new ElementEllipsoid();

  public BallInteractionApp() {
    ball.addInteractionListener( this); // sends interactions to this
    // enables interactions that change positions
    ball.getInteractionTarget(Element.TARGET_POSITION)
                          .setEnabled(true);
    panel.addElement(ball);
    frame.setDefaultCloseOperation(javax.swing.JFrame.EXIT_ON_CLOSE);
    frame.setVisible(true);
  }

  public void interactionPerformed(InteractionEvent _event) {
    switch(_event.getID()) {
    case InteractionEvent.MOUSE_DRAGGED:
      System.out.println("Ball x="+ball.getX());
      System.out.println("Ball y="+ball.getY());
      ball.setZ(0); // always set z to zero
      break;
    default:
      break;
    }
  }

  public static void main(String[] args) {
    new BallInteractionApp();
  }
}
```

Interaction Source

An object that implements the InteractionSource interface is an object that can generate interaction events in response to user actions and report these events to a list of registered interaction listeners. The implementing class typically keeps an internal list of the listeners that will be informed whenever an action event takes place. A listener only appears once in this list even if it has been added multiple times.

A source can contain one or more targets, that is, hot spots that trigger interaction events (see InteractionTarget for details). For example, an arrow has targets that enable a listener to move the arrow when you drag its tail and to resize the arrow when you drag its head. The InteractionEvent contains information about which target was selected by the user.

InteractionSource defines the following methods:

```
public interface InteractionSource {
  InteractionTarget getInteractionTarget (int target);
```

```
void addInteractionListener (InteractionListener listener);
void removeInteractionListener (InteractionListener listener);
}
```

The first method takes an integer that identifies the access to one of the targets. Sources should document their targets and provide integer numbers to reference them. The second and third methods add and remove interaction listeners to the source.

Notice that it is customary that source interactions are disabled by default. This helps avoid undesired interaction events. Notice also that adding a listener doesn't enable the target(s) of a source. The user must activate the target explicitly using its setEnable method.

Interaction Listener

InteractionListener is an interface that defines a single method. Classes implementing this interface need to register themselves with an InteractionSource using the addInteractionListener method.

```
public interface InteractionListener {
  void interactionPerformed(InteractionEvent event);
}
```

A source will call the interactionPerformed method whenever an interaction occurs. The provided event includes relevant information about the interaction that can be used by the listener to determine what to do in response.

Interaction Event

An InteractionEvent passes information to an InteractionListener that describes what interaction has taken place with the InteractionSource and which target has been hit. This class has six public named constants and two public methods.

```
public class InteractionEvent extends ActionEvent {

  static public final int MOUSE_PRESSED;
  static public final int MOUSE_DRAGGED;
  static public final int MOUSE_RELEASED;
  static public final int MOUSE_ENTERED;
  static public final int MOUSE_EXITED;
  static public final int MOUSE_MOVED;

  public Object getInfo();
  public MouseEvent getMouseEvent();
}
```

InteractionEvent is a subclass of java.awt.event.ActionEvent. It adds an object that contains information about the target that has been hit to the event. Sources should document how the information in the passed object should be accessed.

The InteractionEvent superclass defines the getID method. Elements set the identifier to one of the named constants, such as TARGET_POSITION, with obvious meanings. Notice how the identifier is used in BallInteractionApp.

Interaction Target

The `InteractionTarget` interface is the most technical of the interfaces in this subsection because sources can have more that one target and because sources can be nested within groups. A target is a hotspot in the source that is sensible to user interaction and is identified by a named constant such as `TARGET_POSITION` or `TARGET_SIZE`.

The `InteractionTarget` interface differentiates what target in the source was hit and helps the target affect the underlying source during the interaction. This interface is not listed here because we do not show you how to define your own interactive targets. The `setEnabled` method is the most important. This method enables us to activate a target within an `Element` so that the object can respond to mouse actions.

```
element.getInteractionTarget(Element.TARGET_POSITION).setEnabled(true);
```

The `setAffectsGroup` method sets the scope of an interaction when an element belongs to a group.

```
element.getInteractionTarget(Element.TARGET_POSITION).
    setAffectsGroup(true);
```

If the user drags the element, then the entire group will move. You can, for example, build a cart consisting of a block and two wheels and then move the entire cart by dragging the block.

Interaction Example

Users will typically deal with interaction by implementing the `InteractionListener` interface in one or more objects and adding these listeners to `InteractionSources`. Interaction sources are `DrawingPanel3D` and `Element` objects within the panel. The elements within a panel will report mouse actions that hit any of their possible targets and `DrawingPanel3D` reports actions that take place on its empty space. The `getInfo` method returns null if the user rotates or zooms the camera by dragging the mouse and holding the Control or Shift keys, respectively. The user can, however, select a point in 3D space by holding the Alt key, in which case the `getInfo` method returns a `double[3]` array identifying the point.

Listing 11.5 shows how a listener can discriminate between interaction on the different parts of a scene. Notice, in particular, how the listener distinguishes between the panel and the objects within the panel.

Listing 11.5 `Interaction3DApp` demonstrates the OSP 3D interaction framework.

```
package org.opensourcephysics.manual.ch11;
import org.opensourcephysics.display3d.core.interaction.*;
import org.opensourcephysics.display3d.simple3d.*;
import org.opensourcephysics.frames.*;

public class Interaction3DApp implements InteractionListener {
  Element particle = new ElementCircle();
  ElementArrow arrow = new ElementArrow();

  Interaction3DApp() {
    Display3DFrame frame = new Display3DFrame("3D Interactions");
    frame.setPreferredMinMax(-2.5, 2.5, -2.5, 2.5, -2.5, 2.5);
```

```
        particle.setSizeXYZ(1, 1, 1);
        // enables interactions that change positions
        particle.getInteractionTarget(Element.TARGET_POSITION)
                                 .setEnabled(true);
        // accepts interactions from the particle
        particle.addInteractionListener(this);
        frame.addElement(particle); // adds the particle to the panel
        // enables interactions that change the size
        arrow.getInteractionTarget(Element.TARGET_SIZE).setEnabled(true);
        // accepts interactions from the arrow
        arrow.addInteractionListener(this);
        frame.addElement(arrow); // adds the arrow to the panel
        // enables interactions with the 3D Frame
        frame.enableInteraction(true);
        // accepts interactions from the frame's DrawingPanel3D
        frame.addInteractionListener(this);
        frame.setDefaultCloseOperation(javax.swing.JFrame.EXIT_ON_CLOSE);
        frame.setVisible(true);
    }

    public void interactionPerformed(InteractionEvent evt) {
        Object source = evt.getSource();
        // check for a particular mouse action
        if(evt.getID()==InteractionEvent.MOUSE_PRESSED) {
            System.out.println("Mouse Pressed");
        }
        if(source==particle) { // check for a particular element
            System.out.println("A particle has been hit");
        }
    }

    static public void main(String args[]) {
        new Interaction3DApp();
    }
}
```

11.7 ■ VECTORS

Vector operations listed in Table 11.2 can be performed in more than one way in Java. A straightforward way is to represent the components of an n-dimensional vector using an n-component array of numbers. Another approach is to define a class that stores components as instance variables. Both representations are useful. The first is easy to use, while the second provides better encapsulation and is slightly faster when doing arithmetic because instance variables are accessed more quickly than array elements. The array-based implementation uses static methods and is defined in the VectorMath class and the object-based implementation uses instance variables and is defined in the Vec3D class. The Vec3D constructor simply copies the components and is coded as follows:

```
double x, y, z;  // instance variables

public Vec3D(double x, double y, double z) { // constructor
    this.x = x;
    this.y = y;
    this.z = z;
}
```

Table 11.2 Vector operations for 3D modeling are defined in the numerics package.

org.opensourcephysics.numerics.VectorMath org.opensourcephysics.numerics.Vec3D	
cross2d	Computes the cross product of two 2D vectors. The value returned is the component of the vector that is perpendicular to the given vectors.
cross3d	Computes the cross product of two 3D vectors.
dot	Computes the dot product of two vectors.
magnitude	Computes the magnitude (length) of a vector.
magnitudeSquared	Computes the magnitude squared of a vector.
normalize	Normalizes a vector.
perp	Computes the perpendicular part of one vector with respect to another vector.
plus	Adds two vectors.
project	Computes the projection of one vector onto another vector.

The dot (Euclidian inner) product of n-dimensional vectors **a** and **b** is a scalar c such that

$$c = \mathbf{a} \cdot \mathbf{b} = |a|\,|b|\cos\theta, \tag{11.2}$$

where θ is the angle between the vectors. This expression is equivalent to

$$c = \sum_{i=0}^{n-1} a_i b_i, \tag{11.3}$$

where each vector is expressed using components $\mathbf{a} = (a_0, a_1, \ldots, a_{n-1})$ in a Cartesian basis set. The static dot method in the VectorMath class computes the dot product between arrays of numbers.

```
static public double dot(double[] a, double[] b) {
  int aLength = a.length;
  if(aLength!=b.length) {
    throw new UnsupportedOperationException("ERROR:
            Vectors must be of equal dimension in dot product.");
  }
  double sum = 0;
  for(int i = 0; i<aLength; i++) {
    sum += a[i]*b[i];
  }
  return sum;
}
```

Because objects (other than arrays) are not involved, the VectorMath class is easy to use. A dot product, for example, is computed as follows:

```
double[] a = new double[] {1,2,3};
double[] b = new double[] {4,5,6};
double c = VectorMath.dot(a,b);
```

The `magnitude` method computes the vector magnitude (length) by taking the square root of the sum of the squares of the vector components:

$$|a| = \left(\sum_{i=0}^{n-1} a_i^2 \right)^{1/2}. \tag{11.4}$$

There is also a `magnitudeSquared` method that returns the dot product of a vector with itself. This method is useful when doing comparison between lengths. Computing the square root is computationally expensive and often unnecessary because $|a| > |b|$ if $a^2 > b^2$.

The following code fragment computes a dot product using the `Vec3D` class.

```
Vec3D a = new Vec3D(1,2,3);
Vec3D b = new Vec3D(4,5,6);
double c = a.dot(b);
```

Because this is conceptually similar to code shown for the `VectorMath` class, we will usually only show code for the `VectorMath` class.

The projection of vector **a** onto vector **b** is a vector with magnitude **a** · **b** in the direction of **b**:

$$\mathbf{a}_{\|b} = \frac{\mathbf{a} \cdot \mathbf{b}}{|b|^2} \mathbf{b}. \tag{11.5}$$

The component of **a** that is perpendicular to vector **b** is useful and can be computed by subtracting $\mathbf{a}_{\|b}$ from **a**:

$$\mathbf{a}_{\perp b} = \mathbf{a} - \frac{\mathbf{a} \cdot \mathbf{b}}{|b|^2} \mathbf{b}. \tag{11.6}$$

These operations are implemented in the `project` and `perp` methods in the `VectorMath` class.

The cross (vector) product of three-dimensional vectors **a** and **b** is another vector **c** that is perpendicular to **a** and **b** and has a magnitude c given by

$$c = |\mathbf{a} \times \mathbf{b}| = |a|\,|b| \sin \theta, \tag{11.7}$$

where θ is again the angle between the two vectors. The cross product can be evaluated as follows using vector components:

$$\mathbf{a} \times \mathbf{b} = (a_y b_z - b_y a_z,\ a_z b_x - b_z a_x,\ a_x b_y - b_x a_y). \tag{11.8}$$

The cross product is often used to create a new vector that is perpendicular (orthogonal) to two others.

11.8 ■ TRANSFORMATIONS

A function whose domain is an n-dimensional space and whose range is an m-dimensional space is known as a *transformation*. In the context of three-dimensional simulation and visualization, the domain and the range are arrays of numbers representing coordinates along a set of spacial axes (basis set). This is expressed in standard mathematical notation as $\mathcal{T} : \mathcal{R}^n \rightarrow \mathcal{R}^m$. In the current context, we transform a 3D point into another 3D point so $m = n = 3$.

The Transformation interface in the numerics package abstracts these mathematical concepts using one-dimensional arrays of numbers.

```
package org.opensourcephysics.numerics;
public interface Transformation extends Cloneable {
  public Object clone ();
  public double[] direct (double[] point);
  public double[] inverse (double[] point) throws
      UnsupportedOperationException;
}
```

The direct and clone methods are guaranteed to succeed, but the inverse method may fail and then throw an UnsupportedOperationException. The direct method transforms the values and returns the given array object; the clone method creates a copy of the Transformation object.

Creating a new object using the clone method is an important Java concept. The clone method must insure that the new object's internal fields (variables) contain the same values as the original object and that objects within the original object are also cloned. If an object's clone method satisfies this criterion, then it can declare itself to be Cloneable. The following code fragment shows how the clone method is used to create a copy of a point and return transformed values in that point.

```
// transformation implements the Transformation interface;
// array is of type double[]
double[] transformedArray = transformation( (double[]) array.clone());
```

Translation class shown in Listing 11.6 is a concrete example of a Transformation that adds an offset to each coordinate value:

$$\mathcal{T}(x, y, z) \rightarrow (x + a, y + b, z + x). \tag{11.9}$$

The main method applies this transformation to a 3D box.

Listing 11.6 An implementation of the Transformation interface.

```
package org.opensourcephysics.manual.ch11;
import org.opensourcephysics.display3d.simple3d.*;
import org.opensourcephysics.numerics.Transformation;

public class Translation implements Transformation {
  double a, b, c;

  public Translation(double a, double b, double c) {
    this.a = a;
    this.b = b;
    this.c = c;
  }

  public Object clone() {
    return new Translation(a, b, c);
  }

  // assumes point is double[3]
  public double[] direct(double[] point) {
    point[0] += a;
```

```
      point[1] += b;
      point[2] += c;
      return point;
   }

   public double[] inverse(double[] point)
        throws UnsupportedOperationException {
      point[0] -= a;
      point[1] -= b;
      point[2] -= c;
      return point;
   }

   static public void main(String args[]) {
      Element box = new ElementBox();
      Transformation translation = new Translation(2, -4, 1);
      box.setTransformation(translation);
      System.out.println("box x="+box.getX());
   }
}
```

Run the Translation program and notice that the x-coordinate of the box has not changed the default value of zero. The transformation changes the drawing geometry but not the position and size of the Element. OSP 3D elements clone the given transformation whenever their setTransformation method is invoked. Thus, changing the original transformation has no effect unless a new setTransformation is invoked again.

Although the concept of a transformation is very general, we need it most often to rotate and translate points and vectors in three-dimensional space. These transformations are examples of *linear transformations* because the transformations associate with addition $\mathbf{v}_1 + \mathbf{v}_2$ and commute with scalar multiplication:

$$\mathcal{T}(\mathbf{v}_1 + \mathbf{v}_2) = \mathcal{T}(\mathbf{v}_1) + \mathcal{T}(\mathbf{v}_2) \tag{11.10a}$$

$$\mathcal{T}(a\mathbf{v}_1) = a\mathcal{T}(\mathbf{v}_1), \tag{11.10b}$$

where $\mathbf{v}_1$ and $\mathbf{v}_2$ are points or vectors in three-dimensional space.

Three-dimensional transformations can be described in many different ways including spoken language: "Rotate the cube by thirty degrees about the y-axis." Physicists and mathematicians prefer a mathematical description of this process. Two common implementations are rotation matrices and quaternions, and the OSP numerics package contains both implementations. Users should select whichever implementation fits their model. Because many physics texts solve dynamics problems using a sequence of rotations about three Cartesian axes (Euler angles), we'll study rotation matrices first.

11.9 ■ MATRIX TRANSFORMATIONS

A matrix can be used to implement the Transformation interface using the usual rules for matrix multiplication:

$$\begin{bmatrix} \bar{x} \\ \bar{y} \\ \bar{z} \end{bmatrix} = \begin{bmatrix} m_{0,0} & m_{0,1} & m_{0,2} \\ m_{1,0} & m_{1,1} & m_{1,2} \\ m_{2,0} & m_{2,1} & m_{2,2} \end{bmatrix} \begin{bmatrix} x \\ y \\ z \end{bmatrix}, \tag{11.11}$$

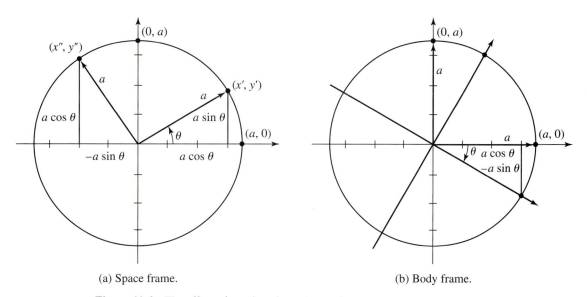

(a) Space frame.   (b) Body frame.

Figure 11.6   The effect of rotation about the z-axis can be derived using simple trigonometry. The figure shows two points being rotated through an angle θ as seen in the (a) space and (b) body frames.

where

$$\begin{bmatrix} \bar{x} \\ \bar{y} \\ \bar{z} \end{bmatrix} = \begin{bmatrix} xm_{0,0} + ym_{0,1} + zm_{0,2} \\ xm_{1,0} + ym_{1,1} + zm_{1,2} \\ xm_{2,0} + ym_{2,1} + zm_{2,2} \end{bmatrix}. \tag{11.12}$$

This can be written compactly as

$$\bar{\mathbf{v}} = \mathcal{M}\mathbf{v}, \tag{11.13}$$

where $\mathcal{M}$ represents the entire matrix, and $\mathbf{v}$ is again a point or a vector.

Rotation transformations can be expressed in matrix form using trigonometry as shown in Figure 11.6. Rotating an object through an angle θ (with respect to the origin) transforms a point $(a, 0)$ into a new point (x', y'). The coordinates of this point are now $(a \cos \theta, a \sin \theta)$ with respect to the original axes. Similarly, a point on the y-axis $(0, a)$ transforms into a point (x'', y'') with coordinates $(-a \sin \theta, a \cos \theta)$.

An alternative view of this process is to think of the transformation as a change of coordinate system rather than as a rotation of the object. The frame of reference attached to the object $(\bar{x}, \bar{y})$ is referred to as the body frame and the original coordinate system (x, y) is referred to as the space frame. The coordinates of a given point of the body (such as $(a, 0)$) do not change in the body frame, but they change (in this case to $(a \cos \theta, -a \sin \theta)$) with respect to the space frame.

The transformation from the space frame to the body frame due to a rotation about the z-axis can be expressed as a matrix as follows:

$$\mathcal{R}_z(\theta) = \begin{bmatrix} \cos \theta & -\sin \theta & 0 \\ \sin \theta & \cos \theta & 0 \\ 0 & 0 & 1 \end{bmatrix}, \tag{11.14}$$

where

$$\bar{\mathbf{v}} = \mathcal{R}_z \mathbf{v}. \tag{11.15}$$

The inverse transformation (body to space) is almost the same except that the direction of rotation (sign of θ) changes thereby changing the sign of the $\sin \theta$ terms:

$$\mathcal{R}_z^{-1}(\theta) = \begin{bmatrix} \cos \theta & \sin \theta & 0 \\ -\sin \theta & \cos \theta & 0 \\ 0 & 0 & 1 \end{bmatrix}. \tag{11.16}$$

The $\mathcal{R}_z$ matrix has a number of very important properties that will be true for all rotation matrices. These properties are:

- The dot product of any row or column with itself is one. Matrices that have this property are said to be normalized.

- The dot product of any row (column) with a different row (column) is zero. Matrices that have this property are said to be orthogonal.

- Every column of $\mathcal{R}_z$ represents the coordinates of a unit vector along a rotated coordinate axis. In other words, column one is the transformed unit vector $(1, 0, 0)$, column two is the transformed unit vector $(0, 1, 0)$, and column three is the transformed unit vector $(0, 0, 1)$. Note that the unit vector along the z-axis (column three) remains unchanged because the rotation is about the z-axis.

- The off-diagonal elements of the matrix change signs if the row and column indices are reversed, $m_{i,j} = m_{j,i}$ where $i \neq j$.

- The transpose of the rotation matrix $\mathcal{R}_z^t$ is the matrix inverse so that $\mathcal{R}_z \mathcal{R}_z^t = 1$. This property is easy to prove using the first two properties and the usual rules of matrix multiplication.

Additional rotation matrices about the other axes can be derived. The rotation about the x-axis through an angle θ is

$$\mathcal{R}_x(\theta) = \begin{bmatrix} 1 & 0 & 0 \\ 0 & \cos \theta & -\sin \theta \\ 0 & \sin \theta & \cos \theta \end{bmatrix}, \tag{11.17}$$

and a rotation about the y-axis through an angle θ is

$$\mathcal{R}_y(\theta) = \begin{bmatrix} \cos \theta & 0 & \sin \theta \\ 0 & 1 & 0 \\ -\sin \theta & 0 & \cos \theta \end{bmatrix}. \tag{11.18}$$

11.10 ■ ANGLE-AXIS REPRESENTATION

Euler showed that the orientation of a rigid body can be expressed as a rotation by an angle about a fixed axis. This construction of a rotation is known as the *angle-axis* representation.

Although rotations about one of the coordinate axes are easy to derive and can be combined using the rules of linear algebra to produce such an arbitrary rotation, the derivation of the rotation matrix for a rotation about the origin by an angle θ around an axis (unit length direction vector) with arbitrary orientation $\hat{\mathbf{r}}$ is not hard to derive.

The strategy is to decompose the vector $\mathbf{v}$ (or point P) that will be rotated into components that are parallel and perpendicular to the direction $\hat{\mathbf{r}}$ as shown in Figure 11.7. The parallel part $\mathbf{v}_\parallel$ does not change, while the perpendicular part $\mathbf{v}_\perp$ is a two-dimensional rotation in a plane perpendicular to $\hat{\mathbf{r}}$. The parallel part is the projection of $\mathbf{v}$ onto the unit vector $\hat{\mathbf{r}}$:

$$\mathbf{v}_\parallel = (\mathbf{v} \cdot \hat{\mathbf{r}})\hat{\mathbf{r}}, \tag{11.19}$$

and the perpendicular part is what remains of $\hat{\mathbf{v}}$ after we subtract the parallel part:

$$\mathbf{v}_\perp = \mathbf{v} - (\mathbf{v} \cdot \hat{\mathbf{r}})\hat{\mathbf{r}}. \tag{11.20}$$

To calculate the rotation of $\mathbf{v}_\perp$, we need two perpendicular basis vectors in the plane of rotation. If we use $\mathbf{v}_\perp$ as the first basis vector, we can then take the cross product of $\mathbf{v}_\perp$ with $\hat{\mathbf{r}}$ to produce a vector $\mathbf{w}$ that is guaranteed to be perpendicular to $\mathbf{v}_\perp$ and $\hat{\mathbf{r}}$:

$$\mathbf{w} = \hat{\mathbf{r}} \times \mathbf{v}_\perp = \hat{\mathbf{r}} \times \mathbf{v}. \tag{11.21}$$

The rotation of $\mathbf{v}_\perp$ can now be expressed in terms of this new basis:

$$\mathbf{v}' = \mathcal{R}(\mathbf{v}_\perp) = \cos\theta \mathbf{v}_\perp + \sin\theta \mathbf{w}. \tag{11.22}$$

The final result is the sum of this rotated vector and the parallel part that does not change:

$$\mathcal{R}(\mathbf{v}) = \mathcal{R}(\mathbf{v}_\perp) + \mathbf{v}_\parallel \tag{11.23a}$$

$$= \cos\theta \mathbf{v}_\perp + \sin\theta \mathbf{w} + \mathbf{v}_\parallel \tag{11.23b}$$

$$= \cos\theta [\mathbf{v} - (\mathbf{v} \cdot \hat{\mathbf{r}})\hat{\mathbf{r}}] + \sin\theta (\hat{\mathbf{r}} \times \mathbf{v}) + (\mathbf{v} \cdot \hat{\mathbf{r}})\hat{\mathbf{r}} \tag{11.23c}$$

$$= [1 - \cos\theta](\mathbf{v} \cdot \hat{\mathbf{r}})\hat{\mathbf{r}} + \sin\theta (\hat{\mathbf{r}} \times \mathbf{v}) + \cos\theta \mathbf{v}. \tag{11.23d}$$

Equation (11.23d) is known as the *Rodrigues formula* and provides a way of constructing rotation matrices in terms of the direction of the axis of rotation $\hat{\mathbf{r}} = (r_x, r_y, r_z)$, the cosine of the rotation angle $c = \cos\theta$, and the sine of the rotation angle $s = \sin\theta$. If we expand the vector products in (11.23d), we obtain the matrix

$$\mathcal{R} = \begin{bmatrix} tr_x r_x + c & tr_x r_y - sr_z & tr_x r_z + sr_y \\ tr_x r_y + sr_z & tr_y r_y + c & tr_y r_z - sr_x \\ tr_x r_z - sr_y & tr_y r_z + sr_x & tr_z r_z + c \end{bmatrix}, \tag{11.24}$$

where $t = 1 - \cos\theta$.

The `Matrix3DTransformation` class in the numerics package provides a ready-to-use implementation of `Transformation` using (11.24). This class also defines static convenience methods that create the rotation matrices $\mathcal{R}_x$, $\mathcal{R}_y$, and $\mathcal{R}_z$ introduced in Section 11.9.

Listing 11.7 uses the `Matrix3DTransformation` to rotate an ellipsoid about the $(0.5, 0.5, 1)$ axis. The direction specifying the rotation need not be normalized.

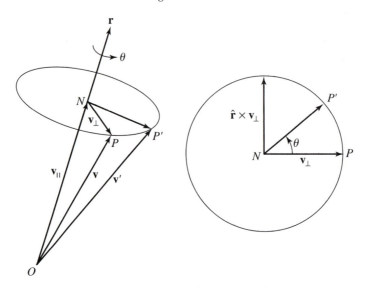

Figure 11.7 A rotation of a vector **v** about an axis $\hat{\mathbf{r}}$ to produce $\mathbf{v}'$ can be decomposed into parallel $\mathbf{v}_{\parallel}$ and perpendicular $\mathbf{v}_{\perp}$ components.

Listing 11.7 `MatrixRotationApp` rotates an ellipsoid using the axis-angle representation.

```
package org.opensourcephysics.manual.ch11;
import org.opensourcephysics.display3d.simple3d.*;
import org.opensourcephysics.frames.Display3DFrame;
import org.opensourcephysics.numerics.*;

public class MatrixRotationApp {
  public static void main(String[] args) {
    Display3DFrame frame = new Display3DFrame("Axis-angle Rotation");
    frame.setDecorationType(VisualizationHints.DECORATION_AXES);
    Element ellipsoid = new ElementEllipsoid();
    Element arrow = new ElementArrow();
    ellipsoid.setSizeXYZ(0.4, 0.4, 1.0);
    frame.addElement(arrow);
    frame.addElement(ellipsoid);
    frame.setVisible(true);
    frame.setDefaultCloseOperation(javax.swing.JFrame.EXIT_ON_CLOSE);
    double theta = 0;
    double[] axis = new double[] {0.5, 0.5, 1}; // rotation axis
    arrow.setSizeXYZ(axis);
    while(true) { // animate until the program exits
      try {
        Thread.sleep(100);
      } catch(InterruptedException ex) {}
      theta += Math.PI/40;
      Transformation transformation =
          Matrix3DTransformation.rotation( theta, axis);
      ellipsoid.setTransformation(transformation);
      frame.render();
    }
  }
}
```

Suppose that we wish to orient an object such that a direction within the object points in the direction of another vector **b**. How do we create a transformation that performs this rotation? Because the axis of rotation must be perpendicular to both vectors, we use the cross product to create an axis array. The cosine of the rotation angle is easily computed by normalizing the two vectors and taking the dot product. The static createAlignmentTransformation method in the Matrix3DTransformation class constructs the needed transformation. It is implemented as follows:

```
public static Matrix3DTransformation
    createAlignmentTransformation(double[] v1, double v2[]){
    v1 = VectorMath.normalize((double[]) v1.clone());
    v2 = VectorMath.normalize((double[]) v2.clone());
    double theta = Math.acos(VectorMath.dot(v1, v2));
    double[] axis = VectorMath.cross3D(v1, v2);
    return Matrix3DTransformation.Rotation(theta, axis);
}
```

AlignmentApp uses the createAlignmentTransformation method to align the z-axis of a cylinder with a vector at an angle of 45° in the xy-plane. This program is available on the CD although it is not shown here because it is similar to Listing 11.7.

11.11 ■ EULER ANGLE REPRESENTATION

Euler angles are generally described in physics texts as a group of three rotations about a set of body frame axes. An object is created with the body frame aligned with the world's Cartesian coordinate system. The first rotation is about the body frame's y-axis; the second rotation is about the new z-axis, and the third rotation is about the new y-axis. Other definitions of Euler angles are possible. For example, the Java 3D API defines Euler angles as three rotations about a fixed set of axes. All possible positions of an object can be represented using either of these conventions.

In order to construct a rotation using Euler angles, we need to create transformations that rotate an object about the x-, y-, and z-axes. The following analysis shows the steps required. A full rotation matrix A can be obtained by writing the product of three individual rotation matrices:

$$A = R_1(\psi)R_2(\theta)R_3(\phi) . \tag{11.25}$$

The first rotation is about the y-axis through an angle ϕ. The second rotation is about the z-axis through an angle θ. The final rotation is a rotation about the new y-axis through an angle ψ. Thus, the third rotation and first rotation matrices are similar, but because matrix multiplication does not commute, the result can be quite complicated. This sequence of rotations gives the final transformation matrix using the physics convention for Euler angles:

$$R_1(\psi)R_2(\theta)R_3(\phi) =$$

$$\begin{bmatrix} -\sin\phi\sin\psi + \cos\phi\cos\theta\cos\psi & -\sin\theta\cos\phi & \sin\phi\cos\psi + \sin\psi\cos\phi\cos\theta \\ \sin\theta\cos\psi & \cos\theta & \sin\theta\sin\psi \\ -\sin\psi\cos\phi - \sin\phi\cos\theta\cos\psi & \sin\phi\sin\theta & \cos\phi\cos\psi - \sin\phi\sin\psi\cos\theta \end{bmatrix}.$$

$$\tag{11.26}$$

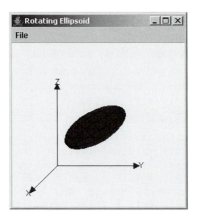

Figure 11.8   The orientation of an object can be set using quaternions to represent rotations.

Unfortunately, the matrix in (11.26) is singular when $\sin \theta = 0$. Because we must invert the above matrix to solve problems in rigid-body dynamics, the Euler angle approach becomes unstable whenever θ approaches 0 or π. A better approach for numerical computation is to abandon Euler angles and to use *quaternions*.

11.12 ■ QUATERNION REPRESENTATION

Quaternions were invented by Hamilton as an elegant extension of complex numbers and were later used by Cayley to describe rotations as shown in Figure 11.8. Although quaternions may be unfamiliar, they are easy to use and can be related to the axis-angle representation introduced in Section 11.10.

A quaternion can be represented in terms of real and *hypercomplex* numbers i, j, and k as

$$\hat{q} = q_0 + iq_1 + jq_2 + kq_3 = (q_0, q_1, q_2, q_3), \tag{11.27}$$

where the hypercomplex numbers obey Hamilton's rules:

$$i^2 = j^2 = k^2 = ijk = -1. \tag{11.28}$$

Like imaginary numbers, the quaternion conjugate is defined as

$$\hat{q}^* = q_0 - iq_1 - jq_2 - kq_3 = (q_0, -q_1, -q_2, -q_3). \tag{11.29}$$

Unlike imaginary numbers, quaternion multiplication does not commute and obeys the rules:

$$ij = k, \quad jk = i, \quad ki = j, \quad ji = -k, \quad kj = -1, \quad ik = -j. \tag{11.30}$$

Although it would be a mistake to identify hypercomplex numbers with unit vectors in three-dimensional space (just as it would be a mistake to identify the imaginary number i

with the y direction in a two-dimensional space), it is convenient to think of a quaternion as the sum of a scalar q_0 and a vector $\mathbf{q}$:

$$\hat{q} = q_0 + \mathbf{q} = (q_0, \mathbf{q}). \tag{11.31}$$

The quaternion is said to be *pure* if the scalar part is zero.

By using the above definitions for multiplication of hypercomplex numbers, the product of two quaternions $\hat{p}$ and $\hat{q}$ can be shown to be

$$\hat{p}\hat{q} = q_0 p_0 - \mathbf{q} \cdot \mathbf{p} + q_0 \mathbf{p} + p_0 \mathbf{q} + \mathbf{p} \times \mathbf{q}, \tag{11.32}$$

where $\mathbf{p}$ and $\mathbf{q}$ are the vector part of the quaternions $\hat{p}$ and $\hat{q}$, respectively.

As with matrices, the inverse $\hat{q}^{-1}$ of a quaternion is another quaternion such that $\hat{q}^{-1}\hat{q} = (1, 0, 0, 0)$. It is easy to show that the inverse of $\hat{q} = (q_0, q_1, q_2, q_3)$ is $(q_0, -q_1, -q_2, -q_3)$ by carrying out the multiplication using (11.32).

Cayley discovered that if he normalized a quaternion so that $q_0^2 + q_1^2 + q_2^2 + q_3^2 = 1$, it could be used to describe pure rotations. The quaternion representation of rotation through an angle θ about a normalized axis vector (n_x, n_y, n_z) can be written as

$$\hat{q} = \cos\frac{\theta}{2} + (in_x + jn_y + kn_z)\sin\frac{\theta}{2}. \tag{11.33}$$

We can use this quaternion to rotate a vector $\mathbf{v}$ by constructing a pure quaternion $\hat{v}$ using the vector as the quaternion's q_1, q_2, and q_3 components. The quaternion product

$$\hat{v}' = \hat{q}\hat{v}\hat{q}^{-1} \tag{11.34}$$

produces a new pure quaternion $\hat{v}'$ whose vector part is the rotated 3D vector. This expression is closely related to the axis-angle representation and is used in the `Quaternion` class to implement the `Transformation` interface.

The arithmetic operations described in this section have been implemented in the `Quaternion` class and are listed in Table 11.3. This class stores a quaternion using instance variables q0, q1, q2, and q3 and contains methods that perform quaternion multiplication and other operations. The following code fragment uses this class to orient a 3D element.

```
double[] n = new double[]{0.5, 0.5, 1}; // rotation axis
double theta = Math.PI/30;              // rotation angle
double cos = Math.cos(theta/2), sin = Math.sin(theta/2);
Transformation transformation =
     new Quaternion(cos, sin*n[0], sin*n[1], sin*n[2]);
ellipsoid.setTransformation(transformation);
```

`QRotationApp` uses this code. It is available on the CD although it is not shown here because it is similar to Listing 11.7.

11.13 ■ RIGID-BODY DYNAMICS

A solid object that does not deform is known as a *rigid body*. It has a natural coordinate system known as the *body frame* that has the important property that the body does not wobble if it is spun about one of these axes. The moments of inertia about the body frame

Table 11.3   Quaternion instance methods.

org.opensourcephysics.display3d.numerics.Quaternion	
`add`	Adds the components of another quaternion to this quaternion.
`angle`	Gets the angle between this quaternion and another quaternion.
`clone`	Instantiates an exact copy of this quaternion.
`conjugate`	Changes the sign of the hypercomplex components q_1, q_2, and q_3.
`direct`	Transforms the given point by the direct quaternion rotation.
`getRotationMatrix`	The matrix that performs an equivalent rotation.
`inverse`	Transforms the given point by the inverse quaternion rotation.
`magnitude`	Gets the magnitude of this quaternion.
`magnitudeSquared`	Gets the magnitude squared of this quaternion.
`multiply`	Multiplies this quaternion by another quaternion.
`normalize`	Normalizes this quaternion.
`setCoordinates`	Sets the quaternion coordinates q_0, q_1, q_2, and q_3.
`setOrigin`	Sets the origin about which the rotation will be transformed.
`subtract`	Subtracts the components of another quaternion from this quaternion.

axis (principal moments) are numbers that we label I_1, I_2, and I_3. For symmetric bodies, such as boxes or ellipsoids, the body frame axes are the axes of symmetry of the object. It turns out that the differential equations that describe the rotation of a rigid body are much simpler in the body frame than in a reference frame that is fixed in space. The disadvantage is that the body frame is noninertial and that all vectors must be transformed into this frame when solving the equations. A spinning top simulation, for example, must transform the constant force of gravity which is down in the space frame into a time-varying vector that can point up, down, or sideways as seen from the top's body frame. Fortunately, `Elements` have `toBodyFrame` and `toSpaceFrame` methods that make the transformation between frames easy.

The rotational equation of motion in the body frame can be written as

$$\mathbf{N} = \frac{d\mathbf{L}}{dt} + \boldsymbol{\omega} \times \mathbf{L} \tag{11.35a}$$

$$\mathbf{L} = \mathcal{I}\boldsymbol{\omega}, \tag{11.35b}$$

where $\mathbf{N}$ is the torque on the rigid body, $\mathbf{L}$ is the angular momentum, and ω is the angular velocity. The principal moments of inertia are the diagonal elements of the inertia matrix $\mathcal{I}$. Equation (11.35) is Euler's equation for the motion of a rigid body. Using the three principal moments of inertia, this equation may be written in component form as

$$N_1 = I_1\dot{\omega}_1 + (I_3 - I_2)\omega_3\omega_2 \qquad (11.36a)$$

$$N_2 = I_2\dot{\omega}_2 + (I_1 - I_3)\omega_1\omega_3 \qquad (11.36b)$$

$$N_3 = I_3\dot{\omega}_3 + (I_2 - I_1)\omega_2\omega_1, \qquad (11.36c)$$

and

$$L_1 = I_1\omega_1 \qquad (11.37a)$$

$$L_2 = I_2\omega_2 \qquad (11.37b)$$

$$L_3 = I_3\omega_3, \qquad (11.37c)$$

where all vector components are in the body frame.

An application of Euler's equation for rigid bodies is the computation of the torque that must be applied to a shaft to keep an out-of-balance mass from wobbling. Assume that the angular velocity ω is constant in the space frame and lies along the z-axis. It is easy to rotate the mass-shaft body in the 3D scene by creating a group and transforming the group using a z-axis rotation matrix. In order to compute the angular momentum using (11.37a), we must transform the angular velocity vector from the space frame $(0, 0, \omega_z)$ to the body frame. The `toBodyFrame` method in the `Element` class in Listing 11.8 does just that. It shows the torque on a rectangular sheet rotating on a fixed shaft with uniform angular velocity ω. The `SimulationControl` allows the mass to be tilted on the shaft to produce an out-of-balance configuration that would cause a real system to shake unless a torque is applied to the shaft. The program displays the angular momentum vector and the torque vector using color-coded arrows.

Listing 11.8   A mass rotating on a fixed shaft with
uniform angular velocity ω.

```
package org.opensourcephysics.manual.ch11;
import org.opensourcephysics.controls.*;
import org.opensourcephysics.display3d.simple3d.*;
import org.opensourcephysics.frames.Display3DFrame;
import org.opensourcephysics.numerics.*;

public class TorqueApp extends AbstractSimulation {
  Display3DFrame frame = new Display3DFrame(" Rotation Test");
  Element body = new ElementBox();          // shows rigid body
  Element shaft = new ElementCylinder();    // shows shaft
  Element arrowOmega = new ElementArrow();  // angular velocity of shaft
  Element arrowL = new ElementArrow();      // angular momentum of body
  Element arrowTorque = new ElementArrow();// torque on shaft
  // shaftGroup contains shaft and arrowOmega
  Group shaftGroup = new Group();
  // bodyGroup contains body, arrowL, arrowTorque
  Group bodyGroup = new Group();
  double theta = 0, omega = 0.1, dt = 0.1;
  double Ixx = 1.0, Iyy = 1.0, Izz = 2.0; // moments of inertia

  public TorqueApp() {
    frame.setDecorationType(VisualizationHints.DECORATION_AXES);
    body.setSizeXYZ(1.0, 1.0, 0.1); // thin rectangle
```

```
      shaft.setSizeXYZ(0.1, 0.1, 0.8);
      arrowL.getStyle().setLineColor(java.awt.Color.MAGENTA);
      arrowTorque.getStyle().setLineColor(java.awt.Color.CYAN);
      bodyGroup.addElement(body);
      bodyGroup.addElement(arrowTorque);
      bodyGroup.addElement(arrowL);
      shaftGroup.addElement(bodyGroup);
      shaftGroup.addElement(arrowOmega);
      shaftGroup.addElement(shaft);
      frame.addElement(shaftGroup);
   }

   void computeVectors() {
      // convert omega to body frame
      double[] omega =
         body.toBodyFrame(new double[] {0, 0, this.omega});
      // convert L in body frame
      double[] angularMomentum =
         new double[] {omega[0]*Ixx, omega[1]*Iyy, omega[2]*Izz};
      double[] torque =
         VectorMath.cross3D(omega, angularMomentum);
      arrowL.setSizeXYZ(angularMomentum);
      arrowTorque.setSizeXYZ(torque);
      // position torque arrow at tip of angular momentum
      arrowTorque.setXYZ(angularMomentum);
   }

   public void initialize() {
      omega = control.getDouble("omega");
      arrowOmega.setSizeXYZ(0, 0, omega);
      double tilt = control.getDouble("tilt");
      bodyGroup.setTransformation(
         Matrix3DTransformation.rotationX(tilt));
      computeVectors();
   }

   public void reset() {
      control.setValue("omega", "pi/4");
      control.setValue("tilt", "pi/5");
   }

   protected void doStep() {
      theta += omega*dt;
      shaftGroup.setTransformation(
         Matrix3DTransformation.rotationZ(theta));
      computeVectors();
   }

   public static void main(String[] args) {
      SimulationControl.createApp(new TorqueApp());
   }
}
```

Notice how TorqueApp instantiates a shaft group and a body group in its constructor. The shaft group contains visual representations of the shaft, the angular velocity vector, and the body group. The body group contains representations of a thin rectangular sheet, the angular velocity, and the torque. The body group is rotated about the x-axis in the `initialize` method and the shaft group is rotated about the z-axis in the `doStep` method. Because the body group is within the shaft group, the body group also rotates. The `computeVectors` method displays the angular momentum and torque vectors by computing their values in the rotating (body) frame of reference. The torque is computed using a cross product of the body frame angular velocity and the body frame angular momentum according to the following kinematic formula:

$$\frac{d\mathbf{L}_{\text{space}}}{dt} = \frac{d\mathbf{L}_{\text{body}}}{dt} + \omega \times \mathbf{L}_{\text{body}} \ . \tag{11.38}$$

Rigid-body dynamics is a rich topic that is covered in most intermediate level mechanics textbooks. Unfortunately, its development is beyond the scope of this Guide. The best way to approach these types of problems is to write the equation of motion of the body in terms of quaternion components and their derivatives. The derivation of these equations can be found in *An Introduction to Computer Simulation Methods* by Gould, Tobochnik, and Christian. The `SpinningTopApp` program from this book is available on the CD. The application of quaternions to molecular dynamics simulations can be found in books by Rapaport or Allen.

11.14 ■ JOGL

Glenn Ford has developed a JOGL implementation of OSP 3D API and this implementation is located in the jogl package `org.opensourcephysics.display3d.jogl`. Jogl package code and the examples are available on the CD in the optional `osp_jogl.zip` archive. All that is required to use the jogl package is to install the operating system specific JOGL library and import the OSP jogl package.

```
import org.opensourcephysics.display3d.jogl.*;
```

The JOGL library is available for Windows 2000 or later, OS X 10.3 or later, Solaris, and Linux. It can be downloaded from `https://jogl.dev.java.net/`. Libraries and installation instructions vary depending on the operating system because JOGL directly accesses video hardware. For example, on Windows the library is distributed in a zip file and you must copy the enclosed DLL files into the system folder (usually `C:\windows` or something similar) and the JOGL jar files into Java VM's class path (usually `lib\ext`) subdirectories in the JDK and the JRE.

Using the OSP jogl package not only produces more realistic images, but it also allows programmers to customize the scene in ways that the simple 3D package does not allow. The jogl package implementation of the `DrawingPanel3D` interface has three methods that can be overridden to access underlying OpenGL library: `additionalGLInit`, `preCameraGL`, and `postCameraGL`. The `additionalGLInit` method is invoked just after the standard initialization of the OSP JOGL display, the `preCameraGL` method is invoked just before the camera rotates the scene, and the `postCameraGL` method is called just after the camera rotates the scene. The `additionalGLInit` and `postCameraGL` methods are used to add lights to a scene in the `SpotLightApp` program shown in Listing 11.9.

Listing 11.9   A bouncing ball simulation with a spotlight that shines on a surface.

```java
package org.opensourcephysics.manual.ch11;
import java.awt.*;
import net.java.games.jogl.*;
import org.opensourcephysics.controls.*;
import org.opensourcephysics.display3d.jogl.*;
import org.opensourcephysics.numerics.*;

public class SpotLightApp extends AbstractSimulation {
  private static boolean original_light = true, spot_light = true;
  Element ball, floor;
  double velocity = 0, time = 0;
  Vec3D lightPosition;
  DrawingPanel3D panel;
  DrawingFrame3D frame;

  public SpotLightApp() {
    // bouncing ball
    ball = new ElementSphere();
    ball.setXYZ(0, 0, 9.0);
    ball.setSizeXYZ(2, 2, 2);
    ball.getStyle().setFillColor(java.awt.Color.YELLOW);
    ball.getStyle().setResolution(new Resolution(70, 70, 0));
    // floor
    floor = new ElementBox();
    floor.setXYZ(0.0, 0.0, 0.0);
    floor.setSizeXYZ(7.0, 7.0, 1.0);
    floor.getStyle().setResolution(new Resolution(70, 70, 2));
    // light will go above the ball and circle around
    lightPosition = new Vec3D(1, 0, 10);
    // Create the panel
    panel = new LightTestPanel();
    panel.getComponent().setBackground(Color.black);
    panel.getVisualizationHints().setDecorationType(
        VisualizationHints.DECORATION_NONE);
    panel.getVisualizationHints().setAllowQuickRedraw(false);
    // Add the objects to panel
    panel.addElement(ball);
    panel.addElement(floor);
    frame = new DrawingFrame3D(panel);
    frame.setDefaultCloseOperation(javax.swing.JFrame.EXIT_ON_CLOSE);
    frame.setSize(300, 300);
    frame.setVisible(true);
  }

  protected void doStep() {
    double z = ball.getZ(), dt = 0.05;
    time += dt;
    z += velocity*dt;   // moves the ball
    velocity -= 9.8*dt; // acceleration changes the velocity
    if((z<=1.0)&&(velocity<0)) {
      velocity = -velocity;
    }
    ball.setZ(z);
    // make the light circle around the scene
```

```
      lightPosition.x = Math.cos(time)*1.5;
      lightPosition.y = Math.sin(time)*1.5;
    }

    public static void main(String[] args) {
      AbstractSimulation s = new SpotLightApp();
      s.startSimulation(); // starts the thread
    }

    private class LightTestPanel extends DrawingPanel3D {
      public void additionalGLInit(GLDrawable drawable) {
        initSpot(drawable.getGL());
      }

      public void postCameraGL(GLDrawable drawable) {
        GL gl = drawable.getGL();
        if(original_light) {
          gl.glEnable(GL.GL_LIGHT0);
        } else {
          gl.glDisable(GL.GL_LIGHT0);
        }
        if(spot_light) {
          gl.glEnable(GL.GL_LIGHT2);
          updateSpotProperties(gl);
        } else {
          gl.glDisable(GL.GL_LIGHT2);
        }
      }

      private void initSpot(GL gl) {
        // white light
        float ambient[] = {1.0f, 1.0f, 1.0f, 1.0f};
        // properties of the light
        gl.glLightfv(GL.GL_LIGHT2, GL.GL_AMBIENT, ambient);
        // Spot properties
        // define the spot direction and cutoff
        updateSpotProperties(gl);
        // exponent properties define the concentration of the light
        gl.glLightf(GL.GL_LIGHT2, GL.GL_SPOT_EXPONENT, 70.0f);
        // light attenuation (default values used here:
        // no attenuation with distance)
        gl.glLightf(GL.GL_LIGHT2, GL.GL_CONSTANT_ATTENUATION, 0.0f);
        gl.glLightf(GL.GL_LIGHT2, GL.GL_LINEAR_ATTENUATION, 0.001f);
        gl.glLightf(GL.GL_LIGHT2, GL.GL_QUADRATIC_ATTENUATION, 0.03f);
        gl.glEnable(GL.GL_LIGHT2);
      }

      private void updateSpotProperties(GL gl) {
        // set the light position
        float position[] = {(float) lightPosition.x,
            (float) lightPosition.y, (float) lightPosition.z, 1.0f};
        gl.glLightfv(GL.GL_LIGHT2, GL.GL_POSITION, position);
        // our light will always face down
        float direction[] = {0, 0, -1};
        // spot direction
        gl.glLightfv(GL.GL_LIGHT2, GL.GL_SPOT_DIRECTION, direction);
```

```
            // angle of the cone light emitted by the spot:
            // value between 0 and 180
            gl.glLightf(GL.GL_LIGHT2, GL.GL_SPOT_CUTOFF, 10);
        }
    }
}
```

Elements in the OSP jogl package can be customized by overriding the `additionalGL` method. This method is called just before the element is filled and edgelines are drawn. The panel's `preCameraGL` method and the element's `additionalGL` method are used to add surface texture in the `TextureApp` program shown in Listing 11.10.

Listing 11.10   A sphere with a striped texture map.

```
package org.opensourcephysics.manual.ch11;
import net.java.games.jogl.GL;
import net.java.games.jogl.GLDrawable;
import org.opensourcephysics.display3d.jogl.*;

public class TextureApp {
  DrawingPanel3D panel;
  DrawingFrame3D frame;

  public TextureApp() {
    panel = new TextureTestPanel();
    Element ball = new TexturedSphere(makeStripeImage1());
    ball.setSizeXYZ(4, 4, 4);
    ball.setXYZ(0, -3, 0);
    ball.getStyle().setResolution(new Resolution(30, 30, 0));
    panel.addElement(ball);
    ball = new TexturedSphere(makeStripeImage2());
    ball.setSizeXYZ(2, 2, 2);
    ball.setXYZ(0, 3, 0);
    ball.getStyle().setResolution(new Resolution(30, 30, 0));
    panel.addElement(ball);
    frame = new DrawingFrame3D(panel);
    frame.setDefaultCloseOperation(javax.swing.JFrame.EXIT_ON_CLOSE);
    frame.setSize(300, 300);
    frame.setVisible(true);
  }

  public static void main(String[] args) {
    new TextureApp();
  }

  private class TextureTestPanel extends DrawingPanel3D {

    private float xequalzero[] = {1.0f, 0.0f, 0.0f, 0.0f};
    private float slanted[] = {1.0f, 1.0f, 1.0f, 0.0f};
    private float currentCoeff[];
    private int currentPlane;
    private int currentGenMode;
    // Set all of the proper flags for the textures we want to use

    public void preCameraGL(GLDrawable drawable) {
      GL gl = drawable.getGL();
```

```
        gl.glEnable(GL.GL_DEPTH_TEST);
        gl.glShadeModel(GL.GL_SMOOTH);
        gl.glTexEnvf(GL.GL_TEXTURE_ENV, GL.GL_TEXTURE_ENV_MODE,
                     GL.GL_MODULATE);
        currentCoeff = xequalzero;
        currentGenMode = GL.GL_OBJECT_LINEAR;
        currentPlane = GL.GL_OBJECT_PLANE;
        gl.glTexGeni(GL.GL_S, GL.GL_TEXTURE_GEN_MODE, currentGenMode);
        gl.glTexGenfv(GL.GL_S, currentPlane, currentCoeff);
        gl.glEnable(GL.GL_TEXTURE_GEN_S);
        gl.glEnable(GL.GL_TEXTURE_1D);
        gl.glEnable(GL.GL_CULL_FACE);
        gl.glEnable(GL.GL_LIGHTING);
        gl.glEnable(GL.GL_LIGHT0);
        gl.glEnable(GL.GL_AUTO_NORMAL);
        gl.glEnable(GL.GL_NORMALIZE);
        gl.glCullFace(GL.GL_BACK);
        gl.glMaterialf(GL.GL_FRONT, GL.GL_SHININESS, 64.0f);
    }
}

private static final int stripeImageWidth = 32;
// Make a striped texture that's opaque

private byte[] makeStripeImage1() {
    byte[] stripeImage = new byte[stripeImageWidth*4];
    for(int j = 0;j<stripeImageWidth;j++) {
        if(j<=4) {
            stripeImage[j] = (byte) 255;
        } else {
            stripeImage[j] = (byte) 60;
        }
        if(j>4) {
            stripeImage[j+stripeImageWidth] = (byte) 255;
        } else {
            stripeImage[j+stripeImageWidth] = (byte) 60;
        }
        stripeImage[j+2*stripeImageWidth] = (byte) 60;
        stripeImage[j+3*stripeImageWidth] = (byte) 255;
    }
    return stripeImage;
}

// Make a translucent gradient-striped texture
private byte[] makeStripeImage2() {
    byte[] stripeImage = new byte[stripeImageWidth*4];
    for(int j = 0;j<stripeImage.length;j++) {
        stripeImage[j] = (byte) (j*5);
    }
    return stripeImage;
}

private class TexturedSphere extends ElementSphere {
    private byte stripeImage[] = new byte[stripeImageWidth*4];

    public TexturedSphere(byte[] stripeImage) {
```

```
            super();
            this.stripeImage = stripeImage;
        }

        // Set the texture before rendering
        public void additionalGL(GLDrawable drawable) {
            GL gl = drawable.getGL();
            gl.glPixelStorei(GL.GL_UNPACK_ALIGNMENT, 1);
            gl.glTexParameterf(GL.GL_TEXTURE_1D, GL.GL_TEXTURE_WRAP_S,
                GL.GL_REPEAT);
            gl.glTexParameterf(GL.GL_TEXTURE_1D, GL.GL_TEXTURE_MAG_FILTER,
                GL.GL_LINEAR);
            gl.glTexParameterf(GL.GL_TEXTURE_1D, GL.GL_TEXTURE_MIN_FILTER,
                GL.GL_LINEAR);
            gl.glTexImage1D(GL.GL_TEXTURE_1D, 0, 4, stripeImageWidth, 0,
                GL.GL_RGBA, GL.GL_UNSIGNED_BYTE, stripeImage);
        }
    }
}
```

11.15 ■ PROGRAMS

The following examples are in the `org.opensourcephysics.manual.ch11` package.
JOGL examples can be found in the `org.opensourcephysics.manual.ch11_jogl`
package.

AlignmentApp

`AlignmentApp` uses a quaternion to align a 3D `Element` with a vector. See Section 11.10.

BallInteractionApp

`BallInteractionApp` uses the `InteractionTarget` API to enable changes in position.
See Section 11.6.

BounceApp

`BounceApp` presents a 3D visualization of a bouncing ball as described in Section 11.3.

Box3DApp

`Box3DApp` creates a 3D box using the simple 3D package and sets it properties. See Section 11.2.

CameraApp

`CameraApp` creates a 3D scene and sets the camera's properties. See Section 11.4.

ElementApp

`ElementApp` demonstrates the appearance and properties of 3D elements. See Section 11.5.

FreeRotationApp

`FreeRotationApp` models the torque-free rotation of a spinning object. See Section 11.13.

GroupApp

`GroupApp` demonstrates how a group of objects can be positioned as a single object as described in Section 11.5.

Interaction3DApp

`Interaction3DApp` demonstrates how to add and handle actions in a `Display3DFrame`. See Section 11.6.

LorenzApp

`LorenzApp` models the Lorenz attractor by extending `AbstractSimulation` and implementing the `doStep` method. See Section 11.5.

MatrixRotationApp

`MatrixRotationApp` demonstrates how to use the `Matrix3DTransformation` class to rotate a 3D element. See Section 11.10.

QRotationApp

`QRotationApp` demonstrates how to use the `Quaternion` class to rotate an `Element`. See Section 11.12.

SHO3DApp

`SHO3DApp` creates a 3D harmonic oscillator simulation by extending `AbstractSimulation` and implementing the `doStep` method.

SpinningTopApp

`SpinningTopApp` models the dynamics of a spinning top using quaternions. See Section 11.13.

TorqueApp

`TorqueApp` models Euler's equations to show the torque on an object that is spinning about a fixed axis. See Section 11.13.

CHAPTER

12

XML Documents

by Doug Brown and Wolfgang Christian

Open Source Physics defines an xml framework that can be used to store, retrieve, and display structured documents.

12.1 ■ BASIC XML

Extensible Markup Language (xml) is a significant development in information technology because it allows us to define a portable language for data that organizes and categorizes the data in a consistent and self-describing manner. It is used by OSP applications such as Launcher (Chapter 15) to organize curricular material and Tracker (Chapter 16) to store video analysis data. If you are unfamiliar with the xml language, don't worry. Although xml can be an unforgiving language to write using a text editor, it is highly structured and therefore not very difficult to program. The OSP library provides all the tools that you need to create simple xml documents. We start by studying the xml representation of an ideal gas shown in Listing 12.1.

Listing 12.1   An ideal gas model described using xml.

```
<?xml version="1.0" encoding="UTF-8"?>
<object class="java.lang.Object">
  <property name="comment" type="string">An xml description of an ideal
      gas.</property>
  <property name="pressure" type="double">100000.0</property>
  <property name="volume" type="double">2.0</property>
  <property name="temperature" type="double">273.0</property>
</object>
```

XML documents have a very simple structure. Like hypertext markup language (html) documents they are written in plain text and consist of one or more named *elements* that store data between their opening and closing tags. Listing 12.1 starts with a header that identifies it as an xml document. The document has a single root element that contains all xml content. The root of an Open Source Physics xml document is an `<object>` element that stores data in child `<property>` elements with *name* and *type* attributes. The *value* of the property is written between the opening and closing property tags.

Unlike html, anyone can define their own xml tags, their rules, and their meaning using a Document Type Definition (DTD). The Open Source Physics DTD is located in the resources package and defines only two element types, `<object>` and `<property>`. The `<object>` element stores the object's Java class as an attribute and the object's properties in child `<property>` elements.

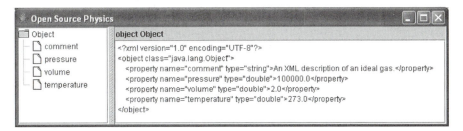

Figure 12.1 A graphical view of a basic xml document describing an ideal gas. When a node is selected, the corresponding xml `<object>` or `<property>` element is displayed in the right-hand panel.

Our xml framework is consistent with the control API described in Chapter 5. Programs instantiate a subinterface of `Control` named `XMLControl` and then set and get values from this control. The OSP implementation of `XMLControl` is the `XMLControlElement` class. Listing 12.2 uses the `XMLControlElement` to set values for an ideal gas model and to write these values to a file named `gas_data.xml`.

Listing 12.2 An `XMLControl` can be used to write data to a file.

```
package org.opensourcephysics.manual.ch12;
import org.opensourcephysics.controls.XMLControl;
import org.opensourcephysics.controls.XMLControlElement;

public class WriteXMLApp {
  static String fileName = "gas_data.xml";

  public static void main(String[] args) {
    XMLControl xml = new XMLControlElement();
    xml.setValue("comment", "An XML description of an ideal gas.");
    xml.setValue("pressure", 1.0E5);
    xml.setValue("volume", 2.0);
    xml.setValue("temperature", 273.0);
    xml.write(fileName);
  }
}
```

Because the `gas_data.xml` file is a text file, it can be examined in any text editor. XML aware editors will often provide a tree view of the content as shown in Figure 12.1.

An `XMLControl` supports standard OSP control methods such as `getDouble` and `getString` to retrieve values of common data types. The `ReadXMLApp` program shown in Listing 12.3 uses these methods to read the `gas_data.xml` data file.

Listing 12.3 An `XMLControl` can be used to read data from an xml file.

```
package org.opensourcephysics.manual.ch12;
import org.opensourcephysics.controls.XMLControl;
import org.opensourcephysics.controls.XMLControlElement;

public class ReadXMLApp {
  public static void main(String[] args) {
    XMLControl xml = new XMLControlElement("gas_data.xml");
    System.out.println(xml.getString("comment"));
```

```
        System.out.println(xml.getDouble("pressure"));
        System.out.println(xml.getDouble("volume"));
        System.out.println(xml.getDouble("temperature"));
    }
}
```

Listing 12.4 shows how to display data in an xml property tree (see Figure 12.1) using an `XMLTreePanel`. When an object or property node is selected in the tree, the corresponding xml `<object>` or `<property>` element is displayed. Primitive and string property types are directly editable in the text field.

> **Listing 12.4**   An `XMLTreePanel` is used to display an `XMLControl` in an xml property tree.

```
package org.opensourcephysics.manual.ch12;
import org.opensourcephysics.controls.*;
import org.opensourcephysics.display.OSPFrame;
import javax.swing.JFrame;

public class ShowXMLApp {
    public static void main(String[] args) {
        XMLControl xml = new XMLControlElement("gas_data.xml");
        // display xml data in a tree view
        JFrame frame = new OSPFrame(new XMLTreePanel(xml));
        frame.setSize(650, 550);
        frame.setDefaultCloseOperation(JFrame.EXIT_ON_CLOSE);
        frame.setVisible(true);
    }
}
```

12.2 ■ OBJECT PROPERTIES

Unlike basic OSP controls, xml controls can set and get objects in addition to primitive data types. Because child elements can be Java arrays, collections, and objects as well as primitive data types, property elements are often nested within an xml document. An `XMLControl` can, for example, store arrays and interactive shapes as shown in Listing 12.5. Note the various data types. The `InteractiveShape` object, for example, is a child element that has double, boolean, and color properties.

> **Listing 12.5**   Objects in an xml document are nested within other objects.

```
package org.opensourcephysics.manual.ch12;
import org.opensourcephysics.controls.*;
import org.opensourcephysics.display.*;
import javax.swing.JFrame;

public class WriteShowNestedXMLApp {
    public static void main(String[] args) {
        // create the XMLControl
        XMLControl xml = new XMLControlElement();
        xml.setValue("name", "Test Object");
        xml.setValue("comment", "A test of XML.");
        xml.setValue("temperature", 273.4);
```

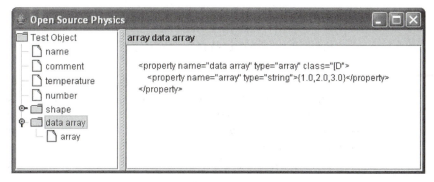

Figure 12.2   An xml property tree displays nested xml properties.

```
xml.setValue("number", 200);
xml.setValue("shape",
        InteractiveShape.createRectangle(0.0, 1.0, 2.0, 5.0));
xml.setValue("data array", new double[] {1, 2, 3});
// write the xml document
xml.write("nested_data.xml"); // save to file
// display the xml document in a tree view
JFrame frame = new OSPFrame(new XMLTreePanel(xml));
frame.setSize(650, 550);
frame.setDefaultCloseOperation(JFrame.EXIT_ON_CLOSE);
frame.setVisible(true);
  }
}
```

Listing 12.5 writes an xml document to a file and displays the document in the xml tree shown in Figure 12.2. When a node contains an object, the object's properties are displayed in a subtree and the corresponding xml description is displayed.

An xml element's `name` property has a special significance for `Objects`. If the `name` property is set, then the value of that property will be displayed as the node name in the xml inspector.

```
xml.setValue("name","Test Object");
```

If the name property is not set, the inspector uses the class name when displaying the node in the tree.

Listing 12.6 shows how arrays and objects are retrieved from an `XMLControl`. Arrays and collections of primitive data types are instantiated within the `XMLControl` when the `getObject` is invoked. The program creates the objects and prints their properties in the system console after reading the `nested_data.xml` data file.

Listing 12.6   XML reads structured data from a file.

```
package org.opensourcephysics.manual.ch12;
import org.opensourcephysics.controls.XMLControl;
import org.opensourcephysics.controls.XMLControlElement;

public class ReadNestedXMLApp {
  static String fileName = "nested_data.xml";

  public static void main(String[] args) {
```

```
      XMLControl xml = new XMLControlElement();
      xml.read(fileName);
      System.out.println(xml.getString("comment"));
      System.out.println(xml.getDouble("temperature"));
      System.out.println(xml.getInt("number"));
      double[] array = (double[]) xml.getObject("data array");
      for(int i = 0, n = array.length;i<n;i++) {
        System.out.println(array[i]);
      }
      System.out.println(xml.getObject("shape"));
   }
}
```

Because the `getObject` method returns a generic Java `Object`, the value returned must be cast to the correct data type. The `getObject` method will instantiate any Java class that defines a no argument constructor or an object loader that implements the `XML.ObjectLoader` interface. This interface makes xml-based reading and writing possible and is described in the next section. Note that the `InteractiveShape` class and many other OSP classes have predefined *object loaders* that implement this interface.

An `XMLControl` instance stores name-value pairs, but what happens if we attempt to read a value that has not been stored? If there is no `boolean`, then `getBoolean` returns `false`; if there is no `double`, then `getDouble` returns `Double.NaN`; if there is no `int`, then `getInt` returns `Integer.MIN_VALUE`; and if there is no `Object`, then `getObject` returns `null`.

```
XMLControl xml = new XMLControlElement();
Color color = (Color) xml.getObject("background color");
// color will be null because a name-value pair has not been set
```

12.3 ■ OBJECT LOADERS

Objects save and load xml documents using an xml *object loader* that implements the `XML.ObjectLoader` interface shown in Listing 12.7.

Listing 12.7   The `XML.ObjectLoader` interface.

```
// See the XML class definition in the controls package for
// additional ObjectLoader documentation
public interface ObjectLoader {
  // Saves data from an object in an XMLControl
  public void saveObject(XMLControl control, Object obj);

  // Creates an object using data in an XMLControl
  public Object createObject(XMLControl control);

  // Loads an object with data from an XMLControl
  public Object loadObject(XMLControl control, Object obj);
}
```

Implementations of the `XML.ObjectLoader` interface create, save, and load objects for a single Java class. Although loaders can be defined and implemented by any class, it is usually convenient and efficient to create them using an inner class. Listing 12.8 defines an `IdealGas` class that uses an anonymous inner class as a loader. The loader is created within the argument to the `XML.setLoader` method. Only one loader is needed for all `IdealGas` objects and this loader is accessed automatically from within the xml framework as needed.

Listing 12.8   A model of an ideal gas that defines an `XML.ObjectLoader`.

```
package org.opensourcephysics.manual.ch12;
import org.opensourcephysics.controls.*;

public class IdealGas {
  static final double R = 8.31; // gas constant J/mole/K
  double p, v, t; // pressure, volume, and temperature in MKS units

  public IdealGas() {
    this(101.3e3, 22.4e-3, 273);
  }

  public IdealGas(double pressure, double volume, double temperature) {
    p = pressure;
    v = volume;
    t = temperature;
  }

  public double getMoles() {
    return p*v/R/t;
  }

  public String toString() {
    return "ideal gas model: P="+p+"  V="+v+"  T="+t;
  }

  static {
    XML.setLoader(IdealGas.class, new XML.ObjectLoader() {
      public void saveObject(XMLControl control, Object obj) {
        IdealGas gas = (IdealGas) obj;
        control.setValue("pressure", gas.p);
        control.setValue("volume", gas.v);
        control.setValue("temperature", gas.t);
      }

      public Object createObject(XMLControl control) {
        return new IdealGas(); // creates a gas model at STP
      }

      public Object loadObject(XMLControl control, Object obj) {
        IdealGas gas = (IdealGas) obj;
        gas.p = control.getDouble("pressure");
        gas.v = control.getDouble("volume");
        gas.t = control.getDouble("temperature");
        return gas;
      }
    });
  }
}
```

The `IdealGas` explicitly registers its loader using the static method `XML.setLoader(classtype, loader)`. The `XML.setLoader` method adds the loader to the xml *loader registry*. Registering a loader for a given class allows objects of that class to be saved and loaded automatically. Another way for a class to register an xml loader is to define a public static getLoader() method within the class itself that returns the loader. Most Open Source Physics classes use the getLoader() method because the loader is only created as needed. The xml framework first looks for a registered loader; if none, it looks for a static getLoader method in the object class; if still none, then it uses a default loader. Classes with a registered loader can save object data in an xml document using the saveObject method as shown in Listing 12.9.

Listing 12.9   Loaders can be used to save object data.

```
package org.opensourcephysics.manual.ch12;
import org.opensourcephysics.controls.XMLControl;
import org.opensourcephysics.controls.XMLControlElement;

public class IdealGasSaveApp {
  public static void main(String[] args) {
    XMLControl xml = new XMLControlElement();
    // create and save data for an ideal gas with the given state
    xml.saveObject(new IdealGas(1.0E5, 2.0, 300));
    xml.write("gas_object.xml"); // save to file
  }
}
```

Object loaders can also load xml data into an existing object. The IdealGasLoadApp program (Listing 12.10) creates an IdealGas in its default state at standard temperature and pressure. It then reads data into an XMLControl and loads the data into the gas object by invoking the loadObject method.

Listing 12.10   Loaders can be used to set properties of existing objects.

```
package org.opensourcephysics.manual.ch12;
import org.opensourcephysics.controls.*;

public class IdealGasLoadApp {
  public static void main(String[] args) {
    IdealGas gas = new IdealGas(); // ideal gas at STP
    // creates an XML control and reads the xml data
    XMLControl xml = new XMLControlElement("gas_object.xml");
    // load gas object with data from xml control
    xml.loadObject(gas);
    // print a string representation of the gas
    System.out.println(gas);
  }
}
```

Object loaders can be used to instantiate Java objects. The IdealGasCreateApp shown in Listing 12.11 first reads a data file into an XMLControl. Because the control's loadObject method is passed a null argument, the control invokes the loader's createObject method and passes the created object to the loader's loadObject method.

Listing 12.11   Loaders can be used to create new objects.

```
package org.opensourcephysics.manual.ch12;
import org.opensourcephysics.controls.XMLControl;
import org.opensourcephysics.controls.XMLControlElement;
```

```
public class IdealGasCreateApp {
  public static void main(String[] args) {
    XMLControl xml = new XMLControlElement("gas_object.xml");
    // creates object with stored data
    IdealGas gas = (IdealGas) xml.loadObject(null);
    System.out.println(gas);
  }
}
```

Object loaders provide a convenient way to implement the `Cloneable` interface. An `XMLControlElement` is constructed using the current object. Invoking the control's `loadObject` method (see Listing 12.12) with a null argument instantiates a new object using control's data.

Listing 12.12   Object loaders can be used to clone objects.

```
// implementation of Cloneable interface using XML
public Object clone(){
  XMLControl control = new XMLControlElement(this);
  return control.loadObject(null);
}
```

12.4 ■ OSP APPLICATIONS

XML is used throughout the Open Source Physics library to store a program's data, to create menus, to inspect objects, and to transfer data between programs. For example, many Open Source Physics controls have menu items to read, save, and inspect an object's state using xml. Because a program's state depends on data stored in the control and on data stored in the model, we define a class named `OSPApplication` that contains references to both these objects. The application's loader creates an xml tree that has the `OSPApplication` object as the root document. The first element is named *control* and contains the control's variable names and values. The second element is named *model* and contains the model's data. If an object such as the model does not have a loader, the object's class name is the only saved data.

Listing 12.13 defines a simple model that creates particles at random locations. Because these particles are interactive, they can be dragged within the display frame.

Listing 12.13   A model that creates particles at random locations.

```
package org.opensourcephysics.manual.ch12;
import org.opensourcephysics.controls.*;
import org.opensourcephysics.display.InteractiveShape;
import org.opensourcephysics.frames.DisplayFrame;

public class ParticleApp extends AbstractCalculation {
  DisplayFrame frame = new DisplayFrame("x", "y", "Particles");

  public void calculate() {
    frame.clearDrawables(); // remove old cirlces
    int n = control.getInt("number of particles");
    double r = control.getDouble("radius");
    for(int i = 0;i<n;i++) {
```

```
        double x = -10+20*Math.random();
        double y = -10+20*Math.random();
        frame.addDrawable(InteractiveShape.createCircle(x, y, r));
    }
}

public void reset() {
  control.setValue("number of particles", 10);
  control.setValue("radius", 0.5);
}

public static XML.ObjectLoader getLoader() {
  return null;
  // uncomment the next line to enable the loader
  // return new ParticleAppLoader();
}

public static void main(String[] args) {
  // creates the program and reads xml data using
  // command line arguments
  CalculationControl.createApp(new ParticleApp(), args);
  // command line arguments are optional
  // use args[0] = "default_particles.xml" to read the example
}
}
```

Run `ParticleApp` and inspect the application's xml tree using the menu item under the control's File menu. Note that the control node contains its name-value pairs but that the model does not because the `getLoader` method in `ParticleApp` does not return a loader. We next discuss the loader that saves the model's data so that the program's entire configuration can later be recreated. Listing 12.14 shows how such a loader is written.

Listing 12.14 An `XML.ObjectLoader` for the `ParticleApp`.

```
package org.opensourcephysics.manual.ch12;
import org.opensourcephysics.controls.XML;
import org.opensourcephysics.controls.XMLControl;

public class ParticleAppLoader implements XML.ObjectLoader {
  public Object createObject(XMLControl control) {
    ParticleApp model = new ParticleApp();
    return model;
  }

  public void saveObject(XMLControl control, Object obj) {
    ParticleApp model = (ParticleApp) obj;
    control.setValue("frame", model.frame);
  }

  public Object loadObject(XMLControl control, Object obj) {
    ParticleApp model = (ParticleApp) obj;
    XMLControl childControl = control.getChildControl("frame");
    if(childControl==null) {
      return obj;
    }
    childControl.loadObject(model.frame);
    model.calculate(); // calculate with the new values
```

```
      model.frame.repaint();
      model.frame.setVisible(true);
      return obj;
  }
}
```

The `ParticleApp` loader shown defines methods to create, load, and save the model. Because the frame already defines an object loader that stores particle data, we only need to store and load the frame. Objects with xml loaders are saved in an object property using the `setValue` method. The following statement shows how the display frame's data are saved.

```
control.setValue("frame", model.frame);
```

Loading child xml data into an existing object is a two-step process. First the child `XMLControl` representing the named object is retrieved from the parent `XMLControl`. The child's `loadObject` method is then passed the object to be loaded. The passed object must, of course, have an xml loader.

```
XMLControl childControl = control.getChildControl("frame");
childControl.loadObject(model.frame);
```

OSP graphical user interfaces such as `CalculationControl` and `AnimationControl` are not xml controls. They do, however, use xml controls internally to load and save data. First enable the use of the `ParticleAppLoader` by uncommenting the line containing the loader in the `ParticleApp` program's `getLoader` method. Run the program, open the inspector in the `CalculationControl` File menu, and expand the model's xml tree. Observe how the program's control and model are nested in the inspector. Select the `radius` parameter in the control node. Enter a new value in the inspector's text field and note that the value displayed within the graphical user interface changes. Select a particle and change its x- and y-coordinates in the inspector. Note that the particle repositions itself on the screen. Close the inspector and save the application's state in a file named `default_particles.xml`. Change the program's state and then read the data file back into the program to restore the saved configuration.

XML data can be read into our graphical user interface controls using the `loadXML` method.

```
// defined for standard graphical user interface controls
control.loadXML("default_particles.xml");
```

We can also read xml data during program instantiation by passing xml data to the user interface control using the `main` method's command line arguments.

```
public static void main(String[] args) {
  CalculationControl.createApp(new ParticleApp(), args);
}
```

Run `ParticleApp` with the `default_particles.xml` command line parameter and note that the saved data determines the program's default conditions. The most recently accessed xml data file will be loaded into a program when the Reset button is pressed in any of the standard OSP controls. The ability to set a program's default parameters with an xml data file is designed for curriculum authors wishing to distribute a program with multiple initial conditions using the `Launcher` as described in Chapter 15.

12.5 ■ CLIPBOARD DATA TRANSFER

The ability to store and create objects using xml allows users to easily pass data between applications using the clipboard. The user copies an xml description onto the clipboard and pastes this description into another program. The following code fragment shows how the system clipboard is used.

```
protected void copyAction(XMLControlElement control) {
  StringSelection data = new StringSelection(control.toXML());
  Clipboard clipboard =
      Toolkit.getDefaultToolkit().getSystemClipboard();
  clipboard.setContents(data, null);
}
```

The mechanism for cutting and pasting clipboard contents to and from a `DrawingFrame` is, in fact, already in place. To examine how this mechanism works, run `ParticleApp` and `ScratchPadApp` simultaneously. The `ScratchPadApp` is shown in Listing 12.15.

Listing 12.15   A program with a `DrawingFrame` that has been enabled to accept clipboard data.

```
package org.opensourcephysics.manual.ch12;
import org.opensourcephysics.display.*;
import javax.swing.JFrame;

public class ScratchPadApp {
  public static void main(String[] args) {
    PlottingPanel panel = new PlottingPanel("x", "y", "Scratch Pad");
    panel.setPreferredMinMax(-10, 10, -10, 10);
    DrawingFrame frame = new DrawingFrame(panel);
    frame.setEnabledPaste(true);
    frame.setEnabledReplace(true);
    frame.setVisible(true);
    frame.setDefaultCloseOperation(JFrame.EXIT_ON_CLOSE);
  }
}
```

Create a random distribution of particles using `ParticleApp` and select the Copy item under the Edit menu. This action copies the particle xml data to the clipboard. Select the Paste item under the `ScratchPadApp` File menu to paste the particles into this second program. Repeat.

12.6 ■ PROGRAMS

The following examples are in the `org.opensourcephysics.manual.ch12` package.

IdealGasCreateApp

`IdealGasCreateApp` creates a new ideal gas model using xml data. See Section 12.3.

IdealGasLoadApp

`IdealGasLoadApp` loads an ideal gas model using xml data. See Section 12.3.

IdealGasSaveApp

`IdealGasSaveApp` saves an ideal gas model using xml data. See Section 12.3.

ParticleApp

`ParticleApp` demonstrates how to transfer xml data between programs as described in Section 12.4.

ReadNestedXMLApp

`ReadNestedXMLApp` reads objects into an xml document from an xml file as described in Section 12.2.

ReadXMLApp

`ReadXMLApp` reads primitive data types into an xml document from an xml file as described in Section 12.1.

ScratchPadApp

`ScratchPadApp` demonstrates how to cut and paste drawables using the clipboard as described in Section 12.5.

ShowXMLApp

`ShowXMLApp` reads an xml document and displays the document in an xml tree view. See Section 12.1.

WriteShowNestedXMLApp

`WriteShowNestedXMLApp` writes an xml document and displays the document in an xml tree view as described in Section 12.2.

WriteXMLApp

`WriteXMLApp` writes an xml document to a file as described in Section 12.1.

CHAPTER
13

Video

by Doug Brown and Wolfgang Christian

The OSP media API is defined in the *media* package (`org.opensourcephysics.media`) and contains classes for playing, recording, and analyzing digital videos. Concrete implementations of this API are available in subpackages for *animated gif* images and *QuickTime* video formats. The *Tracker* application described in Chapter 16 uses the API described in this chapter to build a full-featured cross-platform video analysis program.

13.1 ■ OVERVIEW

Because videos of real-world phenomena or of simulations of phenomena are now common, video tools are playing an increasingly important role in physics education software. In order to meet this need, the Open Source Physics library contains an API for the recording and playing of video. This API is defined in the `org.opensourcephysics.media.core` package. Implementations of this API for *animated gif* images and for *QuickTime* movies are available in the gif `org.opensourcephysics.media.gif` and the quicktime `org.opensourcephysics.media.quicktime` subpackages of the media package, respectively. Because the media API is often used with the Apple QuickTime plug-in and this plug-in may be unavailable on some computers, the media framework is distributed as an optional code library `osp_media.zip` from the OSP website.

13.2 ■ ANIMATED GIFS

Creating a simple animated gif is straightforward using the gif media package as shown in Listing 13.1. This example will run on any platform because the animated gif format is supported in the standard Java library.

> **Listing 13.1**   A `GifVideoRecorder` creates *animated gif* images from a sequence of images.

```
package org.opensourcephysics.manual.ch13;
import java.io.*;
import java.awt.image.*;
import org.opensourcephysics.display.*;
import org.opensourcephysics.media.gif.*;

public class GifRecorderApp {
  public static void main(String[] args) {
```

```
String fileName = "Rotates.gif";
// create a drawing panel, frame, and content
DrawingPanel panel = new DrawingPanel();
// a buffered panel automatically generates an image
panel.setBuffered( true);
DrawableShape rectangle =
    DrawableShape.createRectangle(0, 0, 1.5, 15);
panel.addDrawable(rectangle); // add rectangle to panel
DrawingFrame frame =
    new DrawingFrame("Creating \""+fileName+"\"", panel);
frame.setVisible(true);
frame.setDefaultCloseOperation(javax.swing.JFrame.EXIT_ON_CLOSE);
// create a gif video recorder
GifVideoRecorder recorder = new GifVideoRecorder();
try {
  // create the new video and set the frame duration
  recorder.createVideo();
  recorder.setFrameDuration(100); // 10 fps
  // get gif encoder to set optional repeat in browser
  // (default is 1, 0 repeats continuously)
  recorder.getGifEncoder().setRepeat(0);
  // gif encoder can also set transparent color or quality
  recorder.getGifEncoder().setTransparent(panel.getBackground());
  // add animation frames to the video
  for(int i = 0;i<20;i++) {
    rectangle.setTheta(i*Math.PI/10);
    // generates an image because panel is buffered
    BufferedImage image = panel.render();
    // use the displayed image as the video frame
    recorder.addFrame(image);
    try {
      Thread.currentThread().sleep(100);
    } catch(InterruptedException ex1) {}
  }
  // save the completed video file
  recorder.saveVideo(fileName);
} catch(IOException ex) {}
panel.setMessage("Animated gif saved.");
}
}
```

The GifVideoRecorder class is instantiated and its properties are set to create an animated gif with a 100-millisecond frame duration that plays continuously in a video browser. Video frames are images, and one way to create these images is to invoke the render method in a DrawingPanel. Notice that the DrawingPanel must be buffered to use this technique. If the buffered option is not set, the DrawingPanel does not generate an image and the render method returns null. Display3D panels, on the other hand, are always buffered and the render method will always return an image.

Playing an animated gif is even easier as shown in Listing 13.2. The GifVideo constructor takes a file name, and the images in this file are read into an array. These images are then played by the VideoPanel at the frame rate encoded into the animated gif file. VideoPanel is a subclass of InteractivePanel with a video player control as shown in Figure 13.1. A video panel always draws the video as a background behind other drawable objects.

Figure 13.1   A `VideoPanel` can play animated gifs and QuickTime video formats.

Listing 13.2   `GifVideoApp` plays *animated gif* images.

```
package org.opensourcephysics.manual.ch13;
import java.io.*;
import org.opensourcephysics.display.*;
import org.opensourcephysics.media.core.*;
import org.opensourcephysics.media.gif.*;

public class GifPlayerApp {
  public static void main(String[] args) {
    String fileName = "Rotates.gif";
    DrawingPanel panel = new VideoPanel();
    DrawingFrame frame = new DrawingFrame(panel);
    frame.setVisible(true); // will also work with a hidden frame
    frame.setDefaultCloseOperation(javax.swing.JFrame.EXIT_ON_CLOSE);
    try { // create and draw a gif video
      GifVideo animatedGif = new GifVideo(fileName);
      panel.addDrawable(animatedGif);
      frame.setTitle("Playing \""+fileName+"\"");
    } catch(IOException ex) {}
  }
}
```

13.3 ■ PLAYING QUICKTIME VIDEO

The Apple QuickTime player plays the most popular video formats, and Apple distributes a QuickTime for Java library with this player that allows us to access the native API. On Windows we recommend that you install QuickTime 7.0 or later (after installing Java) because it automatically installs the necessary library.[1] The latest version of QuickTime may be downloaded from

[1]QuickTime 6.4 or 6.5 can also be used if you select a custom installation and install all options.

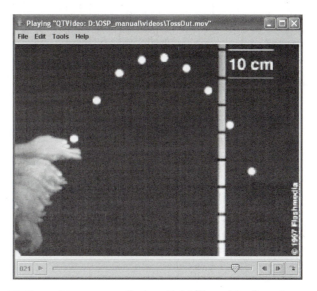

Figure 13.2 The `QTPlayerApp` program displays QuickTime video formats.

```
http://www.apple.com/quicktime/download/
```

The shared video API allows QuickTime videos to be recorded and played simply by replacing `GifVideoRecorder` and `GifVideo` with `QTVideoRecorder` and `QTVideo` in Listings 13.1 and 13.2. The `QTPlayerApp`, shown in Listing 13.3 produces the output shown in Figure 13.2. This program has menu items for loading QuickTime videos with a chooser and for verifying proper installation of Apple's QuickTime library `QTJava.zip`. If the `verifyQuickTime` method does not find `QTJava.zip` in the Java VM's external library folder (path shown on dialog), search for the file and copy it into this folder.[2] Note that when you update the Java VM, the `QTJava.zip` library may need to be copied from the old Java VM to the new Java VM. This library and other Java extension libraries are usually in the Java VM's `lib/ext` folder.

Listing 13.3 `QTVideo` plays video files that are supported by QuickTime.

```
package org.opensourcephysics.manual.ch13;
import org.opensourcephysics.display.*;
import org.opensourcephysics.media.core.*;
import org.opensourcephysics.media.quicktime.*;
import javax.swing.*;
import java.awt.event.*;
import java.io.File;

public class QTPlayerApp {
  VideoPanel panel = new VideoPanel();
  DrawingFrame frame = new DrawingFrame("QT Video Player", panel);
```

---

[2]On Windows computers, the QuickTime installer sometimes places the `QTJava.zip` library into the `C:/windows/system32` directory. The most recent version of QuickTime 7 places this library into a subdirectory in the `C:/Program Files` folder. *Caveat emptor.* Microsoft, Apple, and Sun need to get their act together.

```
QTPlayerApp() {
  frame.setVisible(true); // will also work with a hidden frame
  frame.setDefaultCloseOperation(JFrame.EXIT_ON_CLOSE);
  addMenuItems();
  loadVideo();
}

public void addMenuItems() {
  JMenu menu = frame.getMenu("File"); // gets menu from toolbar
  menu.addSeparator();
  JMenuItem videoItem = new JMenuItem("Load Video");
  videoItem.addActionListener(new ActionListener() {
    public void actionPerformed(ActionEvent e) {
      loadVideo();
    }
  }); // end of addActionListener
  menu.add(videoItem);
  JMenuItem verifyItem = new JMenuItem("Verify QuickTime");
  verifyItem.addActionListener(new ActionListener() {
    public void actionPerformed(ActionEvent e) {
      verifyQuickTime();
    }
  }); // end of addActionListener
  menu.add(verifyItem);
}

public void loadVideo() {
  panel.clear(); // removes all drawables including the old video
  try {
    QTVideo video = new QTVideo(); // displays file chooser
    panel.addDrawable(video);
    frame.setTitle("Playing \""+video.toString()+"\"");
  } catch(java.io.IOException ex) {
    System.out.println("Error creating QTVideo: "+ex.toString());
  }
}

public void verifyQuickTime() {
  File dir = new File(System.getProperty("java.ext.dirs"));
  File file = new File(dir, "QTJava.zip");
  String s = file.exists()
             ? "QTJava.zip found" : "QTJava.zip not found";
  JOptionPane.showMessageDialog(frame,
      s+" in "+dir.getAbsolutePath(), "Verify QuickTime",
      JOptionPane.INFORMATION_MESSAGE);
}

public static void main(String[] args) {
  new QTPlayerApp();
}
}
```

If the QTVideo class is instantiated without an argument, it requests a file name using a chooser dialog. A wide variety of file types can be selected from the chooser, including *mpeg* developed by the motion picture experts group, *avi* developed by Microsoft, and *mov* developed by Apple Computer.

13.4 ■ CONTROLLING VIDEO DISPLAY

After a video has been instantiated, its properties are set using accessor methods.

```
video.setLooping(true);      // turns on looping
video.setRate(0.4);          // play() will be in slow motion
video.setAngle(Math.PI/6);   // rotate counterclockwise
video.setRelativeAspect(2);  // stretch the video horizontally
video.setWidth(1.3);         // set width in world units to scale the video
```

Video frames can be modified before they are displayed by adding filters to the video filter stack. The `GhostFilter`, for example, overlays images to produce a "live" motion diagram.

```
video.getFilterStack().addFilter(new GhostFilter());
```

Search the core media package for files matching `*Filter` to find other implementations of the video `Filter` interface such as `DeinterlaceFilter`, `BrightnessFilter`, and `GrayScaleFilter`.

A `VideoPanel` can draw videos and other objects in either *image-space* (pixel-space) or *world-space*. When drawing in image-space, the image reference frame (that is, the image itself) is fixed. When drawing in world-space, the world reference frame is fixed. The image reference frame defines positions in pixel units relative to the upper left corner of a video image—that is, the upper left corner of a 320×240 video is at $(0.0, 0.0)$ and the lower right corner is at $(320.0, 240.0)$. The `DrawingSpacesApp` available on the CD demonstrates the relationship between image-space and world-space.

A `VideoPanel` contains a `VideoPlayer` object with play control buttons, a slider, and a frame readout. A player's display properties can be set as follows:

```
VideoPanel vidPanel = new VideoPanel(video);
VideoPlayer player = vidPanel.getPlayer();
player.setInspectorButtonVisible(true);
player.setReadoutType("step"); // sets readout type
```

The `VideoPlayer` class contains a `VideoClip` object that breaks a video into segments called *clips*. A clip has a start frame and an end frame that can be set. You can also skip frames within a clip when the player's Step button is pressed.

```
VideoClip clip = player.getVideoClip();
clip.setStartFrameNumber(3); // also indirectly sets video start frame
clip.setStepSize(4);         // also indirectly sets video end frame
```

The `VideoPropertiesApp` program output is shown in Figure 13.3. This program uses these code snippets and is available on the CD.

13.5 ■ VIDEO CAPTURE TOOL

The `VideoCaptureTool` in the tools package (Figure 13.4) provides a user interface for creating and saving animation videos in both gif image and QuickTime movie formats. `CaptureAnimationApp`, shown in Listing 13.4, illustrates its use.

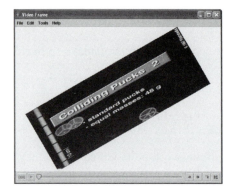

Figure 13.3   The `VideoPropertiesApp` program sets video display properties and rotates the video clip.

Figure 13.4   The `VideoCaptureTool` can be added to any buffered drawing panel. The tool captures frames when the panel's `render` method is invoked.

Listing 13.4   `CaptureAnimationApp` uses a `VideoCaptureTool` to automatically capture video frames from a drawing panel.

```
package org.opensourcephysics.manual.ch13;
import org.opensourcephysics.display.*;
import org.opensourcephysics.controls.*;
import org.opensourcephysics.tools.*;

public class CaptureAnimationApp extends AbstractSimulation {
  // create a drawing panel and frame and add an arrow to animate
  DrawingPanel panel = new DrawingPanel();
  DrawingFrame frame = new DrawingFrame(panel);
  Arrow arrow = new Arrow(0, 0, 5, 0);
  double theta = 0, dtheta = Math.PI/30;

  public CaptureAnimationApp() {
    panel.addDrawable(arrow);
    panel.setBuffered(true); // video capture requires buffering
```

```
      frame.setAnimated(true);
      frame.setVisible(true);
      frame.setDefaultCloseOperation(javax.swing.JFrame.EXIT_ON_CLOSE);
    }

  protected void doStep() {
    theta += dtheta;
    arrow.setXlength(5*Math.cos(theta));
    arrow.setYlength(5*Math.sin(theta));
    panel.setMessage("theta="+decimalFormat.format(theta));
  }

  // creates and shows the video capture tool
  public void showVideoCaptureTool() {
    if(panel.getVideoCaptureTool()==null) {
      panel.setVideoCaptureTool(new VideoCaptureTool());
    }
    panel.getVideoCaptureTool().setVisible(true);
  }

  public static void main(String[] args) {
    (SimulationControl.createApp(new CaptureAnimationApp(),
        args)).addButton("showVideoCaptureTool", "Capture");
  }
}
```

Here a `DrawingPanel` is used to draw a simple animation using an animation control. When a `VideoCaptureTool` is created and assigned to a buffered `DrawingPanel`, the `VideoCaptureTool` will add a frame to the video whenever the panel's `render` method is invoked and the Capture check box is selected. The animation must be visible on the screen or the render method will skip drawing a new image. This is done for efficiency as there is usually no reason to draw a window if it is iconified or hidden. Captured videos are saved with a standard Save As dialog.

13.6 ■ IMAGE AND VIDEO ANALYSIS

A `VideoPanel` can display common still-image formats as shown in Figure 13.5. The `LineProfileApp`, shown in Listing 13.5, opens a jpg file showing hydrogen and helium spectra. A cross-section of this image is then displayed using a `TLineProfile`.

Listing 13.5 `LineProfileApp` reads brightness values across a row of pixels in an image.

```
package org.opensourcephysics.manual.ch13;
import java.io.*;
import java.awt.*;
import javax.swing.*;
import org.opensourcephysics.display.*;
import org.opensourcephysics.media.core.*;

public class LineProfileApp {
  VideoPanel vidPanel;
  Dataset dataset = new Dataset();
```

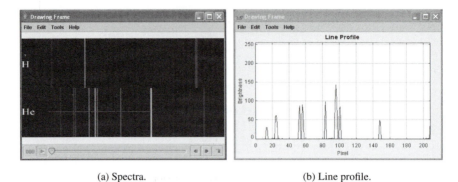

(a) Spectra.   (b) Line profile.

Figure 13.5   The LineProfileApp program shows an image of hydrogen and helium spectra and a cross-section of the image intensity.

```
PlottingPanel plotPanel =
     new PlottingPanel("Pixel", "Brightness", "Line Profile");
TLineProfile profile = new TLineProfile(100, 150, 310, 150);

public LineProfileApp() {
  Video video = null;
  try {
    video = new ImageVideo("videos/Spectra.jpg");
  } catch(IOException ex) {}
  // create the video panel that will also plot the data
  vidPanel = new VideoPanel(video) {
    // overrides paintEverything to plot data
    protected void paintEverything(Graphics g) {
      super.paintEverything(g);
      plotData();
    }
  }; // end of VideoPanel constructor
  vidPanel.setShowCoordinates(false);
  vidPanel.setDrawingInImageSpace(true);
  // add the line profile tool
  profile.setColor(Color.red);
  vidPanel.addDrawable(profile);
  // create the drawing frame
  DrawingFrame frame = new DrawingFrame(vidPanel);
  frame.setDefaultCloseOperation(JFrame.EXIT_ON_CLOSE);
  frame.setVisible(true);
  // create the plotting panel for the profile
  createPlots();
}

private void plotData() {
  dataset.clear();
  int[] data = profile.getProfile();
  for(int n = 0;n<data.length;n++) {
    dataset.append(n, data[n]);
  }
  plotPanel.repaint();
}
```

```
private void createPlots() {
  // make a plot panel to display data
  plotPanel.setAutoscaleX(true);
  plotPanel.setPreferredMinMaxY(0, 255);
  DrawingFrame drawingFrame = new DrawingFrame(plotPanel);
  // make a dataset and set its display properties
  dataset.setConnected(true);
  dataset.setMarkerShape(Dataset.NO_MARKER);
  plotPanel.addDrawable(dataset);
  drawingFrame.setVisible(true);
}

public static void main(String[] args) {
  new LineProfileApp();
}
}
```

TLineProfile is one of many specialized interactive objects created for the *Tracker* video analysis program. It reads image pixels along a path specified by beginning and ending points. These pixel values are obtained using the getProfile method.

The line profile is plotted whenever the TLineProfile is moved. One way to do this is to dig into the OSP library and notice that a DrawingPanel draws its list of drawable objects by invoking the paintEverything method. Because the panel is redrawn after an object is moved, we can obtain an up-to-date line profile by overriding the paintEverything method and plotting the data after the superclass implementation is invoked. This is done in LineProfileApp by creating an anonymous inner class that extends VideoPanel.

The VideoAnalysisApp program uses other Tracker objects to create a small video analysis program that draws graphs of x vs. t and y vs. t for points that are marked by clicking on the video. It is not shown here but available on CD. Although you can adapt this code if your application requires video analysis, a more robust (and easier) solution may be to distribute your program together with Tracker in a Launcher package as described in Chapters 15 and 16.

13.7 ■ PROGRAMS

The following examples are in the org.opensourcephysics.manual.ch13 package.

CaptureAnimationApp

CaptureAnimationApp uses a VideoCaptureTool to automatically capture video frames from a drawing panel as described in Section 13.5.

DrawingSpacesApp

DrawingSpacesApp draws a video in image-space and world-space as described in Section 13.4.

GifPlayerApp

GifPlayerApp, described in Section 13.2, plays *animated gif* images.

GifRecorderApp

`GifRecorderApp`, described in Section 13.2, records an *animated gif*.

LineProfileApp

`LineProfileApp` shows an image of hydrogen and helium spectra and a cross-section of the image intensity as described in Section 13.6.

QTPlayerApp

`QTPlayerApp`, described in Section 13.3, displays QuickTime video formats.

VideoAnalysisApp

`VideoAnalysisApp` draws graphs of x vs. t and y vs. t for points that are marked by clicking on the video as described in Section 13.6.

VideoPropertiesApp

`VideoPropertiesApp` sets video display properties and rotates the video clip as described in Section 13.4.

CHAPTER

14

Utilities

Almost every development project has a small collection of utilities and tricks that save time by providing quick and easy access to common programming tasks. This chapter describes utilities that we have found useful in various OSP projects.

14.1 ■ RESOURCE LOADER

A program often needs access to *resources* such as text, images, or sounds, and these resources can be located almost anywhere. Although Java provides a variety of tools for reading files, the API depends on the type of file and on the location of the file. During development a resource file may be in a directory on a local hard drive, but this same resource may be placed on a Web server or packaged in a jar file after deployment. Unfortunately, the resource must be read differently from each of these locations. The ResourceLoader class in the tools package solves this problem by providing a standard API for reading common file types and by searching multiple locations in a predictable search order. We can use the ResourceLoader to obtain the image shown in Figure 14.1 by invoking the static getResource method.

```
String name = "earth.gif";
Resource res = ResourceLoader.getResource(name);
Image image = res.getImage();
```

We can use this same syntax to obtain a text file.

```
String name = "orbit_help.html";
Resource res = ResourceLoader.getResource(name);
String txt = res.getString();
```

Or we can obtain a reference to the resource that can be used later.

```
String name = "http://www.opensourcephysics.org/help.html";
Resource res = ResourceLoader.getResource(name);
URL page = res.getURL();
```

Listing 14.1 shows a program that uses a ResourceLoader to read and print the data.txt file located in the directory from which the program was executed.

Listing 14.1 GetTextApp loads a text file.

```
package org.opensourcephysics.manual.ch14;
import org.opensourcephysics.tools.*;

public class GetTextApp {
  public static void main(String[] args) {
```

253

Figure 14.1   An image resource can be loaded from almost anywhere.

```
  // get string from resource
  String name = "data.txt";
  Resource res = ResourceLoader.getResource(name);
  if(res==null) {
    System.out.println("Resource not found.  Name="+name);
    return;
  }
  System.out.println("path="+res.getAbsolutePath());
  System.out.println(res.getString());
  }
}
```

When the getResource method is invoked from a Java application, the ResourceLoader searches for the resource in the following order: in directories and subdirectories, on the Web, and inside jar archives. The search starts in the directory from which the program was executed. If the resource is not found, it then assumes the given name refers to a url and searches the Web. If the resource is again not found, it searches within the jar file from which the program was launched. The code remains the same if the getResource method is invoked from within an applet. The search begins in directories starting at the the applet's codebase and then searches within the jar file(s) specified in the applet's archive tag. The resource name can include a relative path to another directory, such as ./images/earth.gif, or even to an external site via a url.

The getResource method has a second signature that makes it easy to place resources in directories relative to a class definition. A program can store its resources in the same package (directory) as the byte code (class file) that uses the resource and then pass an appropriate class to the loader. The GetImageApp program shown in Listing 14.2 uses this syntax to read the earth.jpg image from this chapter's code package ch14.

CAUTION: Java compilers and Java development environments may not copy resources from a project's code directory (src) to a project's output directory (bin or classes). Most development environments have a setting that determines which files are copied, and you must enable this setting for the resource file types.

Table 14.1   Resource accessor methods.

org.opensourcephysics.tools.Resource	
getAbsolutePath	The location from where the resource was loaded.
getAudioClip	The audio resource.
getBufferedImage	The image resource converted into a `BufferedImage`.
getFile	The file associated with this resource.
getIcon	The image resource converted into an `ImageIcon`.
getImage	The image resource.
getString	The text resource.

Listing 14.2   `GetImageApp` loads an image from the same location as the `GetImageApp` class.

```
package org.opensourcephysics.manual.ch14;
import org.opensourcephysics.display.*;
import org.opensourcephysics.frames.*;
import org.opensourcephysics.tools.*;

public class GetImageApp {
  public static void main(String[] args) {
    Resource res =
        ResourceLoader.getResource("earth.jpg", GetImageApp.class);
    MeasuredImage mi =
        new MeasuredImage(res.getBufferedImage(), -1, 1, -1, 1);
    DisplayFrame frame = new DisplayFrame("Earth");
    frame.addDrawable(mi);
    frame.setVisible(true);
    frame.setDefaultCloseOperation(javax.swing.JFrame.EXIT_ON_CLOSE);
  }
}
```

The loader can cache resources in memory so that they can load quickly in response to subsequent requests. Caching is disabled by default and is enabled as follows:

```
ResourceLoader.setCacheEnabled(true);
```

The loader defines utility methods such as `getImage` and `getString` that obtain a resource of a given type using a single method (see Table 14.1).

```
// gets an image from the Web using a url
String s = "http://www.cabrillo.edu/~dbrown/images/doug48.gif";
//ResourceLoader gets the resource and converts it to an image
Image image = ResourceLoader.getImage(s);
```

Consult the resource loader code and the documentation to learn about additional features such as the ability to add search paths.

Text files such a html and *rich text format* (rft) files are resources that can be displayed in a `JEditorPane`. We have combined this Swing component with a OSP resource loader to create the `TextFrame` class. Listing 14.3 uses this frame to display an html page that is stored in this chapter's code package `ch14`.

Listing 14.3 TextFrameApp loads and displays an html page containing the GNU GPL license.

```
package org.opensourcephysics.manual.ch14;
import org.opensourcephysics.display.*;

public class TextFrameApp {
  public static void main(String[] args) {
    TextFrame frame = new TextFrame("gpl.html", TextFrameApp.class);
    frame.setVisible(true);
    frame.setDefaultCloseOperation(javax.swing.JFrame.EXIT_ON_CLOSE);
  }
}
```

14.2 ■ CUSTOM MENUS

Subclasses of OSPFrame such as ControlFrame and DrawingFrame instantiate menu bars in order to provide easy access to common tasks such as saving files. A JMenu bar contains various menus, and these menus are identified by names such as File, Edit, and Help which are listed on the bar. Pressing a name shows a drop-down list of menu items some of which lead to additional submenus. The default menu bar may not be suitable for your application and can be replaced. An alternative to replacement is to add or remove selected items from a menu bar that already exist.

The OSPFrame class provides utility methods to delete menus and menu items from a frame using text labels to identify the object.

```
DisplayFrame frame = new DisplayFrame("Empty Frame");
frame.removeMenu("Tools");              // deletes an entire menu
frame.removeMenuItem("File", "Inspect"); // deletes an item from a menu
```

The OSPFrame class provides methods to retrieve a menu from the menu bar for customization.

```
JMenu menu = frame.getMenu("Help");
```

If the requested menu does not exist, a new menu with the given label is created and added to the menu bar. The JMenu object that is returned can now be modified by adding menu items and separators. Display panels also have a pop-up menu and this menu can also be customized. A reference to the pop-up menu is obtained from the display panel and a menu item is added. Listing 14.4 creates two menu items and adds them to the menu bar and the pop-up menu. Each item has an action listener that invokes the showHelp method.

Listing 14.4 CustomHelpApp adds a menu item to the Help menu.

```
package org.opensourcephysics.manual.ch14;
import org.opensourcephysics.display.*;
import java.awt.event.*;
import javax.swing.*;

public class CustomHelpApp {
  DrawingFrame frame = new DrawingFrame(new DrawingPanel());

  public CustomHelpApp() {
    // custom help added to Help menu
    JMenu menu = frame.getMenu("Help");
```

```
      JMenuItem helpItem = new JMenuItem("Custom Help");
      helpItem.addActionListener(new ActionListener() {
        public void actionPerformed(ActionEvent e) {
          showHelp();
        }
      }); // end of addActionListener
      menu.addSeparator();
      menu.add(helpItem);
      // custom help added to pop-up menu
      JPopupMenu popup = frame.getDrawingPanel().getPopupMenu();
      helpItem = new JMenuItem("Custom Help");
      helpItem.addActionListener(new ActionListener() {
        public void actionPerformed(ActionEvent e) {
          showHelp();
        }
      }); // end of addActionListener
      popup.add(helpItem);
      frame.setVisible(true); // shows the DrawingFrame
      frame.setDefaultCloseOperation(JFrame.EXIT_ON_CLOSE);
    }

    public void showHelp() {
      JOptionPane.showMessageDialog(frame, "This space for rent.",
          "Program Help", JOptionPane.INFORMATION_MESSAGE);
    }

    public static void main(String[] args) {
      new CustomHelpApp();
    }
}
```

Menus can also be defined and created using xml. An xml document containing *launch nodes* is created as described in Chapter 12 or using `LaunchBuilder` as described in Chapter 15. The xml file name is then passed to the `parseXMLMenu` method in an `OSPFrame` as shown in Listing 14.5. The example adds a Documents menu that contains menu items to launch (execute) the `CustomHelpApp` and `GetImageApp` programs from this chapter's code package.

Listing 14.5   XMLMenuApp adds a menu to a frame's menu bar.

```
package org.opensourcephysics.manual.ch14;
import org.opensourcephysics.display.*;

public class XMLMenuApp {
  public XMLMenuApp() {
    DrawingFrame frame = new DrawingFrame("XML Menus", null);
    // "custom_menu.xml" was created using LaunchBuilder
    // parseXMLMenu uses the ResourceLoader to read the file
    frame.parseXMLMenu("custom_menu.xml", XMLMenuApp.class);
    frame.setSize(300, 150);
    frame.setVisible(true);
    frame.setDefaultCloseOperation(javax.swing.JFrame.EXIT_ON_CLOSE);
  }

  public static void main(String[] args) {
    new XMLMenuApp();
  }
}
```

14.3 ■ CHOOSER

Java provides an easy-to-use file chooser that enables users to browse a disk to open and save files. Users often navigate to a working directory and expect the program to remember this location during subsequent file operations. This behavior is easy to achieve if the same chooser is used over and over because a chooser stores information from session to session.

The `OSPFrame` class instantiates a static file chooser that can be used to access the local file system.

```
JFileChooser chooser = OSPFrame.getChooser();
```

When the `getChooser` method is first invoked, the chooser opens in a directory specified by the value of the static `chooserDir` variable. We initialize this variable to point to the application's starting directory.

```
chooserDir = System.getProperty("user.dir",null);
```

No other action can be taken until the chooser is closed because the chooser is a modal dialog box. If the chooser is closed and `getChooser` is again invoked, the same chooser object is returned. Listing 14.6 uses a chooser to load a resource into a `TextFrame`.

> **Listing 14.6**   `ChooserApp` opens (reads) a resource using a chooser and displays the file in a `TextFrame`.

```java
package org.opensourcephysics.manual.ch14;
import org.opensourcephysics.controls.*;
import org.opensourcephysics.display.*;
import javax.swing.JFileChooser;

public class ChooserApp {
  TextFrame frame = new TextFrame(null);

  public void choose() {
    frame.setVisible(true);
    JFileChooser chooser = OSPFrame.getChooser();
    if(chooser.showOpenDialog(null)==JFileChooser.APPROVE_OPTION) {
      OSPFrame.chooserDir = chooser.getCurrentDirectory().toString();
      String fileName = chooser.getSelectedFile().getAbsolutePath();
      frame.loadResource(fileName, null);
      OSPFrame.chooserDir = chooser.getCurrentDirectory().toString();
    }
  }

  public static void main(String[] args) {
    OSPControl.createApp(new ChooserApp()).addButton("choose",
        "Choose");
  }
}
```

Although the static chooser returned by `getChooser` will open the directory where it was last used, it is sometimes necessary to create custom choosers and these choosers should also open in the directory that was last used. One way to achieve this is to store the current directory in the `chooserDir` variable when a chooser is closed.

```java
// instantiate a chooser so that it opens in OSPFrame.chooserDir
JFileChooser chooser = new JFileChooser(new File(OSPFrame.chooserDir));
```

Table 14.2 Graphical user interface utility methods are defined in the `GUIUtils` class.

org.opensourcephysics.display.GUIUtils

`clearDrawingFrameData`	Clears the data in animated `DrawingFrame` objects and repaints the frame's content.
`closeAndDisposeOSPFrames`	Disposes all `OSPFrame` objects except the given frame.
`enableMenubars`	Enables and disables the menu bars in `DrawingFrame` objects.
`renderAnimatedFrames`	Renders all `OSPFrame` objects whose animated property is true.
`repaintAnimatedFrames`	Repaints all `OSPFrame` objects whose animated property is true.
`showDrawingAndTableFrames`	Shows all drawing and table frames.

```
// set OSPFrame.chooserDir when a chooser is closed
OSPFrame.chooserDir = chooser.getCurrentDirectory().toString();
```

Multiple choosers will not remain "synchronized," but new choosers will at least start in a common directory.

14.4 ■ STATIC METHODS

Open Source Physics simulations typically display drawing frames when an Initialize button is pressed, disable menus when a Run button is pressed, and clear data when a Reset button is pressed. The `GUIUtils` class in the display package defines static methods that perform these and other routine housekeeping chores (see Table 14.2). These methods often use a `static` method in the `Frame` class that returns an array containing frames created by the application.

```
Frame[] frames = Frame.getFrames();
```

Methods in the `GUIUtils` class process elements in the array by performing the requested action. In order to control the scope of these actions, some methods check to see if a property is set. For example, the `renderAnimatedFrames` method is called from within an animation thread, but certain frames may have their `animated` property set to false and these frames will not be rendered to improve performance. Likewise, frames with their `autoclear` property set to false will not loose their data when an Initialize button is pressed.

The `GUIUtils` class also contains methods to create and save images based on the `EpsGraphics2D` package by James Mutton. `EpsGraphics2D` is suitable for creating high-quality encapsulated postscript (EPS) graphics for use in documents and papers and can be used just like a standard `Graphics2D` object.

Many Java programs use `Graphics2D` to draw objects on the screen, and while it is easy to save the output as a png or jpeg file, it is a little harder to export it as an EPS for inclusion in a document or paper.

The `EpsGraphics2D` class makes the whole process extremely easy and you can use it to create eps, jpeg, and png images. If you choose eps, the images that drawn can be resized without leading to any of the jagged edges you may see when resizing pixel-based images such as jpeg and png files.

The `saveImage` method in the `GUIUtils` class is designed to print OSP components. A component such as drawing panel is passed to the `saveImage` method along with an output file and an output file format. There is an optional third parameter (which can be null) that is used by the dialog box. The entire process is extremely simple. Invoking the following method brings up a dialog box that requests a file name. The user enters a file name and an eps image of the drawing panel's contents is created.

```
GUIUtils.saveImage(drawingPanel, "eps", DrawingFrame.this);
```

14.5 ■ INSPECTORS

Open Source Physics components, such as `DrawingPanel` and `ControlFrame`, often have inspectors that provide information about the internal state of an object to the user. Because default inspectors are designed for technical users who are debugging programs or teaching computational physics, they may be inappropriate for a general audience. There are, however, mechanisms for disabling these default inspectors and replacing them with other components. Replacement inspectors are `JFrames` and can therefore be defined using standard Swing or Ejs components as shown in Listing 14.7.

Listing 14.7   Default inspectors can be replaced with custom components.

```
package org.opensourcephysics.manual.ch14;
import org.opensourcephysics.display.*;
import org.opensourcephysics.ejs.control.GroupControl;
import java.awt.Color;
import javax.swing.*;

public class CustomInspectorApp {
  DrawingPanel panel = new DrawingPanel();
  DrawingFrame frame = new DrawingFrame(panel);
  GroupControl control; // reference to the inspector

  public CustomInspectorApp() {
    panel.setPopupMenu(null); // disable the pop-up on right click
    JDialog inspector = getCustomInspector();
    panel.setCustomInspector(inspector); // inspector for right click
    frame.setCustomInspector(inspector); // inspector for file menu
    frame.setVisible(true);               // shows the DrawingFrame
    frame.setDefaultCloseOperation(JFrame.EXIT_ON_CLOSE);
  }

  public void setColor() {
    if(control.getBoolean("isRed")) {
      panel.setBackground(Color.RED);
    } else {
      panel.setBackground(Color.BLUE);
    }
  }

  JDialog getCustomInspector() {
    control = new GroupControl(this);
```

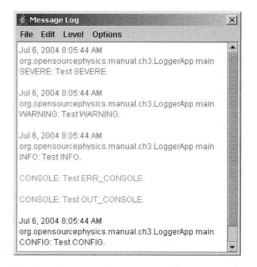

Figure 14.2   An OSPLog displays logger and system console messages in a modal dialog box.

```
    JDialog dialog = (JDialog) control.add("Dialog",
        "name=inspector; title=Inspector; visible=visible;
        location=300,300; size=100,50").getComponent();
    control.add( "Panel",
        "name=controlPanel; position=center; parent=inspector");
    control.add("CheckBox",
        "parent=controlPanel; variable=isRed; text=Red;
        selected=false; action=setColor");
    return dialog;
  }

  public static void main(String[] args) {
    new CustomInspectorApp();
  }
}
```

14.6 ■ LOGGING

Java 1.4 and above provide a very flexible *logging* API that enables a program to record system information, informational messages, and debugging output. This API is far superior to the old standby System.out.println. The OSPLog class uses the logging API to create a frame that displays logger output as well as system console output such as error messages.[1] Listing 14.8 shows how to post messages to the OSP logger using various logging levels. The output is shown in Figure 14.2. The OSPLog displays the logger's output in a window that allows a user to decide the level of detail using a menu bar. The user may also save the output to a disk file for later analysis.

[1]The OSP log cannot capture the system console from an applet because the typical applet security policy does not allow a program to redirect the console stream to another device. Other logger messages continue to be logged in applet mode.

Listing 14.8   A OSPLog test program showing the various log levels.

```
package org.opensourcephysics.manual.ch14;
import org.opensourcephysics.controls.ConsoleLevel;
import org.opensourcephysics.controls.OSPLog;

public class LoggerApp {
  public static void main(String[] args) {
    OSPLog.getOSPLog().setVisible(true); // displays the output window
    OSPLog.setLevel(ConsoleLevel.ALL);
    OSPLog.severe("Test SEVERE.");
    OSPLog.warning("Test WARNING.");
    OSPLog.info("Test INFO.");
    System.err.println("Test ERR_CONSOLE.");
    System.out.println("Test OUT_CONSOLE.");
    OSPLog.config("Test CONFIG.");
    OSPLog.fine("Test FINE.");
    OSPLog.finer("Test FINER.");
    OSPLog.finest("Test FINEST.");
    // Caution: program will keep running after Dialog window is closed
    // cannot set JFrame to EXIT_ON_CLOSE because OSPLog is a dialog
  }
}
```

Logging does not consume any significant amount of time processing messages below the current level. Allowed levels range from SEVERE to FINEST in the order shown in Listing 14.8. The default level shows messages above Level.OUT_CONSOLE so that console and error messages are displayed. A common debugging strategy is to log messages at FINE, FINER, and FINEST levels and then examine the log at the appropriate level of detail in order to determine at what point a program or algorithm failed.

14.7 ■ PROGRAMS

The following examples are in the org.opensourcephysics.manual.ch14 package.

ChooserApp

ChooserApp uses a file chooser to load a resource as described in Section 14.3.

CustomHelpApp

CustomHelpApp demonstrates how to add a menu item to the Help menu as described in Section 14.2.

CustomInspectorApp

CustomInspectorApp demonstrates how to create a custom inspector for a drawing panel as described in Section 14.5.

GetImageApp

GetImageApp tests the ResourceLoader class by reading an image as described in Section 14.1.

GetTextApp

`GetTextApp` reads the `data.txt` file from the root of the classpath as described in Section 14.1.

LoggerApp

`LoggerApp` demonstrates the use of the `OSPLog` class as described in Section 14.6.

RemoveMenuApp

`RemoveMenuApp` removes a menu and a menu item from a `DisplayFrame` menu bar.

TextFrameApp

`TextFrame` loads an html document from the `ch14` package directory as described in Section 14.1.

XMLMenuApp

`XMLMenuApp` adds a menu to a frame's menu bar using data in an xml document as described in Section 14.2.

15

Authoring Curricular Material

by Mario Belloni and Wolfgang Christian

We describe `Launcher` and `LaunchBuilder` and show how these programs are used to author, organize, and run Java-based curricular material.

15.1 ■ OVERVIEW

As you have already seen in this *Guide*, the Open Source Physics project, with its wide variety of packages, libraries, programs, and files, is a large body of work useful for the study of computational physics. Many instructors, however, do not teach (or do research in) computational physics. In order for these instructors to use OSP material in their courses (or in their educational, experimental, or theoretical research), the various physical models already available must be easily accessible, modifiable, and distributable. The paradigm for authoring, organizing, and running curricular material described in this chapter uses the `Launcher` and `LaunchBuilder` programs to accomplish this goal.

The `Launcher` program displays curricular units by using a tree structure to organize the material according to topic, course, etc., as shown in Figure 15.1. Each unit can include html pages, launchable (executable) programs, and parameters stored in xml files. Delivering curricular material in Launcher packages has several advantages. First, the material can be made self-contained and, therefore, easily distributable (see Chapter 18 for more on the distribution mechanism). Second, the material is only dependent on having a Java VM on a local machine and not on the type of operating system or browser. This is important as there are at least three to four standard browsers on each operating system and testing curricular material on these ever-changing configurations has become increasingly problematic and time consuming.

The `LaunchBuilder` program creates and organizes the curricular units in Launcher packages. Although `Launcher` and `LaunchBuilder` were developed primarily for OSP-based curricular materials, they can be used to launch *any* Java program packaged in a jar file. As shown in Section 15.3, `LaunchBuilder` is an easy-to-understand and easy-to-use editor for creating and organizing curricular material.

In this chapter, after we describe how to create and organize curricular material, we present examples from three short topical units: classical mechanics (orbits), electromagnetism (radiation from point charges), and quantum mechanics (time evolution of superpositions of states).

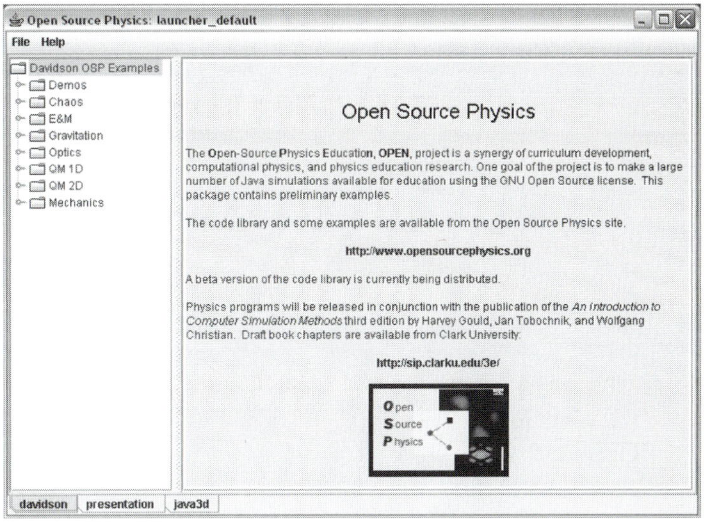

Figure 15.1 Launcher used to display the organizational structure of the ready-to-run Java-based curricular material from the `osp_guide.jar` file.

15.2 ■ LAUNCHER

Launcher enables users to access ready-to-run Java-based curricular material developed by the OSP project and other authors. For a complete listing of the curricular materials distributed with this *Guide*, see Table 1.1 of Chapter 1.

Run the Launcher example file, `osp_demo.jar`, by executing it from the command line or double-clicking it in a GUI file browser. Launcher automatically reads xml data from a file within the jar file and loads the specified programs and resources. A collection of curricular materials is then displayed in a *TabSet* as shown in Figure 15.2 for the `osp_demo.jar`.

A *TabSet* organizes material into one or more tabs, each of which displays an *Explorer* pane and a *Description* pane within Launcher. The *Explorer* pane (left) allows users to easily navigate the tree-based curricular material and launch the associated programs. Selecting a tree node (single click) displays its associated html or text description in the *Description* pane (right). Double-clicking a folder node expands or collapses the contents of that node, while double-clicking a launchable node (depicted with a green arrow) will launch a particular program.

In order to use Launcher, a developer creates a Java archive (a jar file) that designates Launcher as the target using a *manifest* similar to the one shown in Listing 15.1. The manifest is packaged inside the jar file along with compiled Java classes and resources (html pages, xml files, images, and sounds). As the `osp_demo.jar` example shows, all of the resources for a Launcher unit can be packaged inside a single jar file for easy distribution.

Listing 15.1 A Java manifest that executes the `main` method in Launcher.

```
Manifest-Version: 1.0
Built-By: W.Christian
Main-Class: org.opensourcephysics.tools.Launcher
```

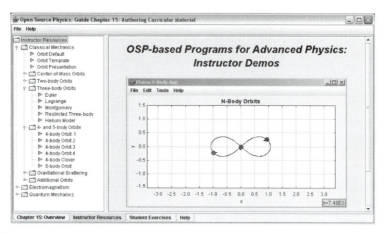

Figure 15.2   An example of how `Launcher` is used to organize the ready-to-run Java-based curricular material which appears in `osp_demo.jar`.

Curriculum authors need not, however, concern themselves with these details. The core OSP library is packaged and distributed in the `osp.jar` file, and this archive already contains a manifest that executes `Launcher`.

`Launcher` reads resources using a text-based xml configuration file, and these resources may be located externally in other jar files, zip files, or directories. Thus, it is easy to author and distribute `Launcher`-based curricular material without recompiling Java code or creating jar files. Although `Launcher` configuration files can be edited using any text editor, they are most easily authored using `LaunchBuilder`, as described in the next section.

15.3 ■ LAUNCHBUILDER

The `osp_demo.jar` file contains over 50 examples of ready-to-run physics curricular material. In order to learn how to author your own material, we will begin with the simpler `osp.jar` file. Run the `osp.jar` file and select the Edit menu item under `Launcher`'s File menu to invoke `LaunchBuilder` as shown in Figure 15.3. `LaunchBuilder` shows the same tree structure as `Launcher`, but a toolbar has been added and the *Description* pane has been replaced with an *Editor* pane with *Display*, *Launch*, and *Metadata* tabs.

Editing Launcher Tabs and Trees

The left-hand side of the `LaunchBuilder` workspace contains the *Explorer* pane with an xml tree. Click on the root node of the tree and then click the Add button to add a node. Each node can be associated with an html document by entering a url in the `URL HTML` text field found in the *Display* tab of the *Editor* pane. If an html document is not available, you may enter an alternate text narrative in the Text Description field near the bottom of the *Display* tab.

As shown in Figure 15.3, in `osp.jar` there is a single `Launcher` tab (displaying the text "Open Source Physics") with a single root node named "Open Source Physics."

Select the root and change its name to "Quantum Mechanics" by entering the name in the *Display* tab `Name` text field (this text field, like all `LaunchBuilder` text fields, turns

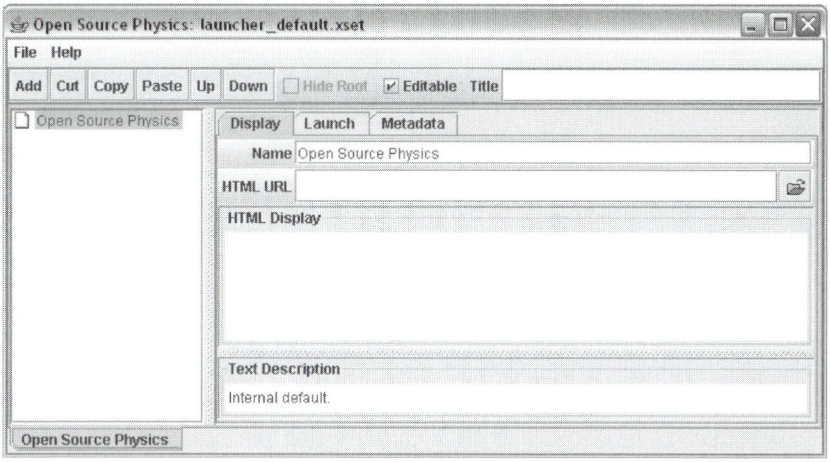

Figure 15.3 Selecting the Edit menu item under `Launcher`'s File menu invokes `LaunchBuilder`. Shown is the basic `LaunchBuilder` workspace that appears when opening the `osp.jar` file.

yellow when editing; hit the Enter key or select another field to apply the change). Note that the `LauncherBuilder` tab label also changes to reflect the new root name.

Use the `Title` text field on the toolbar to give your *TabSet* a title ("My First Launcher"). To see how your *TabSet* will look in `Launcher`, select the *Preview* item from the File menu. Close your preview to return to `LaunchBuilder`.

To save this tab root as an 'independent' `Launcher` node (an xml file), right-click the tab (control-click on Mac) and select the *Save As* item from the pop-up menu as shown in Figure 15.4. In the *Save As* dialog that appears, assign the xml file the name `quantum_mechanics.xml` and save. Notice that the file `quantum_mechanics.xml` now appears in the same directory as the `osp.jar` file.[1]

Independent nodes represent independent content trees that can be referenced by file name and opened as tab roots (*Open*) or imported as branches within larger trees (*Import*). This permits very modular content organization. Independent nodes are identified by a file icon. The color of the icon indicates the type and status of the xml file: yellow is a changed file, white is a saved file, magenta is a jar file, and red is a read-only file.

Now invoke the *Save As* item under the File menu as shown in Figure 15.5. This brings up the file-saver dialog, shown in Figure 15.6, that displays a tree showing the *TabSet* files to be saved. All file names are relative to the base path displayed below the tree. To change a name or base path, click in a field or, for the *TabSet* name and base path, click the Choose button to bring up a file chooser. For our current example, simply accept the default base path and the default *TabSet* file name `launcher_default.xset`.

The saver tree allows you to save all the files defined in the current *TabSet* or to combine multiple files into a single 'self-contained' file by collapsing a tree node. If you collapse the *TabSet* node, then all data will be saved in a single *TabSet* file. File names are not saved.

---

[1]A convenient method for naming a `Launcher` tab involves naming the root and then hiding it. To see how this works, add a child node to the root by clicking the Add button and assigning the name "QM" to it. Now hide the root node by checking the `Hide Root` box on the toolbar. The "QM" node now appears as the root but the tab name is still "Quantum Mechanics."

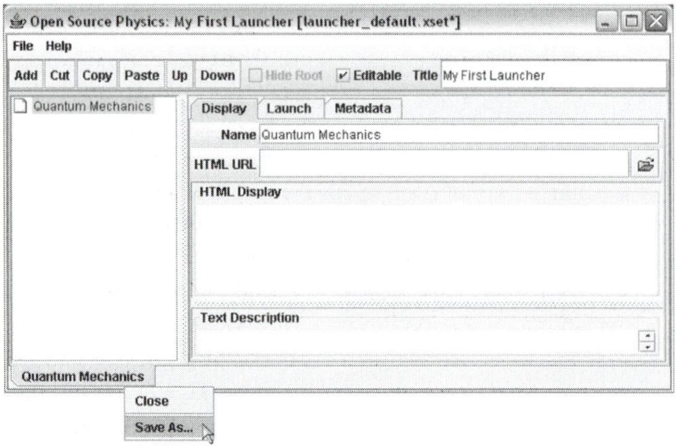

Figure 15.4   The LaunchBuilder showing the *Save As* pop-up menu for saving tabs as xml files.

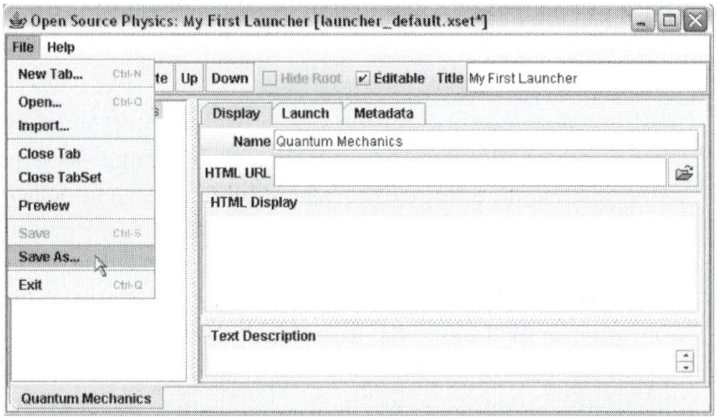

Figure 15.5   The File menu for LaunchBuilder showing the *Save As* menu item.

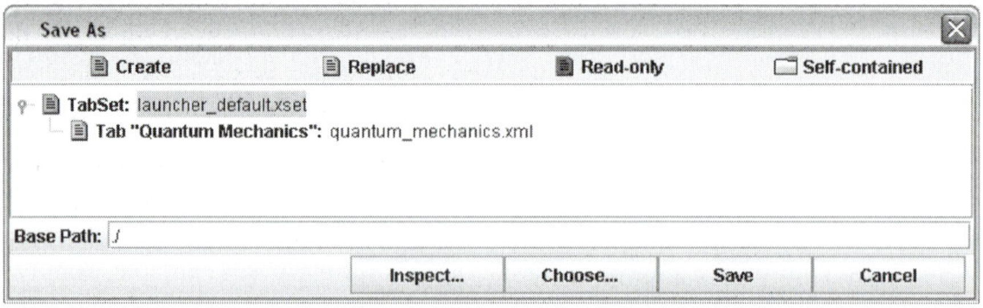

Figure 15.6   The resulting file-saver dialog box from selecting the Save As menu item.

The saver tree icons also give you information about the files you are saving. Green icons indicate new files, while yellow icons indicate that an existing file will be overwritten. Red icons identify read-only files for which you must choose a different name or base path before saving.

Close `Launcher` and `LaunchBuilder` and reopen `osp.jar`. The configuration files, `launcher_default.xset` and `quantum_mechanics.xml`, that were just created where the jar file is located will now load as the default, overriding the internal xml files. On startup both `Launcher` and `LaunchBuilder` search for the default file `launcher_default.xset`, first in the directory containing the jar and then within the jar itself. This search order allows teachers to adapt prepackaged content to their curricular needs by providing edited external files that override the internal ones. Since the `Launcher` will look for these files, moving or deleting them will cause the `Launcher` to load the internal xml files again.

`LaunchBuilder` can be invoked from within `Launcher` only if the *TabSet* is saved with the *Editable* option checked on the toolbar. Unchecking this option protects curricular material from unintended modification. `LaunchBuilder` may also be executed from the command line by creating a separate jar file with a manifest that defines `LaunchBuilder` to be the `Main-Class` or by using Java Web Start.[2]

```
java -classpath osp.jar org.opensourcephysics.tools.LaunchBuilder
```

Although `Launcher` and `LaunchBuilder` accept a command line parameter that can specify any *TabSet* file, it is often useful to use the default name `launcher_default.xset` as in the example. *TabSets* with names other than the default can be opened by selecting the *Open* item from the *File* menu.

Editing Launcher Nodes

The right-hand side of the `LaunchBuilder` workspace contains the *Editor* pane which provides three tabs for setting node properties:

Display **tab** contains fields for specifying the name, html page, and text description of the selected node. This allows authors to name nodes or folders and specify the html page that will be shown when a launch node or folder is selected. A text description can also be specified.

Launch **tab** contains fields for specifying a jar file containing Java applications, a launchable application within that jar, and an optional xml file with parameters to be loaded when the application is launched. This allows authors to associate with a particular node: a jar file, a program within that jar file, and an xml file with preset parameters to load when the node is launched. If an xml file is not specified, the chosen program will use its default parameters when it opens.

Metadata **tab** contains fields for specifying the code and/or curriculum author and other metadata useful for Web-based search engines and resource collections. This allows authors to embed metadata tags within the xml files associated with the curricular material they are creating.

With your newly created *TabSet* open in `Launcher`, select the *Edit* menu item to again enter the `LaunchBuilder` workspace. Select the *Display* tab in the *Editor* pane. Add a child node to the "Quantum Mechanics" node and name it "Harmonic Oscillator" using the `Name`

[2]A Java applet is embedded in an html page and depends on a browser to be executed. A Java application is distributed over the Web from an html page using *Web Start* and is then independent of the Web. The downloaded program has some security restrictions (just like an applet) but is independent of the browser and executes like other programs. In fact, a *Web Start* dialog asks the user if they wish to install an icon on the desktop to access the installed program.

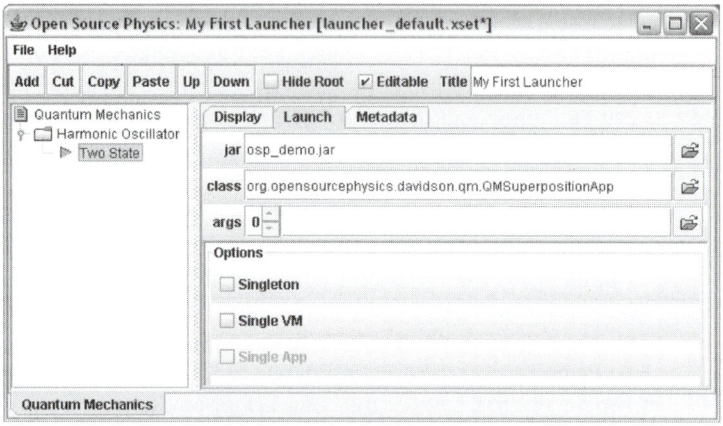

Figure 15.7   Using `LaunchBuilder` to create curricular modules. In "Harmonic Oscillator" a node "Two State" has been added (automatically changing "Harmonic Oscillator" into a folder). With the *Launch* tab selected, `Launcher` is configured to open `osp_demo.jar` and access the program `QMSuperpositionApp` when "Two State" is double-clicked.

text field. You must either hit the Return button on your keyboard or select another text field for this change to take effect.

Now select the "Harmonic Oscillator" node and add a child named "Two State" to it. Note that "Harmonic Oscillator" has become a folder with "Two State" as its sole content and that the frame title now displays an asterisk to indicate that the *TabSet* has changed.

To move a node up or down within a given level of the tree, select the node and click the Up or Down button. Duplicate a node by first copying it to the clipboard (select node and click the Copy button) and then pasting it as a child of a different node (select the parent node and click the Paste button). Remove or move a node using the Cut button.

Select the "Two State" node and associate an html document with it by entering a url in the `HTML URL` field. If no html document is available, enter a text description in the `Text Description` field.

Next we specify a jar file containing the Java applications. These applications need not be OSP programs, although we will assume that is the case in the remainder of this chapter. With the node "Two State" selected (highlighted), select the *Launch* tab in the *Editor* pane. Note the `jar` and `class` text fields which are used for specifying a jar file name and a launch class name, respectively. Also note that the `class` text field is initially disabled since no jar file has been specified. Use the Browse button to the right of the `jar` text field to select the `osp_demo.jar` file. The `class` text field is now enabled and you can now browse the contents of the jar file to select any launchable class (class with a `main` method). Select the `org.opensourcephysics.davidson.qm.QMSuperpositionApp` class. The "Two State" node now displays a green arrow icon to indicate that it executes a Java program as shown in Figure 15.7. Double-click the node to execute the program. Click the Initialize button (turning it into a Start button) in the `control` frame as shown in Figure 15.8 to run the simulation. Save your work.

The `jar` text field allows authors to specify more than one jar file, and child nodes inherit jar files from their parents. In other words, if a jar file is specified at the root of the `Launcher` configuration, that jar file is added to the classpath for all nodes in the tree. Hence, if we

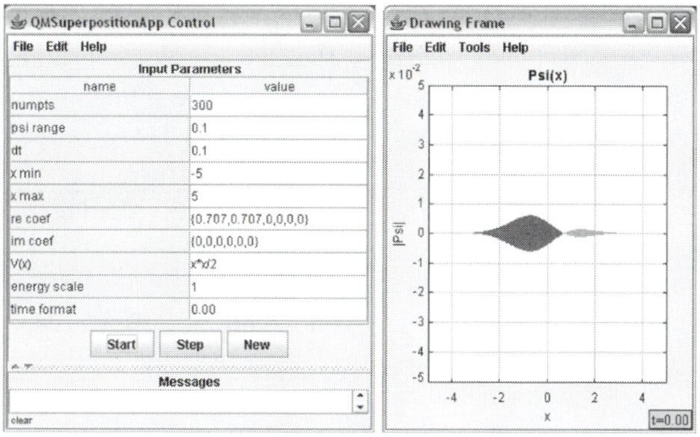

Figure 15.8 The result of launching the node "Two State."

were to populate a folder or collection with programs from the same jar file, it would make sense to add the jar at the root of the collection.

Save all files, exit `LaunchBuilder`, and reopen `Launcher`. Click on the "Two State" node to run the program. We now examine the xml files that were created.

Launcher and XML Files

`Launcher` recognizes both xml and xset file extensions. Both extensions are used to designate a file as an xml document. All `Launcher` xml files use the standard OSP format (see Chapter 12). This format uses an <object> tag with `class` attribute to save a Java object and <property> tags with `name` and `type` attributes to save the object's properties. Since a <property> can be of type <object>, object and property tags are nested.

The `xset` extension signals that the document contains complete *TabSet* data. The `launcher_default.xset` document describing "My First Launcher" is shown in Listing 15.2. To view this file within `Launcher`, open the saver dialog (*File | Save As*), select the *TabSet* node, and click the Inspect button. An xml inspector will then open that shows the xml file organized into a tree structure on the left with the corresponding xml text on the right.

Listing 15.2 An xset document clears the `Launcher` and loads new material.

```
<?xml version="1.0" encoding="UTF-8"?>
<object class="org.opensourcephysics.tools.Launcher$LaunchSet">
  <property name="title" type="string">My First Launcher</property>
  <property name="editor_enabled" type="boolean">true</property>
  <property name="width" type="int">650</property>
  <property name="height" type="int">300</property>
  <property name="divider" type="int">160</property>
  <property name="launch_nodes"
    type="collection" class="java.util.ArrayList">
    <property name="item" type="string">
    quantum_mechanics.xml
    </property>
  </property>
</object>
```

Inspect the *TabSet* xml shown in Listing 15.2. Note that it specifies a collection of launch nodes to be loaded into tabs. Here, there is a single launch node, and the data saved is simply the file name of the independent `Launcher` file that defines the "Quantum Mechanics" tab. Also note the other properties: `title`, `editor_enabled`, `width`, `height`, and `divider`. When Launcher opens a *TabSet*, it first closes any previous tabs, then loads the new tabs, resizes and titles the frame, and sets the divider location based on these values.

Documents with an xml extension describe an 'independent' launch node that can be opened in a tab or attached to an existing tree in `LaunchBuilder`. Inspect the `quantum mechanics.xml` file shown in Listing 15.3. You should recognize the names and other properties of the nodes as well as their parent/child relationships.

With the xml inspector open, again select the *TabSet* node and note how the xml changes when you collapse it, so it is self-contained. All of the data that was previously in the `quantum_mechanics.xml` file is now in the *TabSet* file.

Listing 15.3   An xml document describes an xml tree that can be loaded into a `Launcher`.

```
<?xml version="1.0" encoding="UTF-8"?>
<object class="org.opensourcephysics.tools.LaunchNode">
  <property name="name" type="string">Quantum Mechanics</property>
  <property name="description" type="string"></property>
  <property name="child_nodes" type="collection"
    class="java.util.ArrayList">
  <property name="item" type="object">
  <object class="org.opensourcephysics.tools.LaunchNode">
    <property name="name" type="string">Harmonic Oscillator</property>
    <property name="child_nodes" type="collection"
      class="java.util.ArrayList">
    <property name="item" type="object">
    <object class="org.opensourcephysics.tools.LaunchNode">
      <property name="name" type="string">Two State</property>
      <property name="launch_class" type="string">
        org.opensourcephysics.davidson.qm.QMSuperpositionApp
        </property>
      <property name="classpath" type="string">osp_demo.jar</property>
    </object>
    </property>
    </property>
  </object>
  </property>
  </property>
</object>
```

Saving and Loading Application Data

The *Launch* tab in the `LaunchBuilder` workspace also provides an `args` field for specifying one or more (usually using the `[0]` value) command line arguments passed to the `main` method of the launch class. Generally, the `args[0]` parameter is used to specify the name of an xml document containing initialization parameters. This allows a single Java program to be used with many different parameter sets.

The easiest way to create such an xml file is by using a program's `control` frame. Open `Launcher` and double-click the "Two State" node. Make some changes to the text fields of the `control` frame. Under the `control` frame's *File* menu select the *Save XML* item as shown in Figure 15.9.

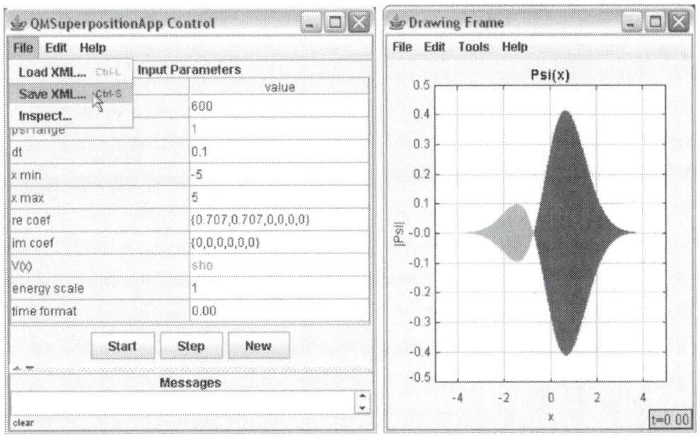

Figure 15.9 The node "Two State" has been launched yielding the following program control (left) and view (right). The default configuration has been altered and is in the process of being saved as an xml file as shown by the *Save XML* dialog.

Save your xml file as `test.xml`. This xml file can be loaded in two ways. First, use the *Load XML* item under the `control` frame's File menu to load the file directly. Second, use the `args[0]` text field (as shown in Figure 15.7) in the *Launch* tab of the `LaunchBuilder` workspace to specify that the parameters saved in the `test.xml` file will automatically load the next time the node "Two State" is selected in `Launcher`. You must, of course, save this change.

In addition, there are several options that control the runtime environment for the program being executed (launched). The *Singleton* option allows only one running instance for a given node. The *Single VM* option instantiates the application in the same Java VM that is executing `Launcher` rather than a separate VM. (Note that because of security restrictions, single VM mode must be selected if `Launcher` is used to distribute curricular material from a server as an applet or using Java Web Start technology.) The *Single App* option allows only one application to run at a time—a running application is terminated when the user double-clicks a node to launch another. The *Show Log* option displays a console-like log window containing error and information messages when the application starts or runs. This option is useful for debugging and for some types of output data. A log level can be specified to log different levels of detail from finest to severe. Launcher nodes inherit runtime options from their parent, so selecting an option at the root will apply to all nodes.

Organizing and Distributing Material

If you have followed the creation of curricular materials using `LaunchBuilder` described thus far, you should have three files in the same directory as your jar files: `launcher_default.xset`, `quantum mechanics.xml`, and `test.xml`. These xml files supersede the internal xml files contained within the jar allowing teachers to author their own material with `LaunchBuilder`. While this is convenient, the issue of organization quickly becomes important. At the moment there are only three xml files to worry about. However, in time there may be tens or hundreds of xml files for *TabSets*, xml files for saved program data, and html files for descriptions. How should we organize them?

The simplest directory structure organizes the html and xml data files in separate subdirectories while leaving the xset file, `launcher_default.xset`, and *TabSets* at the root

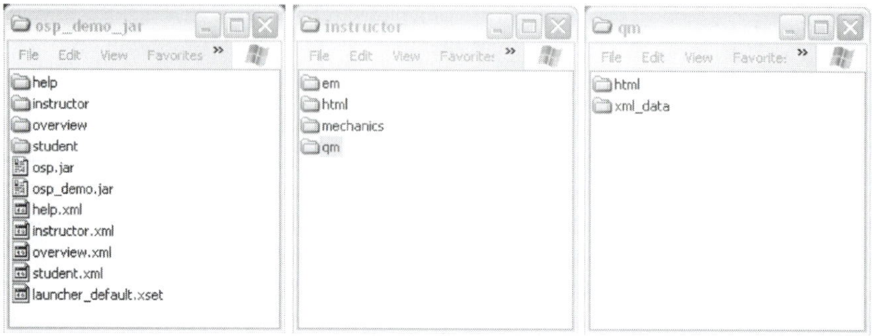

Figure 15.10   The internal file structure of the `osp_demo.jar` file. On the left, the root directory structure is shown; in the center, the subdirectories of the instructor directory showing folders organized according to topic; on the right, the contents of the *qm* directory revealing *xml_data* and *html* subdirectories.

where the jar file is located. Such a directory structure is shown in Figure 15.10. Note that this is the internal file structure of the `osp_demo.jar` file. The xset file at the root and the associated xml files store *TabSet* information that produces the results shown in Figure 15.2. Within the qm directory are the xml_data and html directories associated with the quantum mechanics curricular material.

Figure 15.10 shows a complicated directory structure which often makes distribution difficult. One way to simplify distribution is to compress all xml and xset files into a single `zip` file called `launcher_default.zip`. Subdirectories should be compressed in `launcher_default.zip` as well so that the directory structure is preserved in the zip archive. Figure 15.11 shows the resulting files as a result of compressing xml and xset files. Compressing the files and folders shown in Figure 15.10 would yield the same resulting structure with a different `launcher_default.zip` file. In order to see the prepackaged directory structure of `osp_demo.jar`, which is similar to that shown in Figure 15.10. Expand (unzip) this jar file and look at its contents.

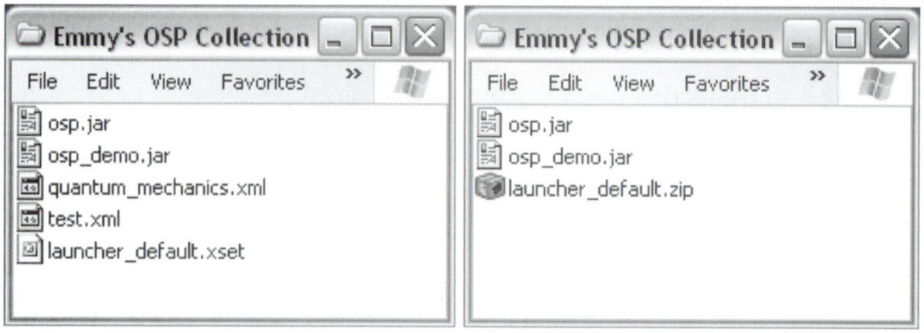

Figure 15.11   Two directories that yield the same result when double-clicking the `osp.jar` file. The directory on the right uses a `launcher_default.zip` archive containing the xml and xset files.

15.4 ■ CURRICULUM DEVELOPMENT OVERVIEW

To accompany this chapter we have created curricular units for classical mechanics (orbits), electromagnetism (radiation from point charges), and quantum mechanics (time evolution of superpositions of states). This material can be accessed by double-clicking the file osp_demo.jar. We briefly begin each section with a review of the theory behind the phenomena we are simulating, then briefly describe the programs used and the curricular unit. A fuller description of the programs (including all of the editable parameters) and the curricular material can be found in the "Overview" and "Help" tabs in the Launcher that opens when accessing osp_demo.jar.

15.5 ■ CLASSICAL MECHANICS: ORBITS

One of the standard topics in classical mechanics at any level is that of Kepler's laws of planetary motion. This topic rests on Newton's law of universal gravitation,

$$\mathbf{F}_{12} = -Gm_1m_2 \frac{\mathbf{r}_{12}}{r_{12}^3}, \tag{15.1}$$

for the gravitational force on mass m_1 due to mass m_2, given a separation $r_{12} = |\mathbf{r}_1 - \mathbf{r}_2| = \sqrt{r_1^2 + r_2^2 - 2\mathbf{r}_1 \cdot \mathbf{r}_2}$, and where $G = 6.67 \times 10^{-11}$ Nm^2/kg^2. The force is attractive and lies along the line separating the two masses.

Kepler's laws can be simply expressed as (1) planets move in ellipses with Sun at one focus, (2) planets sweep out equal areas in equal times as they orbit Sun, and (3) the square of the period of a planet's orbit is proportional to the cube of the semimajor axis a of its orbit: $T^2 = \left(\frac{4\pi^2}{GM}\right) a^3$.

Introductory treatments of Kepler's laws focus on the special cases of two-body systems where the central object is much more massive than the orbiting body, which itself usually orbits in circular motion. More advanced treatments include the general study of elliptical orbits and occasionally cases where the objects are nearly the same mass. Systems with more than two objects are considered less frequently and for good reason. These systems are, in general, not exactly solvable. However, as we shall see, there are special cases in which the three-, four-, and five-body problems can be solved as shown in Figure 15.12.

The simplest Kepler problem involves just two objects with masses m_1 and m_2. These objects are attracted to each other via (15.1), and there is usually some motion of both objects as shown in Figure 15.13. Each object is given a position vector relative to an origin, but it is the *separation vector* or *relative coordinate* $\mathbf{r} = \mathbf{r}_{12} = \mathbf{r}_1 - \mathbf{r}_2$ that appears in the

Figure 15.12   Classical orbit examples from the osp_demo.jar file.

Figure 15.13 Classical orbit examples from the osp_demo.jar file which show three different binary systems' orbits.

gravitational force $\mathbf{F} = -Gm_1m_2\frac{\mathbf{r}}{r^3}$ and the gravitational potential energy $U(r) = -\frac{Gm_1m_2}{r}$ of the system.

We can naively write the Lagrangian for the system:

$$\mathcal{L} = \frac{1}{2}m_1\dot{\mathbf{r}}_1^2 + \frac{1}{2}m_2\dot{\mathbf{r}}_2^2 - U(r), \tag{15.2}$$

which is given in terms of three coordinates, although there are just two degrees of freedom. We use *center-of-mass coordinates*

$$\mathbf{R} = \frac{m_1\mathbf{r}_1 + m_2\mathbf{r}_2}{m_1 + m_2}, \tag{15.3}$$

and the total mass of the system $M = m_1 + m_2$ to write $\mathbf{r}_1 = \mathbf{R} + \frac{m_2}{M}\mathbf{r}$ and $\mathbf{r}_2 = \mathbf{R} - \frac{m_1}{M}\mathbf{r}$, a substitution which allows the Lagrangian to obtain the form

$$\mathcal{L} = \frac{1}{2}M\dot{\mathbf{R}}^2 + \frac{1}{2}\mu\dot{\mathbf{r}}^2 - U(r), \tag{15.4}$$

where $\mu = \frac{m_1m_2}{m_1+m_2}$ is the reduced mass, and $U(r) = -\frac{G\mu M}{r}$. Equation (15.4) suggests that the system behaves as if there were two fictitious particles: one of mass M that moves with the center of mass and another of mass μ that moves with a speed of the relative position as shown in Figure 15.14.

We can use this idea to rewrite the Lagrangian as $\mathcal{L} = \mathcal{L}_{\text{rel}} + \mathcal{L}_{\text{cm}}$, where

$$\mathcal{L}_{\text{cm}} = \frac{1}{2}M\dot{\mathbf{R}}^2 \quad \text{and} \quad \mathcal{L}_{\text{rel}} = \frac{1}{2}\mu\dot{\mathbf{r}}^2 - U(r). \tag{15.5}$$

For the center-of-mass motion $\ddot{\mathbf{R}} = 0$, which means that $\dot{\mathbf{R}} = $ constant, and for the equation of motion of the relative coordinate, we have

$$\mu\ddot{\mathbf{r}} = -\nabla U(r). \tag{15.6}$$

If we choose a frame of reference in which $\dot{\mathbf{R}} = 0$, such as shown in Figure 15.14, the entire Lagrangian is just

$$\mathcal{L} = \frac{1}{2}\mu\dot{\mathbf{r}}^2 - U(r), \tag{15.7}$$

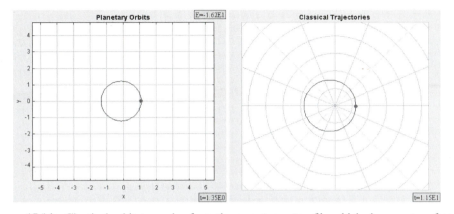

Figure 15.14 Classical orbit examples from the `osp_demo.jar` file which show center-of-mass motion in both Cartesian and polar coordinates.

and the resulting motion will lie on a plane that we can identify as the xy-plane. We can now write the Lagrangian in polar coordinates as

$$\mathcal{L} = \frac{1}{2}\mu\dot{r}^2 + \frac{1}{2}\mu r^2\dot{\phi}^2 - U(r). \tag{15.8}$$

For the ϕ coordinate we find the equation of motion $\mu r^2\ddot{\phi} = 0$ or that $\mu r^2\dot{\phi} = L = $ constant. Here, L is the z-component of the angular momentum and it is a constant.

The radial equation of motion for the r-coordinate gives

$$\mu\ddot{r} = \frac{L^2}{\mu r^3} - \frac{G\mu M}{r^2}. \tag{15.9}$$

The first term on the right-hand side is called the centrifugal force $F_{\text{cf}} = \mu r\dot{\phi}^2 = \frac{L^2}{\mu r^3}$, and the second term on the right-hand side is the gravitational force.

To show that these orbits are, in general, elliptical, we are interested in the *trajectory* plot $r(\phi)$ for an arbitrary orbit. We solve for the time parameter in order to make substitutions that eventually eliminate t. We begin by using the substitution $u = \frac{1}{r}$ which, after taking time derivatives and simplifying, yields the differential equation

$$\frac{d^2u}{d\phi^2} + u = \frac{Gm^2M}{L^2}, \tag{15.10}$$

which has the solution that $u = A\cos(\phi - \phi_0) + \frac{Gm^2M}{L^2}$ or in terms of r

$$r(\phi) = \frac{1}{A\cos(\phi - \phi_0) + \dfrac{Gm^2M}{L^2}}. \tag{15.11}$$

If we set $\phi_0 = 0$ and rewrite, we find that

$$r(\phi) = \frac{1}{A\left[\cos(\phi) + \dfrac{Gm^2M}{AL^2}\right]}, \tag{15.12}$$

Figure 15.15   Classical gravitational scattering examples from the `osp_demo.jar` file which show ever-increasing impact parameters.

where we easily find the maximum (apogee) and minimum (perigee) value of this expression and also find that the eccentricity is $e = \frac{AL^2}{Gm^2M}$.

In addition to closed orbits, we may also consider scattering (unbound) orbits. Particles incident on a cross-sectional area $d\sigma$ will scatter an angle $d\Omega$. The differential cross-section is defined as $\mathcal{D}(\theta) = \frac{d\sigma}{d\Omega}$. In terms of the impact parameter b, we have that $d\sigma = db\, b\, d\phi$ and $d\Omega = \sin(\theta)\, d\theta\, d\phi$ and therefore the differential cross-section becomes

$$\mathcal{D}(\theta) = \frac{d\sigma}{d\Omega} = \frac{b}{\sin(\theta)} \left| \frac{db}{d\theta} \right|. \tag{15.13}$$

We find for the entire cross-section

$$\sigma = \int \frac{d\sigma}{d\Omega}\, d\Omega = \int \mathcal{D}(\theta)\, d\Omega. \tag{15.14}$$

For gravitational scattering, we use the fact that the radial position of the scatterer r is related to angle by (15.11). We begin by evaluating $\frac{1}{r}$ as given by (15.11) and solve for the parameter A using the energy at the particle's extrema (usually $r = -\infty$) in its unbounded orbit. In doing so, after much algebra and trigonometry, we find that the impact parameter b can be written in terms of Θ, the scattering angle (the total angle scattered), as

$$b = \frac{GmM}{2E} \cot\left(\frac{\Theta}{2}\right), \tag{15.15}$$

where E is the energy of the scattered object. Examples of gravitational scattering and the relationship between b and Θ given by (15.15) are shown in Figure 15.15.

Programs and Curricular Materials

The classical mechanics curricular materials use the following OSP programs in the Davidson package `org.opensourcephysics.davidson`. This material is organized as shown in Figure 15.16.

`gravitation.PlanetApp` shows a planet orbiting a massive central star in Cartesian coordinates. The initial position (x, y) and velocity (v_x, v_y) of the orbiting mass can be set.

`gravitation.ClassicalApp` shows a planet orbiting a massive central star in polar coordinates. The initial position (r, ϕ) and velocity (v_r, v_ϕ) of the orbiting mass can be set.

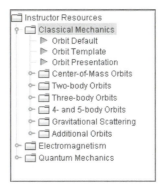

Figure 15.16 The *Explorer* pane of the `Launcher` from the `osp_demo.jar` showing the classical mechanics curricular materials.

`nbody.OrbitApp` shows two or more objects that are gravitationally attracted to each other. Any number of objects can be added with any mass, initial position, and initial velocity.

The curricular material is organized into six units:

Center-of-Mass Orbits shows variations on a planet orbiting a much more massive central star in Cartesian and polar coordinates. The examples show one circular and two elliptical orbits that begin at the same position yet have different initial velocities. It is the difference in initial velocity that accounts for the difference in orbital trajectory around the central star.

Two-Body Orbits shows a variety of two-body orbits where the masses of the two objects are comparable. One set of examples varies the mass of one object, while the other set varies the initial position of one object.

Three-Body Orbits shows four three-body orbits. The *Euler* example shows a special solution determined by Euler in which all three masses always lie on a straight line. The *Lagrange* example shows a special solution determined by Lagrange in which the objects maintain a geometric relationship to each other as vertices of a polygon. In this case, the three objects are always at the vertices of an isosceles triangle (whose sides can change length). The *Montgomery* example shows a stable orbit of three objects that orbit each other in a figure-eight pattern with all three objects tracing out the exact same orbital trajectory. The *Restricted Three-Body* example shows an example of this class of exactly-solvable problems. Two objects with the same mass are in orbit around each other. A third, much less massive, object also orbits, following one object around its orbit.

Four- and Five-Body Orbits shows a variety of unstable orbits with four and five objects all with similar masses with all objects tracing out the exact same orbital trajectory. The four-body *Clover* example is a Lagrange type orbit where the four objects are always at the vertices of a square (whose sides can change length).

Gravitational Scattering shows a variety of examples of unbounded scattering of an object in the gravitational field of a more massive object. The impact parameter varies in each simulation.

Additional Orbits shows center-of-mass orbits of a wider variety than that of the Center-of-Mass Orbits unit.

15.6 ■ ELECTROMAGNETISM: RADIATION

Maxwell's equations are written in MKS units as

$$\nabla \cdot \mathbf{E} = \rho/\epsilon_0 , \qquad \nabla \cdot \mathbf{B} = 0, \tag{15.16}$$

$$\nabla \times \mathbf{E} + \frac{\partial \mathbf{B}}{\partial t} = 0 , \qquad \nabla \times \mathbf{B} - \epsilon_0\mu_0\frac{\partial \mathbf{E}}{\partial t} = \mu_0\mathbf{J}, \tag{15.17}$$

where we choose to write these equations with all of the fields on one side and all the true sources on the other. For a variety of reasons, the calculation of time-dependent fields shown in Figures 15.17–15.19 is easier in the potential formalism. In this formalism, $\mathbf{E} = -\nabla\phi - \frac{\partial \mathbf{A}}{\partial t}$ and $\mathbf{B} = \nabla \times \mathbf{A}$, where ϕ is the electric (scalar) potential and $\mathbf{A}$ is the magnetic (vector) potential. Making these two substitutions in (15.17) using the double-curl rule and simplifying, we are left with only two equations:

$$-\nabla^2\phi - \frac{\partial}{\partial t}(\nabla \cdot \mathbf{A}) = \rho/\epsilon_0, \tag{15.18}$$

and

$$-\nabla^2\mathbf{A} + \nabla(\nabla \cdot \mathbf{A}) + \epsilon_0\mu_0\nabla\left(\frac{\partial \phi}{\partial t}\right) + \epsilon_0\mu_0\frac{\partial^2 \mathbf{A}}{\partial t^2} = \mu_0\mathbf{J}. \tag{15.19}$$

If we choose the Lorentz gauge, $\epsilon_0\mu_0\frac{\partial \phi}{\partial t} + \nabla \cdot \mathbf{A} = 0$, (15.18) and (15.19) reduce nicely to

$$\left[\epsilon_0\mu_0\frac{\partial^2}{\partial t^2} - \nabla^2\right]\phi = \rho/\epsilon_0 \quad \text{and} \quad \left[\epsilon_0\mu_0\frac{\partial^2}{\partial t^2} - \nabla^2\right]\mathbf{A} = \mu_0\mathbf{J}, \tag{15.20}$$

which are wave equations for electromagnetic waves with speed $c = \frac{1}{\sqrt{\epsilon_0\mu_0}} = 3 \times 10^8$ m/s.

Figure 15.17 Electromagnetic radiation examples from the `osp_demo.jar` file which show the electric field vectors (in the xy-plane) of a positively charged particle at increasing constant velocities.

We find that for static cases, the equations in (15.20) reduce to

$$\nabla^2 \phi = -\rho/\epsilon_0 \quad \text{and} \quad \nabla^2 \mathbf{A} = -\mu_0 \mathbf{J}, \tag{15.21}$$

which have the solutions

$$\phi(\mathbf{r}) = \frac{1}{4\pi\epsilon_0} \int \frac{\rho(\mathbf{r}')}{|\mathbf{r} - \mathbf{r}'|} \, dV' \quad \text{and} \quad \mathbf{A}(\mathbf{r}) = \frac{\mu_0}{4\pi} \int \frac{\mathbf{J}(\mathbf{r}')}{|\mathbf{r} - \mathbf{r}'|} \, dV'. \tag{15.22}$$

In the static case, the potentials and fields do not change. However, when charge distributions and current distributions change, the potentials and fields must also change. Where we measure the potentials and fields $\mathbf{r}$ there is, however, a time delay of the signal that the charge and current distributions located at $\mathbf{r}'$ have changed. This signal travels at the speed of light, and hence the relationship between the time associated with the charge and current distributions (the cause) t' and the time when we measure the potentials and fields (the effect) t is

$$t' = t - \frac{|\mathbf{r} - \mathbf{r}'|}{c}, \tag{15.23}$$

where t' is called the retarded time. The time t' is the time at which the signal left, and t is the time at which the signal arrives. The difference is due to the constant velocity c at which the signal travels between $\mathbf{r}'$ and $\mathbf{r}$. For time-dependent charge and current distributions, we must include the retardation effects in (15.22) which yields

$$\phi(\mathbf{r}, t) = \frac{1}{4\pi\epsilon_0} \int \frac{\rho(\mathbf{r}', t')}{|\mathbf{r} - \mathbf{r}'|} \, dV' \quad \text{and} \quad \mathbf{A}(\mathbf{r}, t) = \frac{\mu_0}{4\pi} \int \frac{\mathbf{J}(\mathbf{r}', t')}{|\mathbf{r} - \mathbf{r}'|} \, dV', \tag{15.24}$$

provided one uses the retarded time for t'.

Using $\mathbf{E} = -\nabla\phi - \frac{\partial \mathbf{A}}{\partial t}$ and $\mathbf{B} = \nabla \times \mathbf{A}$ one can, in principle, calculate the electric and magnetic fields for any arbitrary charge and current distribution. These explicit calculations were first worked out by Jefimenko in 1966 and yield

$$\mathbf{E}(\mathbf{r}, t) = \frac{1}{4\pi\epsilon_0} \int \left[\frac{(\mathbf{r} - \mathbf{r}')\rho(\mathbf{r}', t')}{|\mathbf{r} - \mathbf{r}'|^3} + \frac{(\mathbf{r} - \mathbf{r}')\dot{\rho}(\mathbf{r}', t')}{c|\mathbf{r} - \mathbf{r}'|^2} - \frac{\dot{\mathbf{J}}(\mathbf{r}', t')}{c^2|\mathbf{r} - \mathbf{r}'|} \right] dV', \tag{15.25}$$

and

$$\mathbf{B}(\mathbf{r}, t) = \frac{\mu_0}{4\pi} \int \left[\frac{\mathbf{J}(\mathbf{r}', t') \times (\mathbf{r} - \mathbf{r}')}{|\mathbf{r} - \mathbf{r}'|^3} + \frac{\dot{\mathbf{J}}(\mathbf{r}', t') \times (\mathbf{r} - \mathbf{r}')}{c|\mathbf{r} - \mathbf{r}'|^2} \right] dV', \tag{15.26}$$

which reduce to the usual expressions for the electrostatic and magnetostatic fields when $\dot{\rho} = \dot{\mathbf{J}} = 0$.

Radiation due to a point charge q is most directly determined by using the retarded potentials of (15.24) to yield

$$\phi(\mathbf{r}, t) = \frac{1}{4\pi\epsilon_0} \left[\frac{qc}{|\mathbf{r} - \mathbf{r}'|c - (\mathbf{r} - \mathbf{r}') \cdot \mathbf{v}} \right], \tag{15.27}$$

and

$$\mathbf{A}(\mathbf{r}, t) = \frac{\mathbf{v}}{c^2} \phi(\mathbf{r}, t), \tag{15.28}$$

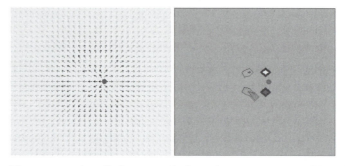

Figure 15.18 Electromagnetic radiation example from the `osp_demo.jar` file which shows the electric field vectors (in the xy-plane) and magnetic field contours (for the field in the z direction) of a positively charged particle changing velocity.

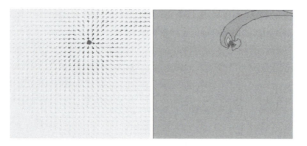

Figure 15.19 Electromagnetic radiation examples from the `osp_demo.jar` file which show the electric field vectors (in the xy-plane) and magnetic field contours (for the field in the z direction) of a positively charged particle in circular motion.

which are the Liénard–Wiechert potentials for moving point charges. Examples of the electric field vectors in the xy-plane and magnetic field contours for the field in the z direction that result from a moving point charge can be seen in Figures 15.17–15.19.

If we consider stationary charges and steady currents (charges moving at a constant velocity), we find that they cause static fields. We may expect from (15.25) and (15.26) that accelerating charges and time-varying currents will create changing electric and magnetic fields and, therefore, the possibility of electromagnetic radiation.

To determine whether or not we will have radiation from a particular moving charge, we look at the Poynting vector $\mathbf{S} = \frac{1}{\mu_0}(\mathbf{E} \times \mathbf{B})$. When we integrate the Poynting vector over a surface, we get the power delivered through that area:

$$P(r) = \oint \mathbf{S} \cdot \mathbf{n} \, da. \tag{15.29}$$

In order to have radiation, there must be some power delivered to infinity, and therefore (15.29) must not vanish as $r \to \infty$. Since the surface area behaves like r^2, for radiating systems the Poynting vector must behave like $\frac{1}{r^2}$, otherwise, the power would vanish at large r. The electric and magnetic fields, therefore, must each fall off like $\frac{1}{r}$ to have radiation. Since static electric and magnetic fields behave like $\frac{1}{r^2}$ at best, the Poynting vector behaves

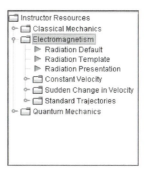

Figure 15.20 The *Explorer* pane of the `Launcher` from the `osp_demo.jar` showing the electromagnetism curricular materials.

like $\frac{1}{r^4}$, and hence point charges associated with these kinds of fields do not radiate. To study radiation, we look for electric and magnetic fields that behave like $\frac{1}{r}$, which we have seen from (15.25) and (15.26).

Qualitatively, one can think about the electric and magnetic fields that are created by point charges in the following way: there are the fields that stay *attached* to the point charge and there are fields that are *thrown off* by the point charge. The fields that are thrown off and make it to infinity are characterized as radiation. Look at Figures 15.17–15.19 again; which ones show radiation?

Programs and Curricular Material

The electromagnetism curricular materials use the following OSP program in the Davidson package `org.opensourcephysics.davidson`. This material is organized as shown in Figure 15.20.

`electrodynamics.RadiationApp` shows a positively charged particle's electric field vectors (in the xy-plane) and magnetic field contours (for the field in the z direction) calculated from the Liénard–Wiechert potentials of (15.27) and (15.28). The trajectory of the particle ($f[x] = x(t)$, $f[y] = y(t)$) can be set, where $c = 1$.

The curricular material is organized into three units:

Constant Velocity shows variations on a positively charged particle moving with constant velocity. The electric and magnetic fields are shown for a particle moving at constant velocities varying from $v = 0$ to $v = 0.99c$.

Sudden Change in Velocity shows the effect of a sudden change in velocity on the electric and magnetic fields associated with a positively charged particle.

Standard Trajectories shows the five standard trajectories of a moving point charge that give rise to radiation. *Bremsstrahlung* shows a moving charge experiencing "braking," a constant deceleration, giving rise to radiation. *SHO x* and *SHO y* show a particle moving in simple harmonic motion in either the x or y direction. There are

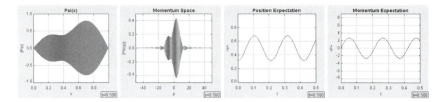

Figure 15.21 Quantum mechanics examples from the `osp_demo.jar` file which show the time evolution of the equal-mix two-state ($n_1 = 1$, $n_2 = 2$) superposition in the infinite square well by displaying the wave function in position and momentum space (shown in color-as-phase representation) and the expectation values of $\hat{x}$ and $\hat{p}$ versus time.

two examples of constant velocity motion of the particle in one direction and a constant acceleration motion in the other direction. *Wiggler x* and *Wiggler y* show a particle moving in simple harmonic motion in either the *x* or *y* direction and a constant velocity in the other direction. A particle following these trajectories (subjected to alternating magnetic fields) is called a wiggler, and such high-intensity radiation from a wiggler can be harnessed in a free electron laser. Finally, there is *Synchrotron* radiation in which the positively charged particle is forced to move in circular motion.

15.7 ■ QUANTUM MECHANICS: SUPERPOSITIONS

Time-dependent quantum-mechanical wave functions are inherently complex (having real and imaginary components) due to the time evolution governed by the Schrödinger equation, which in one-dimensional position space is

$$\left[-\frac{\hbar^2}{2m} \frac{\partial^2}{\partial x^2} + V(x) \right] \psi(x, t) = i\hbar \frac{\partial}{\partial t} \psi(x, t). \tag{15.30}$$

The standard way to visualize the wave functions that solve (15.30) is to either consider just the real part or consider the probability density, approaches that discard all phase information. The `osp_demo.jar` package contains programs for this type of visualization as well as color-as-phase and operator expectation values as shown in Figure 15.21.

As an example, consider the standard problem of a particle of mass *m* confined to a region of length *L* by an infinite square-well potential defined by

$$V(x) = \begin{cases} 0 & \text{for } 0 < x < L \\ \infty & \text{otherwise,} \end{cases} \tag{15.31}$$

which has the position-space energy eigenstates (which are nonzero only within the well):

$$\psi_n(x) = \sqrt{\frac{2}{L}} \sin\left(\frac{n\pi x}{L}\right), \tag{15.32}$$

with energy eigenvalues, $E_n = \frac{\hbar^2 \pi^2 n^2}{2mL^2}$. The first four position-space energy eigenstates are shown in Figure 15.22.

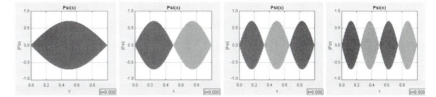

Figure 15.22   Quantum mechanics examples from the osp_demo.jar file which show the first four energy eigenstates in position space (shown in color-as-phase representation) of the infinite square well.

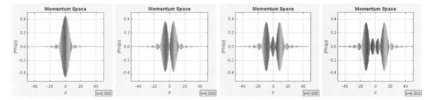

Figure 15.23   Quantum mechanics examples from the osp_demo.jar file which shows the first four energy eigenstates in momentum space of the infinite square well.

The momentum-space energy eigenstates can be determined from the position-space energy eigenstates via the standard Fourier transform:

$$\phi(p) = \frac{1}{\sqrt{2\pi\hbar}} \int_{-\infty}^{+\infty} \psi(x)\, e^{-ipx/\hbar}\, dx, \tag{15.33}$$

which yields

$$\phi_n(p) =$$

$$(-i)\sqrt{\frac{L}{\pi\hbar}} e^{-ipL/2\hbar} \left[e^{+in\pi/2} \frac{\sin[(pL/\hbar - n\pi)/2]}{(pL/\hbar - n\pi)} - e^{-in\pi/2} \frac{\sin[(pL/\hbar + n\pi)/2]}{(pL/\hbar + n\pi)} \right]. \tag{15.34}$$

The first four momentum-space energy eigenstates are shown in Figure 15.23.

In general, the time evolution of wave functions can be written as

$$\psi(x,t) = e^{-i\hat{\mathcal{H}}t/\hbar} \psi(x,0), \tag{15.35}$$

where $\hat{\mathcal{H}}$ is the Hamiltonian of the system. For the position- and momentum-space energy eigenstates of the infinite square well, the time dependence of these states are

$$\psi_n(x,t) = e^{-iE_nt/\hbar}\, \psi_n(x) \quad \text{and} \quad \phi_n(p,t) = e^{-iE_nt/\hbar}\, \phi_n(p), \tag{15.36}$$

where $\psi_n(x)$ and $\phi_n(p)$ are given in (15.32) and (15.34), respectively.

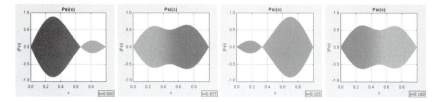

Figure 15.24 Quantum mechanics examples from the `osp_demo.jar` file which shows the first equal-mix two-state superposition ($n_1 = 1$, $n_2 = 2$) of the infinite square well at four different times.

We can consider states with nontrivial time evolution by considering equal-mix two-state superpositions, which is one of the simplest examples of nontrivial time-dependent states.[3] These position- and momentum-space wave functions are

$$\Psi_{n_1 n_2}(x, t) = \frac{1}{\sqrt{2}} \left[\psi_{n_1}(x, t) + \psi_{n_2}(x, t) \right], \tag{15.37}$$

and

$$\Phi_{n_1 n_2}(p, t) = \frac{1}{\sqrt{2}} \left[\phi_{n_1}(p, t) + \phi_{n_2}(p, t) \right]. \tag{15.38}$$

The first equal-mix two-state superposition ($n_1 = 1$, $n_2 = 2$) of the infinite square well is shown in position space at four different times in Figure 15.24.

We can examine the time dependence of more general states by adopting a general description for a time-dependent superposition:

$$\Psi(x, t) = \sum_{n=1}^{\infty} c_n \, e^{-i E_n t / \hbar} \, \psi_n(x), \tag{15.39}$$

where the expansion coefficients satisfy $\sum_n |c_n|^2 = 1$. As an example, consider an initially localized state chosen to be of Gaussian shape and of the form

$$\Psi_{(G)}(x, 0) = \frac{1}{\sqrt{b \sqrt{\pi}}} \, e^{-(x - x_0)^2 / 2b^2} \, e^{i p_0 (x - x_0) / \hbar}. \tag{15.40}$$

The expansion coefficients c_n for this superposition are determined by the integral

$$c_n = \int_0^L [\psi_n^*(x)] \, [\Psi_G(x, 0)] \, dx. \tag{15.41}$$

Such a time-dependent state in the infinite square well experiences so-called quantum-mechanical revivals in which an initially localized wave packet reforms a definite time later. The quantum-mechanical revivals for this system are exact, and any wave packet returns to its initial state after a time T_{rev}. At half this time, $t = T_{\text{rev}}/2$, the wave packet also reforms

---

[3]One of the earliest pedagogical visualizations of the time dependence of such a two-state system is by C. Dean, "Simple Schrödinger Wave Functions Which Simulate Classical Radiating Systems," *Am. J. Phys.* **27**, 161–163 (1959).

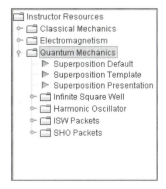

Figure 15.25 The *Explorer* pane of the `Launcher` from the `osp_demo.jar` showing the quantum mechanics curricular materials.

(same shape, width, etc.) but at a location mirrored about the center of the well with 'mirror' (opposite) momentum as well. At various fractional multiples of the revival time pT_{rev}/q the wave packet can also reform as several small copies (sometimes called 'minipackets' or 'clones') of the original wave packet.

The curricular material uses the `QMSuperpositionApp` to show wave functions in color-as-phase representation ($\hbar = 2m = 1$). In color-as-phase representation the amplitude of the wave function is depicted as the distance from the bottom to the top of the wave function at a given point and time. The phase is depicted as the color of the wave function with a map between phase angle and color. In quantum-mechanical time evolution, the phase evolves *clockwise* in the complex plane in time.

The position-space energy eigenstates and their energies are calculated either numerically for any user-defined potential energy function $V(x)$ or calculated analytically for the special cases of the infinite square well, the periodic infinite well, and the simple harmonic oscillator.

The power of `QMSuperpositionApp` is that it allows the author or user to specify a set of expansion coefficients for (15.39) in a comma-delimited list. The program then calculates the sum over states and then automatically evolves the superposition state in time. The use of analytic solutions for each eigenstate allows the simulation to run for a long period of time and yet never accumulate numerical error since the wave function is calculated anew at each time step according to an analytic formula.

Programs and Curricular Material

The quantum mechanics curricular materials use the following OSP programs in the David-son package `org.opensourcephysics.davidson`. This material is organized as shown in Figure 15.25.

qm.QMSuperposition shows the time evolution of a superposition of quantum mechanical wave functions.

qm.QMSuperpositionProbabilityApp adds a view of the position-space probability density.

qm.QMSuperpositionExpectationXApp adds a view of the expectation value of $\hat{x}$, $\langle\hat{x}\rangle$.

`qm.QMSuperpositionExpectationPApp` adds a view of the expectation value of $\hat{p}$, $\langle \hat{p} \rangle$.

`qm.QMSuperpositionCarpetApp` adds a view of the position-space quantum carpet.

`qm.QMSuperpositionMomentumCarpetApp` adds a view of the momentum-space carpet.

`qm.QMSuperpositionFFTApp` adds a view of the momentum-space wave function.

The curricular material is organized into four units:

Infinite Square Well shows the first five energy eigenstates and six two-state superpositions for a particle in an infinite square well. The eigenstates are shown in position and momentum space. The two-state superpositions are time dependent and are shown to evolve in time. The superpositions are shown in position and momentum space. In addition, one can choose to look at the expectation value of $\hat{x}$ and $\hat{p}$.

Harmonic Oscillator shows the first five energy eigenstates and six two-state superpositions for a particle in a harmonic oscillator. The eigenstates are shown in position and momentum space. The two-state superpositions are time dependent and are shown to evolve in time. The superpositions are shown in position and momentum space. In addition, one can choose to look at the expectation value of $\hat{x}$ and $\hat{p}$.

Infinite Square-Well Packets shows four initially localized Gaussian wave packets in an infinite square well. The energy scale, which determines the time scale for the animation, is set at $2/\pi$ which forces $T_{\mathrm{rev}} = 1$. This is a convenient time scale for the study of revivals and fractional revivals which occur at T_{rev} and fractions of T_{rev}, respectively. The time evolution of these states is shown in position and momentum space. In addition, one can choose to look at the expectation value of $\hat{x}$ and $\hat{p}$ and also the position-space quantum carpet.

Harmonic Oscillator Packets shows six different initially localized Gaussian wave packets (with varying initial widths and momenta) in three different harmonic oscillator wells ($\omega = 0.5, 1, 2$).

CHAPTER
16

Tracker

by Doug Brown

Tracker is a video analysis application built using the core OSP library and the optional media framework defined in the `org.opensourcephysics.media` package. Tracker source code is available in the `org.opensourcephysics.cabrillo.tracker` package.

16.1 ■ OVERVIEW

One of the goals of the Open Source Physics project is to develop a Java library that can be used both to teach computational physics and to develop robust high-quality educational software. The Tracker video analysis application described here and the Ejs modeling application described in Chapter 17 are examples of large programs that illustrate and achieve this design goal.

The development of these programs has enabled the OSP project to improve the core library in many ways. The API has been improved and made more consistent, bugs have been found and squashed, and we have learned which tools are useful to the developer community. But most importantly, the development of Tracker and Ejs has resulted in great tools for the physics education community.

16.2 ■ VIDEO ANALYSIS

The value of image and video analysis in physics education is well established. Images, such as the hydrogen spectra shown in Figure 16.1, are familiar to students and contains a wealth of spatial data. The video format adds temporal data (built-in clock) and provides a bridge between direct observations and abstract representations of many physical phenomena. Tracking the positions and colors of features in a video clip and then transforming the resulting data into real-world values and graphical overlays offers many possibilities for building and testing physical models both conceptually and analytically.

Moreover, there are effective and interactive educational video applications that require no feature tracking at all. For example, special effects filters can generate live motion diagrams of moving objects, and theoretical model animations can be overlaid directly on videos of interference patterns. In addition, saving an animation as a video is a useful way to archive, transmit, or display it easily in another application. Often, recording a video of a processor-intensive animation is the most practical way to display it at a reasonable speed.

Combining video analysis with model animations illustrates many features of the experimental method: the videos provide opportunities for objective observation, measurement,

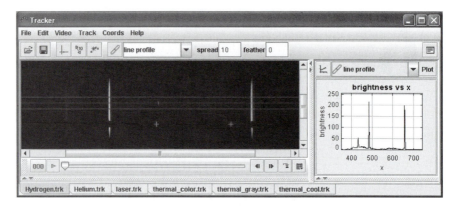

Figure 16.1 Tracker analysis of a hydrogen spectrum.

and model building, and the model animations allow theoretical predictions and comparisons with observed behavior. Tracker has been developed to facilitate this combination by extending the OSP drawing, control, and tools frameworks and thus seamlessly integrating the powerful OSP animation library into video analysis.

As its name implies, Tracker analyzes videos by tracking features of interest in the video image. The examples discussed below illustrate the types of tracks and analysis that are possible. All examples can be found on Tracker's Java Web Start page at

```
http://www.cabrillo.edu/~dbrown/tracker/webstart/
```

16.3 ■ EXAMPLES

Ball Toss

Figure 16.2 shows a tossed ball that has been tracked as a *Point Mass*. The track consists of a series of steps (green diamonds), each corresponding to the ball's position at a single time. In this example, all of the steps are visible simultaneously since "trails" have been turned on. The step associated with the current video frame is highlighted with a bold circle.

A Tracker *Coordinate System*, defined by an origin, angle, and scale, transforms the ball's raw image positions into world coordinates. The red *Axes* shows the current origin and angle, and the yellow *Tape Measure* shows the current scale. A *Plot* view and *Table* view of the ball's world data are displayed on the right.

In Figure 16.3 the Ball Toss video has been magnified and the ball's velocity and acceleration vectors have been displayed with buttons on the *Track Control*. The vectors can be (a) dragged with the mouse for side-by-side comparison or tip-to-tail addition, and (b) multiplied by the mass to display momentum and net force.

There is no limit to how many point masses can be tracked. In addition, a Tracker *Center of Mass* can track a system of masses such as colliding or exploding objects.

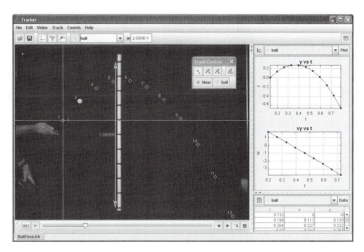

Figure 16.2   Tracker analysis of a tossed ball.

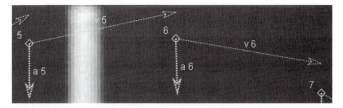

Figure 16.3   Velocity and acceleration vectors in Ball Toss video.

Pendulum

Figure 16.4 uses blue and yellow *Vectors* to track the weight and tension forces acting on a simple pendulum and a cyan *Vector Sum* that shows the time-dependent net force. For comparison, the motion of the pendulum has also been tracked and its acceleration displayed as red arrows. This example shows how Tracker enables students to extend force diagrams beyond static images and to verify their validity visually, thus, reinforcing the conceptual net force-acceleration connection. The numerical data associated with these tracks is, of course, also available for analysis.

Hydrogen Spectrum

Figure 16.1 uses a *Line Profile* track to measure the brightness of the pixels in a video image of a hydrogen spectrum. The pixel positions are converted to wavelength and the resulting spectral intensity data are plotted on the right. A *Calibration Points* track uses red and green laser spots in the spectrum to easily and accurately calibrate the wavelength scale. Students can interactively explore the spectrum in detail by dragging the line profile with the mouse.

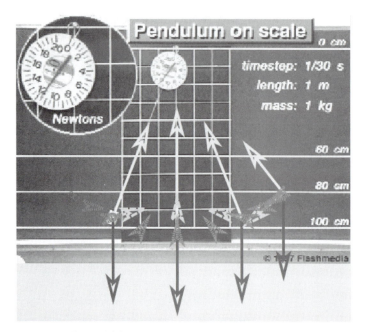

Figure 16.4   Pendulum forces and acceleration.

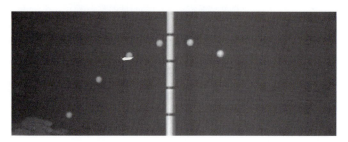

Figure 16.5   Live motion diagram.

Motion Diagram

Figure 16.5 illustrates how Tracker can create a live motion diagram simply by applying a *Ghost Filter* to a video. There are no tracks at all here! Live motion diagrams can be powerful conceptual teaching tools, particularly in interactive lectures.

16.4 ■ INSTALLING AND USING TRACKER

Tracker, like all Open Source Physics programs, requires Java 1.42 or later. In addition, you must install Apple QuickTime.[1] On Windows we recommend QuickTime 7.0 or later (after

[1]Tracker can open jpeg images and animated gifs and these file formats do not require QuickTime.

you install Java) because it automatically installs QuickTime for Java. QuickTime 6.4 or 6.5 can be used if you select custom and install all options. The latest version of QuickTime may be downloaded from

```
http://www.apple.com/quicktime/download/
```

You do not need to purchase QuickTime Pro to use Tracker. For more help refer to the "Installation" help topic on the Tracker website.

```
http://www.cabrillo.edu/~dbrown/tracker/
```

Tracker itself may be launched directly from the Web using Java *Web Start* or downloaded to a desktop. If you download the jar file, double-click `Tracker.jar` to launch Tracker.

Tracker-based curricular material may be organized using Launcher as described in Chapter 15. Launch nodes are associated with the Tracker program, Tracker xml data, and an html page. See Section 16.11 for implementation details.

Once Tracker is launched, a typical analysis session follows the steps listed below. Each of these is described in detail in its corresponding chapter section.

- A video file is opened or imported, video clip properties are set, and video filters are applied, if desired.

- Coordinate system properties are set using the axes, tape measure, and/or one or more calibration points.

- Video features are tracked by creating appropriate tracks and marking the features on each frame of interest.

- Track data are viewed and analyzed in a plot or table view. Data may be copied to the clipboard for pasting into other applications.

- The entire document (video, coordinate system, and tracks) is saved in a Tracker xml data file.

16.5 ■ USER INTERFACE

Every Tracker tab, like the "Untitled" tab shown in Figure 16.6, displays a Tracker document with the following components:

- *Menu Bar* with menus that control most program commands and settings.

- *Toolbar* that offers quick access to frequently used commands, controls, and track data.

- Main *Video View* that displays the video and tracks.

- *Video Player* that controls the video and tracks and provides access to video clip settings.

- *Split Panes* that provide access to additional views.

Figure 16.6   The Tracker user interface.

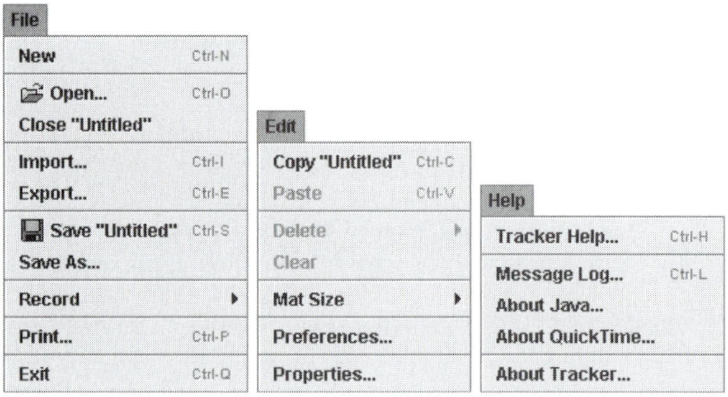

Figure 16.7   Tracker menus.

Menus

The *File*, *Edit*, and *Help* menus are shown in Figure 16.5 and described below. The *Video*, *Track*, and *Coords* menus are discussed in the corresponding chapter sections. The *Window* menu is not shown.

File menu

- *New* creates a new untitled tab with a blank Tracker document.

- *Open* opens a video or Tracker xml file in a new tab.

- *Import* imports a video or selected elements of an xml file into the current tab.

- *Export* saves selected elements of the current tab in an xml file.

- *Save* saves the entire current tab in an xml file.

Figure 16.8   Tracker toolbar.

- *Record* records the current video clip and tracks as a QuickTime movie or animated gif.

Edit menu

- *Copy* copies the selected track to the clipboard. If no track is selected, the entire tab is copied.

- *Clear* deletes all tracks in the current tab.

- *Mat Size* sets the minimum size of the drawing area. The default size is the size of the video.

- *Preferences* displays the preferences dialog as described in Configuring Tracker.

- *Properties* displays the current tab in xml format (i.e., exactly as it would be saved).

Help menu

- *Message Log* displays a message log useful for trouble-shooting.

- *About QuickTime* displays the QuickTime version and verifies correct QuickTime operation.

- *About Java* displays the version of the Java VM being used.

Toolbar

The toolbar always includes the components shown in Figure 16.8 and listed below. When a track is selected, additional data fields may be displayed.

Toolbar components shown from left to right in Figure 16.8 are as follows:

- *Open* button opens a video or Tracker xml file in a new tab.

- *Save* button saves the current tab in an xml file.

- *Axes* button shows and hides the axes.

- *Tape Measure* button shows and hides the tape measure.

- *Track Control* button shows and hides the track control.

- *Footprint* button (shown disabled) sets the color and footprint of the selected track.

- *Selected Track* drop-down selects tracks and displays/edits the name of the selected track.

- *Description* button shows and hides track and document descriptions.

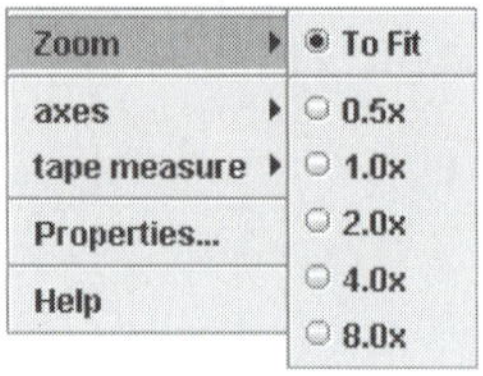

Figure 16.9   The Video pop-up menu.

Figure 16.10   Video player control.

Video View

The video view has a fixed, stable video image. All tracks and the axes and tape measure are marked and/or edited in this view.

To facilitate accurate marking, a pop-up menu (see Figure 16.9) allows you to zoom (magnify) the video image up to 8×. The pop-up menu is displayed by right-clicking the mouse (control-click on Mac); the zoomed image is then centered on the click point. The pop-up menu also provides quick access to track menus and Tracker help.

Video Player

You can drag the video player by the left end to convert it to a floating window as shown in Figure 16.10. Closing the floating window restores the player to its normal position.

Video player components (from left to right) are as follows:

- *Readout* displays the current video time, frame, or step.

- *Play/Pause* button plays, pauses, and resets the video.

- *Slider* enables you to move quickly to a desired frame.

- *Step Forward* and *Step Back* buttons step the video one frame at a time.

- *Loop* button toggles looping (continuous play).

- *Clip Settings* button displays the Clip Inspector (see Clip Inspector).

Split Panes

Split panes allow you to open additional views (see Section 16.10) by moving or clicking a thin divider. By default, the right divider opens a *Plot View* and the bottom divider opens a *World View*. Each of these also contains a split pane for a total of four views. Views can also be opened by selecting them in the *Window* menu.

Figure 16.11   The configuration preferences dialog.

Configuration

Tracker's user interface can be greatly simplified by hiding unwanted features using the preferences dialog shown in Figure 16.11. This is particularly important when introducing students to Tracker for the first time. As they gain familiarity with the program, additional features can be included as needed.

To display the preferences dialog, choose the Edit|Preferences menu item. Selected features will be enabled; unselected ones will be hidden.

There are two ways to save your preferences:

1. Click the Save As Default button. This will immediately save the currently checked items in a default configuration file named `default_config.trk`. New Tracker documents automatically load this file to configure themselves. (Note: The Save As Default button is only available when the *config.save* item is selected in the preferences. If it is hidden and you wish to enable it, select the *config.save* item and click the OK button, then reopen the preferences dialog.)

2. Select the *config.saveWithData* item in the preferences. The preferences will then be included in the xml file whenever the Tracker document is saved. Opening or importing this file will then restore the saved preferences.

16.6 ■ VIDEOS

Tracker can open QuickTime movies, animated gif files, jpeg and gif images, and image sequences. A Tracker document can have only one video opened at a time.

Videos are characterized by their pixel dimensions, frame count, and time intervals between frames (note that there is no requirement for a constant time interval between frames). All videos are displayed in 24-bit RGB format regardless of their original color depth.

It is not necessary to open a video to use Tracker. When there is no video or when the video is hidden, tracks are drawn on a white background.

Video Clips

A *Video Clip* defines a set of equally spaced steps in a video using three parameters: a start frame, a step size, and a step count. The start frame is the frame number of the first step, the

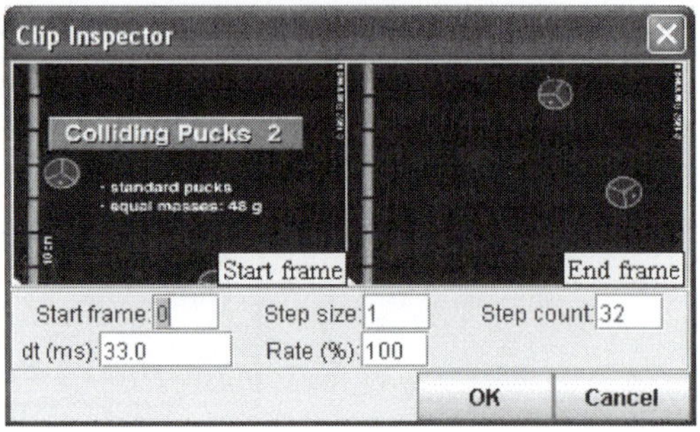

Figure 16.12   Video *Clip Inspector*.

step size is the frame increment between successive steps, and the step count is the number of steps in the clip. For example, a clip with start frame 3, step size 2, and step count 5 would have step numbers 0, 1, 2, 3, and 4 that map to video frame numbers 3, 5, 7, 9, and 11, respectively. Newly opened videos have default start frame 0, step size 1, and step count equal to the video frame count.

Note: point mass tracks work best when their steps are marked on every frame in the clip, so it is useful to set the clip properties before marking. Clip settings may be changed at any time, but previously unmarked frames will then require marking.

A clip is defined for every video and even for no video. For single-frame and null videos, the clip settings apply to tracks, but every step maps to the same video image.

Clip Inspector

The *Clip Inspector* (see Figure 16.12) shows the current video clip settings along with thumbnail images of the start and end frames. In addition, there are fields for setting the mean time *t* between video frames (important for high-speed or time-lapse videos) and the play rate as a percent of normal playback speed. To display the *Clip Inspector*, click the Clip Settings button on the player.

Playing Videos

The player plays only those frames that are video clip steps. By default, all steps are played with no frames dropped even though this often results in slower-than-normal play. If this feature is turned off, a video will attempt to play at the rate specified in the *Clip Inspector* but may drop frames.

The readout can display time in seconds (measured from the clip's start frame), step number, or video frame number. To select a readout type, click the readout and choose from the drop-down list.

Video Filters

It is often useful to process a video image in order to deinterlace, suppress noise, or enhance features of interest. Tracker provides such image processing through the use of video filters. Multiple filters may be applied to a video, in which case the filtered image from each becomes the source image for the next.

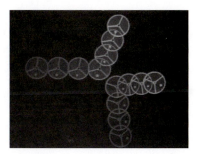

Figure 16.13 The *Ghost Filter* shows ghosts of previous video frames.

Most filters apply to each video image independently, but some, like the *Ghost Filter*, combine multiple images and are apparent only when stepping or playing the video as shown in Figure 16.13.

Tracker currently provides the following filters:

- *Baseline* subtracts a background image from each video frame.

- *Brightness* sets the video brightness and contrast.

- *Deinterlace* selects odd or even fields of interlaced video frames.

- *Ghost* leaves trails of bright objects against a dark background.

- *DarkGhost* leaves trails of dark objects against a bright background.

- *GrayScale* sets the video display to black and white.

- *Negative* reverses the video's colors.

- *Sum* adds or averages video frames.

- *Threshold* turns pixels above the threshold brightness white and below the threshold black.

Most filters have properties that are accessible through a properties dialog. The properties dialog is displayed when a filter is first created and when choosing a filter's Properties menu item.

Video Menu

The *Video* menu (see Figure 16.6) allows a user to import a video into the current tab, close an existing video, toggle the Visible and Play All Steps properties, and create and configure filters. Multiple filters are listed and applied in the order in which they are created. Existing filters can be temporarily disabled or permanently deleted. Choose Video | Filters | Clear to delete all filters.

Figure 16.14   Video menus.

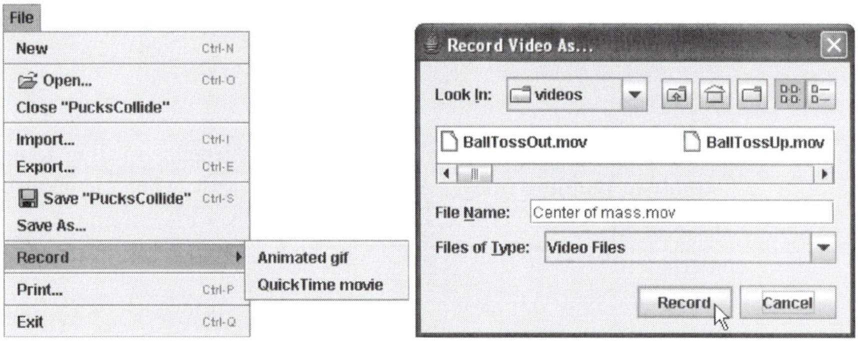

Figure 16.15   Recording menus.

Recording Videos

Tracker can record the current video clip with filters and visible track overlays as a Quick-Time movie or animated gif. The movie or gif will have the dimensions of the current viewport in the main video view. Note: QuickTime movie file sizes can be large since they are not compressed. Animated gif files are always compressed but have only 256 colors.

To record a video, choose the desired format from the File│Record menu shown in Figure 16.6. Assign a file name to the new movie or gif and click the Record button in the chooser dialog.

Tracker will reset the video to the first frame in the clip and ask if you wish to include it in the video. Click Add to record the frame and step forward, Skip to skip it and step forward, or End to end the recording without adding the current frame.

16.7 ■ COORDINATE SYSTEM

Marking a point in a video image uniquely defines its image position. But since a video image is a camera view of the real world, a marked point also has a world position relative to some laboratory reference frame. Transforming image positions into world coordinates is the job of the *Coordinate System*.

A point's image position is its pixel position on the video image, measured from the top left corner. The positive x-axis points to the right and the positive y-axis points down. Image units are like pixels except that they are doubles, not integers. This means that the center of the top left pixel has coordinates (0.5, 0.5); the image position (0.0, 0.0) is the top left corner of the top left pixel! The lower right corner of a 640×480 pixel image is at (640.0, 480.0).

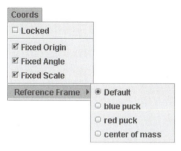

Figure 16.16 Coords menu has items to control the coordinate system.

The world position is the scaled position of the point relative to a specified world reference frame. The reference frame origin may be anywhere on or off the image and the positive *x*-axis may point in any direction. The positive *y*-axis is 90 degrees counterclockwise from the positive *x*-axis.

The coordinate system encapsulates the relationship between image and world coordinates by defining for each frame of the video

- *Origin*: the image coordinates of the world origin.

- *Angle*: the counterclockwise angle in radians from the image *x*-axis to the world *x*-axis.

- *Scale*: the world scale in image units (pixels) per world unit.

With these definitions, the coordinate system constructs a set of affine transformations, powerful Java objects for manipulating points and shapes in 2D space. Locking the coordinate system prevents any changes to the transforms.

Coords Menu

The *Coords* menu (see Figure 16.16) has items for locking the coordinate system, setting the fixed scale, origin, and angle properties of the default coordinate system, and choosing a reference frame. A fixed coordinate system property is the same in every video frame (i.e., the camera view does not change), so setting it in one frame sets it in all. For videos that pan, tilt, or zoom, deselecting an item allows (and requires) the corresponding property to be set independently for every frame in the clip.

The Reference Frame submenu enables you to select reference frames in which the origin moves along with a *Point Mass* or *Center of Mass* track (angle and scale continue to be those of the default coordinate system). Center of mass reference frames are particularly useful when studying collisions.

Axes

Axes provide a convenient way to display and set the origin and angle of the coordinate system. The positive *x*-axis is indicated by a tick mark near the origin. To display the axes, click the Axes button on the toolbar.

To set the origin, select and drag or nudge the origin to the desired location in the main video view as shown in Figure 16.17. To set the angle, click anywhere on the positive *x*-axis

Figure 16.17   The origin and angle of the axes can be set using the mouse or using a text field.

Figure 16.18   The tape measure.

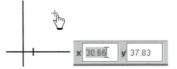

Figure 16.19   The coordinate axes origin can be offset by dragging or using a text box.

and drag or nudge the axis to rotate it about the origin. Hold down the Shift key to restrict angles to five degree increments. The angle is displayed in an angle field on the toolbar. A desired angle may be entered directly in this field.

Tape Measure

The *Tape Measure* shown in Figure 16.7 has a readout that displays the distance between its ends in world units. To display the tape, click the Tape button on the toolbar. Drag the shaft or either end of the tape to move it or to change its length or orientation.

To set the scale (calibrate the video) using the tape measure, first set the ends of the tape at positions that are a known distance apart. Then double-click the tape readout and enter the known distance.

Offset Origin

An *Offset Origin* provides a way to set the position of an origin that is not on the video image. The offset origin has a fixed offset in world units relative to the origin so that when it is dragged the origin moves with it.

To create an offset origin, choose the Track | New | Offset Origin menu item and click on a feature in the video image that has known world coordinates. The offset origin is initially assigned the current world coordinates of the clicked image feature, and the x- and y-components of the offset are displayed on the toolbar. To change the offset, enter the desired values in the toolbar fields. To move an offset origin, select and drag or nudge the origin (see Figure 16.19).

Note that changing the offset value moves the coordinate system origin so that the position of the offset origin remains unchanged. Conversely, moving the offset origin moves the coordinate system origin so that the offset (in world coordinates) remains unchanged.

Figure 16.20 Calibration points set the video frame's scale.

Calibration Points

The *Calibration Points* shown in Figure 16.20 are similar to the offset origin except that there are two points with known world coordinates. When either of the calibration points is dragged, all three coordinate system properties—scale, origin, and angle—change in order to maintain these world coordinates. Calibration points are the easiest way to set coordinate system properties when two features with known world coordinates are visible in all video frames.

To create a set of calibration points, choose the Track | New | Calibration Points menu item and click on two features in the video image that have known world coordinates. The calibration points are initially assigned the current world coordinates of the clicked image features, and their x- and y-components are displayed on the toolbar.

To change the world coordinates of either point, enter the desired values in the toolbar fields. Changing either point's world coordinates changes the scale, origin, and angle so that the image positions of both points remain unchanged.

To move a calibration point, select and drag or nudge the point. Moving a calibration point changes the scale, origin, and angle so that the world coordinates of both points and the image position of the unselected point remain unchanged.

Note: Calibration points are very powerful. It is strongly recommended to display the axes and tape measure then create and "play" with some calibration points to see how they work.

16.8 ■ TRACKS

A *Track* represents a single video feature that evolves over time. All interactive elements in Tracker, including the axes, tape measure, offset origin, and calibration points discussed above, are tracks.

The image position or shape of the feature in a single video frame is known as a step; thus, a track is a series of steps. Each step in turn consists of an array of points that can be selected and moved with the mouse or keyboard. Some steps, like those for point mass tracks, have only a single point, but others, like vector steps, have two endpoints plus a center handle point.

Every track is identified by its name, color, footprint (visible shape), and description. Newly created tracks are assigned default values for the first three properties that depend on the type of track. For example, a point mass might initially be called "mass A" and be drawn as a blue diamond. These are displayed as a toolbar item, as shown in Figure 16.21, when the track is selected.

Figure 16.21   Track name, color, and footprint.

Figure 16.22   Track Control and Track menu.

Track Control

The *Track Control* shown in Figure 16.8 is a compact floating window for creating tracks and controlling the properties of existing tracks. To display the Track Control, click the track control button on the toolbar. The Track menu duplicates some of the track control functionality and provides access to axes and tape measure properties.

Track control buttons (from left to right) are as follows:

- *Trails* button shows and hides all trails.

- *Labels* button shows and hides all labels.

- *Positions* button shows and hides all point mass positions.

- *Velocities* button shows and hides all point mass velocity vectors.

- *Accelerations* button shows and hides all point mass acceleration vectors.

- *Vectors* button shows and hides all vector tracks.

- *Stretch* button stretches all vectors.

- *Dynamics* button multiplies all motion vectors by mass.

To create a new track, select the desired track type from the New button or menu list. A newly created track is automatically selected for identification and marking.

Marking Tracks

Marking a track refers to the process of drawing its steps using the crosshair cursor shown in Figure 16.23. For point mass and vector tracks, this means stepping through the video clip and, on each frame, clicking or dragging the crosshair with the mouse. Offset origin, calibration points, and line profile tracks are marked on only a single video frame and automatically duplicated for all.

Tracks that require marking on multiple video frames have an *Autostep* property (true by default) that, for efficiency, steps the video forward after each frame is marked. Obtaining the crosshair cursor depends on the value of the *Mark By Default* property (false by default): when true, the crosshair is displayed by default on any frame that is not yet marked; when false, the crosshair is displayed only when the user holds down the Shift key.

Figure 16.23   Crosshair cursor.

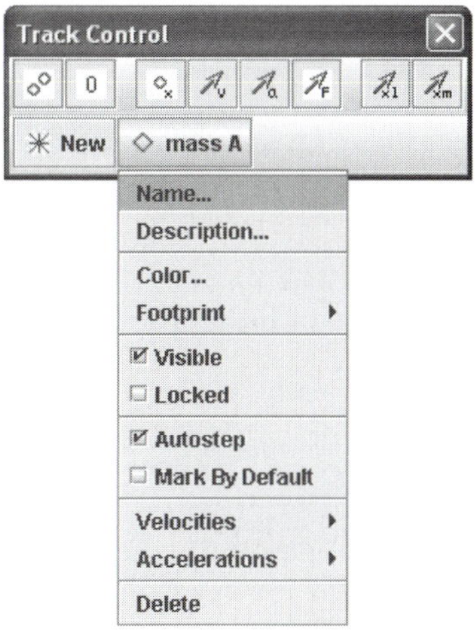

Figure 16.24   Menu for a point mass.

Once a step is marked it can be selected by clicking any of its points. This typically displays data associated with the step in a set of editable fields on the toolbar. A selected step is edited by dragging it with the mouse, nudging it one pixel at a time with the arrow keys, or entering the desired value in a toolbar field. Very fine control is possible at a high zoom level.

Every track has a *Locked* property (false by default). Locking a track prevents the creation, deletion, or modification of its steps, though its identification properties remain editable.

Personal Track Menus

Each track has its own "personal" menu that is added to both the *Track* and *Video* pop-up menus as submenus and to the *Track Control* as a button. Most of a track's properties can be set using its personal menu as shown in Figure 16.24.

16.9 ■ TRACK TYPES

The track types covered in this section are *Point Mass*, *Center of Mass*, *Vector*, *Vector Sum*, and *Line Profile*.

Point Mass

A *Point Mass* track shown in Figure 16.9 represents a mass moving as a point-like object. It is the most fundamental model of a moving inertial object. Point masses are the building

Figure 16.25   Point mass steps and mass field.

Figure 16.26   A velocity vector can be attached to a position.

t	0.10	vx	213.38	vy	316.90	v	382.04	theta	56.0

Figure 16.27   Velocity data are displayed on the toolbar.

blocks from which more complex and realistic models of physical systems are constructed in classical physics.

A newly created point mass is given a default mass of 1.0 (arbitrary units). To change the mass, enter a new value ($m \geq 0$) in the mass field on the toolbar.

Velocity and Acceleration

The instantaneous velocities and accelerations of a point mass are estimated using numerical derivative algorithms. The default algorithm for velocity is $v_n = (x_{n+1} - x_{n-1})/2$ and for acceleration is $a_n = (2x_{n+2} - x_{n+1} - 2x_n - x_{n-1} + 2x_{n-2})/7$. Note that these equations require three sequential positions to determine velocity and five to determine acceleration—if these conditions are not met, velocity and/or acceleration will not be calculated or displayed.

To display velocity or acceleration vectors, click the Velocities or Accelerations button on the track control. The vectors are drawn with dotted lines (Figure 16.26) and are initially attached to their positions (i.e., the tail of the velocity vector for step n is at the step n position).

Note: Some motion vectors, especially accelerations, may be very short. You can artificially stretch them by a factor of 2 or 4 by clicking on the Stretch button on the track control. Zooming in also increases their length.

Select a motion vector by clicking near its center to display its components, magnitude, and direction on the toolbar. The fields will be grayed out since motion vectors are not editable (Figure 16.27).

Drag a vector to detach it from its position and move it around. Drop the vector with its tail near its position to reattach—it will snap to the position. A vector will also snap and attach to the origin when the axes are visible. This is useful for estimating and visualizing its components. Attach all vectors quickly to the origin or their positions with the Tails to Origin or Tails to Position items in a track's personal menu.

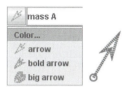

Figure 16.28   The display properties of vectors can be set.

Click the Dynamics button on the toolbar to multiply all velocity and acceleration vectors by their mass. This changes them to momentum and net force vectors, respectively. Tracker changes the labels next to the momentum and net force vectors and draws these vectors with dashed lines to distinguish them from motion vectors.

Change the footprint of a vector by first selecting it, then clicking the Footprint button on the toolbar, and choosing from the list. The "big arrow" footprint (Figure 16.9) is particularly useful for large classroom presentations.

Motion vectors, like all vectors, can be linked tip-to-tail to visually determine their vector sum (see Linking Vectors).

Center of Mass

A *Center of Mass* (cm) track represents the center of mass of a collection of point masses and is itself a point mass with the usual motion vectors. Its mass, however, is not editable, but instead is the sum of its point masses. Similarly, its steps are not marked, but instead are determined by the positions and masses of its point masses. Figure 16.29 shows the green center of mass of a pair of colliding pucks. Center of mass footprints are always solid to distinguish them from independent point masses.

Select the point masses to include in a cm by checking them in the *Select Masses* dialog. The dialog is displayed when the center of mass track is initially created or by choosing Select Masses in its personal menu.

Vector

A *Vector* track can represent any vector but is commonly used as a force in a force diagram. For traditional static force diagrams, the background video can be a still image (photo or drawing) or simply a blank screen. In Tracker, QuickTime videos present new opportunities to study force diagrams that vary with time as illustrated in the earlier Pendulum example.

Figure 16.29   The center of mass of a colliding puck system.

Figure 16.30   Marking and moving a vector.

To mark a vector, click the crosshair cursor at the tail and drag the tip with the pointer hand as shown in Figure 16.9. Vectors are drawn with solid lines to distinguish them from motion vectors. Select any point on a vector to display its components, magnitude, and direction on the toolbar. Enter a desired value in the appropriate field or select and drag/nudge a vector's tip to change the components. Drag or nudge the center of a vector to move it without changing its components.

When the axes are visible, you can drop a vector with its tail near the origin and it will snap to the origin. This is useful for estimating and visualizing its components. Attach all of a vector's steps quickly to the origin with the To Origin item in its personal menu as shown in Figure 16.9.

Linked Vectors

Vectors can be linked tip-to-tail to visually determine their vector sum as shown in Figure 16.9. To link vectors, drag and drop one with its tail near the tip of the other. The dropped vector will snap to the tip when it links. You may continue to link additional vectors in the same way to form a chain.

Note: Tracker makes no attempt to check whether it is mathematically appropriate or physically meaningful to link a given chain of vectors—it simply makes it possible.

When you drag the first vector in a chain of linked vectors, the chain moves as a unit and the vectors remain linked. When you drag a vector further up the chain, it breaks the chain into two smaller chains.

Vector Sum

A *Vector Sum* track represents the sum of a collection of vectors. Its vector steps are not marked, but instead are determined by the components of the vectors in the collection. The green vector sum in Figure 16.9 represents the sum of vector forces **A** and **B**. Vector sums are drawn with a dashed line to distinguish them from independent vectors and motion vectors.

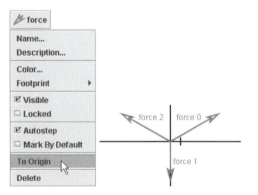

Figure 16.31   Vectors can be set to display at the origin.

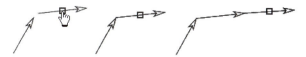

Figure 16.32 Vectors can be linked in order to show a vector sum.

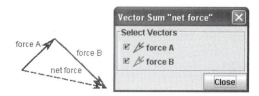

Figure 16.33 A vector sum. Forces **A** and **B** are shown linked but need not be.

Figure 16.34 Line profile tool.

Select the vectors to include in a vector sum by checking them in its Select Vectors dialog. The dialog is displayed when the vector sum track is initially created or by choosing Select Vectors in its personal menu.

Line Profile

A *Line Profile* track (Figure 16.34) is a tool for measuring brightness and other pixel data along a horizontal line on a video image. At each pixel point along the line, it can average the image pixels above and below the line to reduce noise and/or increase sensitivity.

The line profile tool is automatically drawn near the center of the video when first created. Drag either end of the line horizontally to change its length. The pixels covered by the line are the points analyzed by the line profile tool. The maximum length of the line is the pixel width of the video image (for example, 640 pixel point length for a 640×480 image). Drag the center of the line in any direction to position it.

To increase the number of pixels sampled for a smoother average, you can increase the line profile's spread and/or feather. Select the line and enter the value in pixels in the appropriate field on the toolbar as shown in Figure 16.10. The spread pixels, which extend immediately above and below the pixel point, are given the same (full) weight in the average as the pixel point itself. The feather pixels, which extend outside the spread pixels, are given a linearly decreasing weight in the average.

For a given pixel point, the total number of pixels given full weight is $1 + 2s$. The total width of the line profile in pixels is $(1 + 2s + 2f)$. The outline of the line profile shows all spread and feather pixels included in the average.

Figure 16.35   Line profile spread and feather.

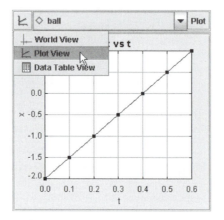

Figure 16.36   Selecting a plot view.

16.10 ∎ VIEWS

Data Views

A *Data View* displays information about the world data associated with tracks as measured by the current coordinate system. Tracker provides three standard data view types in up to four split panes. To open a split pane, drag or click on a divider or select the appropriate Window menu item. Each pane displays the view type selected from a list as shown in Figure 16.36. To display the list, click the View button in the upper left corner of the pane.

Plot Views

The *Plot View* displays one or more graphs of a track's data. The toolbar allows users to select the track. Pressing the Plot button allows a user to choose the plot layout and variables for the selected track. A double layout, for example, allows the user to display two graphs stacked vertically. Both graphs have the same horizontal variable as shown in Figure 16.10.

Graphs include both points and lines. Autoscale is on by default, but right-clicking (control-click on Mac) brings up a pop-up menu for controlling these display properties and accessing print and help menu items. Uncheck the appropriate Auto box in the Scale dialog to set a maximum or minimum value manually. Choose *Scale to Fit* to rescale the graph so all data points are visible when Autoscale is off.

Datatable View

The *Datatable View* (Figure 16.38) displays a table of a track's world data. Like the plot view, it has its own toolbar for selecting the track and choosing the variables. The data displayed in the table can be copied to the clipboard and pasted into a spreadsheet or other application.

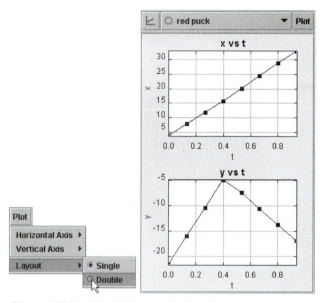

Figure 16.37 A double layout for data from the red puck track.

t	x	y
0	0.034	-0.029
0.133	5.179	-0.074
0.268	10.32	-0.054
0.402	15.487	-0
0.535	20.436	0.002
0.668	25.517	-0.048
0.802	30.534	-0.102
0.935	35.48	-0.099

center of mass ▼ Data

Figure 16.38 A datatable view of the center of mass track.

Select the data columns to be included in the table by clicking the Data button and checking those desired in the dialog displayed.

To copy data to the clipboard, first drag to select the cells of interest, then right-click the data table (control-click on Mac) and select Copy from the pop-up menu. If no cells are selected, all cells will be copied.

World View

A *World View* shown in Figure 16.39 displays the video and tracks in world space, the drawing space used by standard drawing panels. The world view has a fixed x-axis pointing to the right and a fixed scale. All tracks that are visible in the main video view (including the axes and tape measure) are also visible in the world view.

Model animations may be overlaid on top of the video and tracks in a world view enabling a direct visual comparison between theory and experiment. Right-click (control-click on Mac) to display a pop-up menu with print, help, and animation tools.

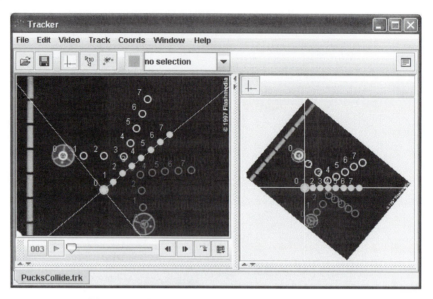

Figure 16.39   Video and world views of a collision.

16.11 ■ TRACKER XML DOCUMENTS

Tracker saves properties of the video clip, coordinate system, and tracks in xml documents with the default extension "trk" (pronounced track). When a saved file is opened, Tracker loads the specified video, sets the clip and coordinate system properties, and recreates the tracks in a new tab that displays the file name. Step positions are saved in image coordinates, so they are not suitable for direct analysis (to access the world data associated with a track, use a datatable view).

A Tracker data file is easily read and edited with any text editor. The xml format conforms to the `doctype` defined in `osp10.dtd` file in the `org.opensourcephysics.resources` package.

Importing and Exporting Data

Videos and tracks can be imported from a Tracker xml file into the current tracker tab or exported from the tab to a file using the File | Import and File | Export menu items (Figure 16.11). When importing or exporting a data file, the available elements are displayed in a dialog that allows the user to select those desired.

Tracker and Launcher

Launcher can be used to open saved xml files in Tracker. Figure 16.41 shows typical launch parameters that open the file `BallToss.trk`. Note that the specified class is TrackerPanel rather than Tracker itself. This loads a new tab into the shared (static) Tracker application, so opening multiple files does not open multiple application windows. The *Single VM* option must be checked for this to work.

Figure 16.40   Importing data from a saved Tracker xml file.

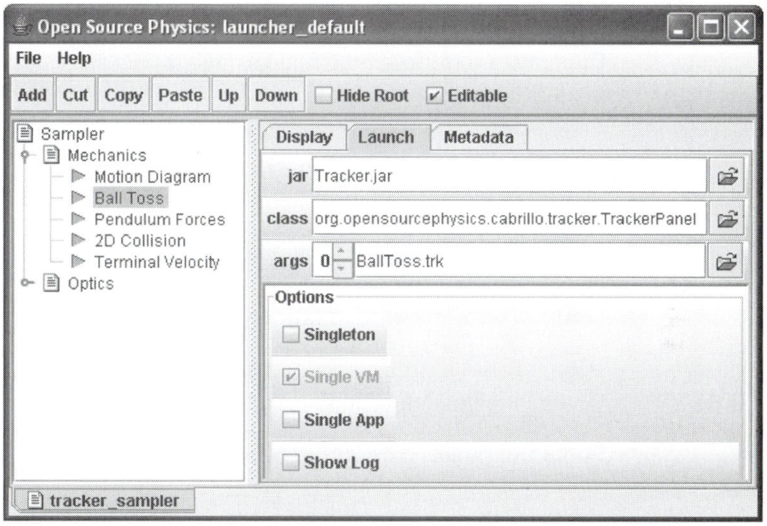

Figure 16.41   Launching Tracker with the xml document BallToss.trk.

16.12 ■ GLOSSARY

Videos

- Video: (noun) a digital file containing one or more images.

- Frame: (noun) a single video image.

- Video Clip: (noun) a subset of frames in a video defined by a start frame, step size (number of frames per step), and step count.

- Step: (noun) a frame included in a video clip.

- Filter: (noun) an image processing algorithm applied to a video image.

Coordinates

- Image Position: (noun) the position of a point measured in pixel units relative to the top left corner of the video image. In a 320×240 pixel image, the upper left corner is at image position (0.0, 0.0) and the lower right is at (320.0, 240.0). Image positions have subpixel resolution.

- World Coordinates: (noun) the position of a point measured in scaled world units relative to a reference frame.

- Coordinate System: (noun) the set of transformations used to convert image positions into world coordinates.

- Reference Frame: (noun) the set of origins and angles used to define the coordinate system. The reference frame can be changed without changing the scale.

- Scale: (noun) the number of image units per world unit.

- Calibrate: (verb) to set the scale using video features with known dimensions.

Tracks

- Track: (noun) a series of steps in time that define the position and orientation of an object or feature in a video; (verb) to follow and/or identify an object's track.

- Step: (noun) (1) the position and orientation of a track at a single time, typically specified by step number (integer ≥ 0) and (2) a video frame included in a video clip; (verb) to move forward in time to the next step number.

- Point: (noun) a single point on the video image. Points are used singularly or in combination to define and/or manipulate a step.

- Footprint: (noun) the visible shape of a step.

- Mark: (verb) to draw a track's steps by clicking and/or dragging the mouse.

17

Easy Java Simulations

by Francisco Esquembre

The Open Source Physics project includes *Easy Java Simulations*, a high-level modeling and authoring tool that can be used both by expert programmers as a fast-prototyping utility and by novices as a simple tool that will help them create their first simulations. This chapter provides an overview of this application.

17.1 ■ INTRODUCTION

The OSP library provides a complete and articulated set of Java classes and utilities for programmers who want to create their own computational physics applications in Java. For these users, the library helps speed up the creation process.

However, OSP is just too powerful to reduce its potential use only to programmers. There are a great number of creative teachers (and students) who may not be fluent enough in Java to use the libraries as provided but who could, if given the chance, develop effective and useful physics simulations.

For this reason, the OSP project aimed, from the very beginning, to find a way for non-programmers to access the concepts and to use the products of the OSP library. *Easy Java Simulations* is the answer to this goal.

Easy Java Simulations (Ejs for short) is a software authoring tool created in Java which sits on top of OSP libraries. It provides a simplified entry point for those who want to create Java applications or applets that simulate physical phenomena.

If you are an experienced Java programmer already or are on the way of becoming one, you may wonder why you need such an authoring tool when you can program directly. Even though it's true that direct programming gives you full control, before you decide to skip this chapter completely, consider that you may want to learn more about a tool that can help you to

- Quickly develop a prototype of an application in order to test an idea or algorithm.

- Boost the creation of sophisticated user interfaces with minimal effort.

- Create simulations whose structure and algorithms other people (especially nonprogrammers) can easily inspect and understand.

- Invite your students or colleagues (who may be new to Java) to create their own simulations.

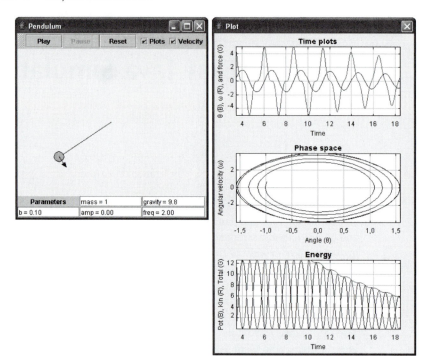

Figure 17.1   A simulation of a simple pendulum created with *Easy Java Simulations*. A damping term was introduced approximately at time $t = 10$ seconds.

- Automate production tasks such as preparing your simulations to be distributed using Web pages or Java Web Start technology.

This chapter provides a general description of *Easy Java Simulations* but does not explain all of its features, such as the installation of the software and the Web distribution of the generated simulations. A detailed manual is, however, available on the OSP CD and on the Ejs website. Instead, this chapter is a short survey that describes how the tool works and how it simplifies the creation of professional OSP simulations.

We will illustrate how to use Ejs by working with one of the most famous physics examples of all time, the simple pendulum. To get acquainted with the main features of Ejs, we will begin by loading, inspecting, and running an existing basic simulation of this physical system. We will then extend the model to include damping and forcing and to improve the visualization of the phenomenon by including phase-space and energy graphs. Figure 17.1 shows the user interface of the simulation we will obtain once we have finished working with it.

17.2 ■ THE MODEL-VIEW-CONTROL PARADIGM MADE SIMPLER

We have mentioned the model-view-control (MVC) paradigm earlier in this Guide (see Section 1.2). This paradigm was introduced as a way of structuring the different parts of

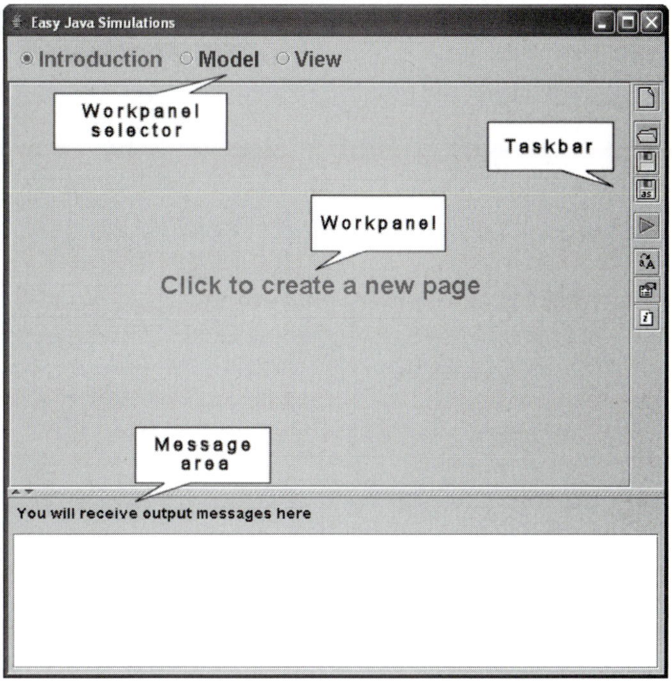

Figure 17.2   *Easy Java Simulations* user interface (with annotations).

a computer program and has proved to be successful in creating versatile applications in a clean, well-defined, and relatively simple way.

Easy Java Simulations uses this same paradigm. It structures a simulation in two main parts: the model and the view. The model is the collection of variables that define the different possible states of the system under study together with the algorithms that describe how these variables change in time or how they respond to user interaction. The view is the generic name that Ejs adopts for both the visualization of the phenomenon and the user interface of the application.

In essence, the Ejs view combines the control and view of the MVC paradigm. This simplifies things for beginners because in modern computer simulations the control is achieved through interaction of the user with the application's graphical user interface.

Because our simulations have mainly a pedagogical purpose, we will add to the model and view a textual (multimedia) part that is designed to contain a short introduction to the simulation or operating instructions for the user. Thus, an Ejs simulation consists of three main parts: the introduction, the model, and the view. Figure 17.2 shows the interface of *Easy Java Simulations* (to which we have added some notes), which reflects this structure.

You are invited to run Ejs while you read this chapter. Although you can find detailed instructions on installing and running it in the manual included in the companion CD, here are some quick guidelines (for those who don't like reading manuals!). To install and run Ejs, follow the next steps.

Install Java 2 JDK.   *Easy Java Simulations* is a Java program that compiles Java programs, hence it requires that you install Java Development Kit (Java 2 JDK) in your computer. The recommended version at the moment of this writing is 1.5.0_04.

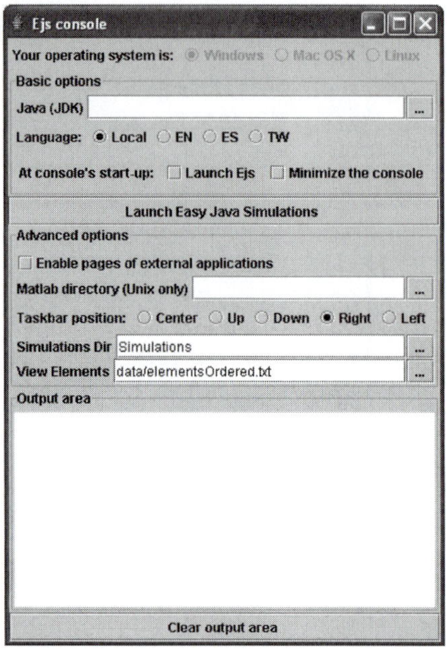

Figure 17.3   The Ejs console that will help you run *Easy Java Simulations*.

Copy Ejs to your hard disk. Installing Ejs requires only copying the **Ejs** directory to your hard disk. This directory is usually distributed as a zip (compressed) file. Just uncompress this file to any suitable directory. (In Unix-like systems, the directory may be uncompressed as read-only. In this case, please enable write permissions for the Ejs directory and subdirectory tree.)

Run the Ejs console. *Easy Java Simulations* is now installed. The distribution includes an `EjsConsole.jar` file that runs a console that lets you do the final configuration for the installation and to start the Ejs program. Run this file by either double-clicking on it (if your system allows you to run it this way) or typing the command

```
java -jar EjsConsole.jar
```

at a system terminal window. You should get the window shown in Figure 17.3.

Tell Ejs where you installed the JDK. (This is only required in Windows and Linux.) Because you may have installed Java 2 JDK in any directory, you will need to enter this directory in the `Java JDK` field of the console.

Run Ejs. Just click on the Launch Easy Java Simulations button of the console.

That's it! You should get the interface displayed in Figure 17.2. Again, you will find more detailed instructions in the CD and also on the Ejs home page `http://fem.um.es/Ejs`.[1]

---

[1]For operating system purists, Ejs installation includes three script files that launch Ejs in the different platforms: `Ejs.bat` for Windows, `Ejs.macosx` for Mac OS X, and `Ejs.linux` for Linux. Before running the file corresponding to your operating system, you may need to slightly edit the script to correctly set the `JAVAROOT` variable (which is defined in the first lines of the script) that points to the directory where you installed the Java JDK.

As you can see in Figure 17.2, the interface of Ejs is rather basic. This was a deliberate design decision. We wanted to make clear from the very beginning that *Easy Java Simulations* is simple. Hence, we avoided providing a large and overwhelming number of icons and menu entries in the Ejs interface. Despite its simplicity, however, the tool has everything that it needs to build controls, models, and views.

We start exploring the interface by looking at the set of icons on the right-hand side taskbar of Figure 17.2. The taskbar provides an icon for each of the main functionalities of Ejs. We will explain the purpose of some of these icons in this chapter, but their meaning and use should be rather natural.

You will also notice in Figure 17.2 that there is a blank area at the lower part of the window with a header that reads, "You will receive output messages here." This is a message area that Ejs uses to display information about the results of the actions we ask it to take.

Finally, the most important part of the interface is the central area of the interface (labeled the `Workpanel` in the figure) and the three radio buttons on top of it, Introduction, Model, and View. These radio buttons are used to display the introduction, the model, and the view in the workpanel.

17.3 ■ INSPECTING AN EXISTING SIMULATION

We can better understand the role of the different parts of a simulation and the utility of the different panels of Ejs by loading and inspecting an existing simulation. We choose for this a basic implementation of the simple pendulum that we created for your perusal and that is included in the distribution of Ejs in the companion CD.

If you click on the `Open` icon of the taskbar, 🖼, a file dialog box will open from which you can load a simulation file (see Figure 17.4).

The file dialog box first displays the contents of the Ejs working directory called `Simulations`. From there you can access any directory or file in the standard way. Open the directory called `_ospguide` and select from it the file called `PendulumBasic.xml`. Click the Open button and Ejs will load a basic version of the simulation of a simple pendulum.

The interface of Ejs will change noticeably. It will load the different parts of the simulation file in the corresponding panels, allowing us to see how the simulation has been designed. It will change its title to include the name of the simulation file, and finally it will also display a message in the message area, reporting that the file has been loaded.

Two new windows will also appear. You can see them in Figure 17.5. They correspond to the view of the simulation, which we will describe a bit later.[2] For the moment, however, we will concentrate on the interface of Ejs itself.

What we see in this interface depends on which part of the simulation was visible when we loaded the file. Because we are interested in learning how to use Ejs to specify a real simulation, we will inspect the different parts of the simulation in turn by browsing the different panels of Ejs.

---

[2]To be more precise, these two windows are, as their titles state, Ejs windows. These Ejs windows enable the author to configure the view. Hence, they are really mock-ups of what the real view will look like, but they are not operative. Notice also that this view is different from the view of the final simulation we will obtain at the end of this chapter, the one displayed in Figure 17.1.

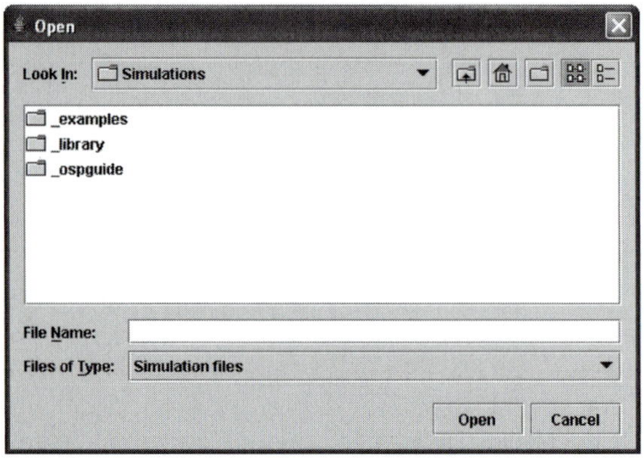

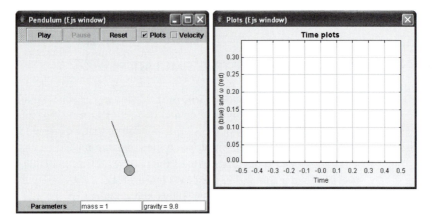

Figure 17.4   File dialog box used to load an existing simulation file.

Figure 17.5   The two windows for the view of the simulation.

The Introduction

The first panel is the introduction of the simulation, shown in Figure 17.6, which consists of an html page with a short introduction to the problem. This part of the interface of Ejs provides a simple editor that can be used to both visualize and edit standard html pages. Right now, we see the editor in its read-only mode. We will learn in Section 17.5 how to set it to edit mode and actually change its contents.

The Model

The second part of the simulation, the model, is more interesting. As you may expect, this is the part where all the physics goes. The physical model of a simple pendulum without friction and without an external driving force is contained in the differential equation

$$\frac{d^2\theta}{dt^2} = -\frac{g}{l}\sin(\theta),\qquad(17.1)$$

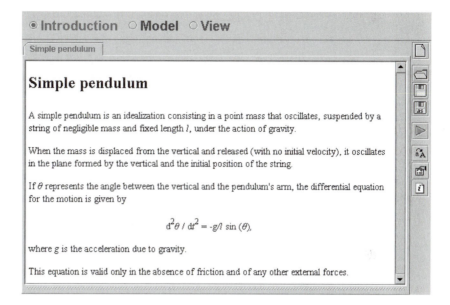

Figure 17.6 The introduction panel for the simulation of a simple pendulum.

where the variable θ corresponds to the angle between the pendulum's arm from the vertical, g is the acceleration due to gravity, and l is the length of the pendulum.

This second-order, nonlinear, ordinary differential equation (ODE) must be solved numerically because there is simply no way to express its solution in terms of elementary functions.[3] To this end, we need to express the second-order differential equation as an equivalent system of two first-order ODEs, introducing the auxiliary variable ω, the angular velocity:

$$\frac{d\theta}{dt} = \omega \tag{17.2}$$

$$\frac{d\omega}{dt} = -\frac{g}{l}\sin(\theta). \tag{17.3}$$

Solving this system of ODEs for the given parameters will give us the evolution of the angular position (and velocity) of the system in time.

The model offers us a set of five subpanels that can be used to accomplish the tasks needed to solve this problem. These five subpanels are: Variables, Initialization, Evolution, Constraints, and Custom. In the Ejs pendulum example, select the Model radio button in order to view the model panel. We now consider the five subpanels one after the other.

Declaring the Variables
In the context of a computer program or simulation, the state of a physical system is determined by the values of a set of variables. Accordingly, to begin modeling our simple

---

[3]To be more precise, the case we consider here in the absence of friction and of any other forces can be explicitly solved using elliptic functions. However, we adopt the numerical approach right from the start in order to be able to deal later with the more general case.

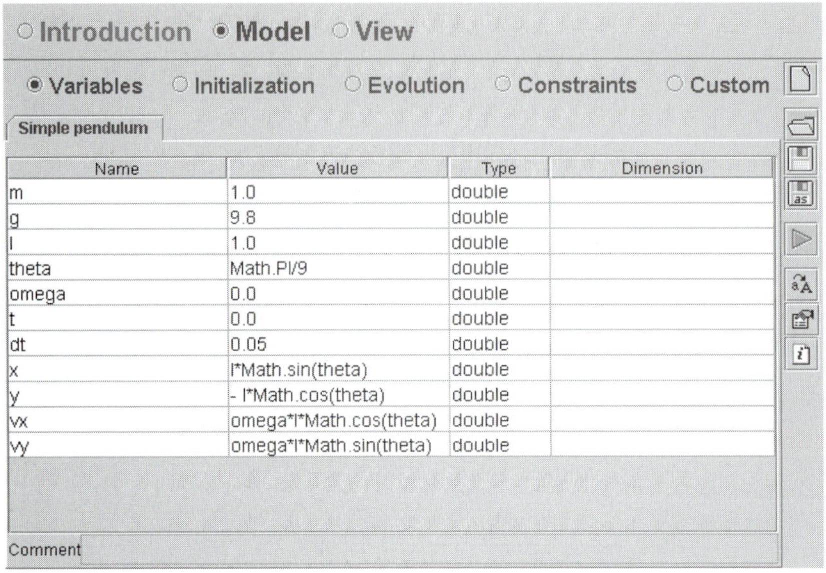

Figure 17.7 Table of variables that describes the state of a simple pendulum.

pendulum, we must provide Ejs with the set of variables that defines the state of the physical system. This is done by editing a simple table of variables in the Variables subpanel of the model part of the interface of Ejs (see Figure 17.7).

In this table we have declared all the variables involved in the differential equations as well as other parameters, such as the mass m (which we don't need for the moment, but that we will need later on) and the time interval dt at which we want to obtain information from the system. Finally, you will notice that we have also declared the variables x, y, vx, and vy that hold the position and the velocity of the pendulum's bob. They will serve to configure the view to display a realistic visualization of the pendulum.

If you want information about the role of a particular variable, you can select it in the table and the comment field at the bottom of the page will display a short description of that variable.

Initializing the Model

The system must be initialized to a valid state before letting the time run. There are two basic ways of initializing variables in Ejs. The first one is using the corresponding cells of the Value column in the table of variables. Just type a constant value, or a simple expression, and it will be assigned to the variable at start-up. This is the option we have used for this simulation.

A second possibility, which is required if your program needs to do some more complex computations to initialize the system, is to use the Initialization subpanel provided by Ejs. In this subpanel you can type the Java code for the algorithm that will compute the correct values for the variables that require these extra computations. Ejs helps to keep this process simple, because you just need to write the Java sentences that contain the algorithm for your computations, and Ejs will automatically wrap these algorithms into a Java method and will take care of calling it at start-up or whenever you reset the simulation.

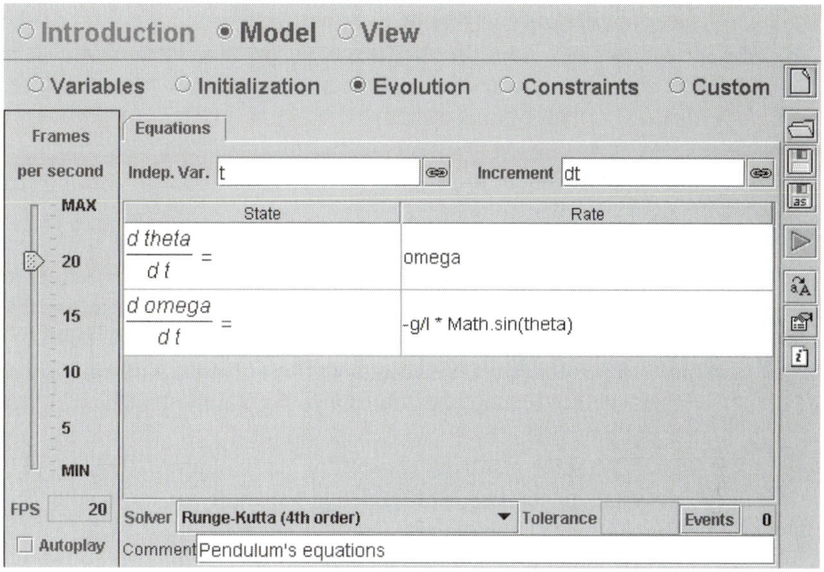

Figure 17.8   Equation editor for our simulation with the pendulum's system of ODEs.

Because the system has been completely initialized in the table of variables, the initialization panel remains empty. We'll have the opportunity to show how to write such a page of Java code when we cover constraints.

The Evolution of the Model

Specifying the evolution of the model consists of providing the algorithms that describe how the state of the system changes in time. That is, what happens to the values of the variables of the system when it evolves in time? In our case, this corresponds to solving numerically the differential equations of the model.

Ejs offers two possibilities for this. The first one is a plain editor for Java code where the author can write directly the numerical algorithm required. You would typically use this option if your main interest is simply numerical algorithms. However, because it is a common situation that the evolution of a system is specified by a system of ODEs, and writing the code to solve this type of problem with accuracy can be tedious (if not difficult), Ejs also offers a dedicated built-in editor. This editor allows us to easily enter the system of ODEs, and it automatically generates the Java code (based on OSP numeric classes) corresponding to some of the most popular numerical algorithms to solve the equations.

This second possibility is the one we chose for our system. If you inspect the Evolution subpanel of Ejs, you will see that it contains our system of ODEs in this specialized editor, as shown in Figure 17.8.

One advantage of this editor is that it displays the system of ODEs in a very familiar way, one that your students can easily recognize and that is easy to understand and modify. The second advantage is that the editor takes care of the worst part of solving the equation: coding the algorithm. Solving differential equations numerically is a sophisticated task, but Ejs (with the help of OSP classes) has automated it in a very convenient way.

You will notice in Figure 17.8 that we have chosen `t` to be the independent variable in the model and `dt` to be the increment for it in each evolution step. This means that the evolution page should solve the equation to move from the current instant of time at `t` to the next instant of time at `t+dt`.

We have chosen the standard fourth-order Runge–Kutta algorithm, as the Solver field immediately below the equations indicates. You will also notice that there are two extra fields, one called Tolerance (that is used only for adaptive algorithms) and one called Events. This last field is used to define and handle state events, such as collisions, that may occur in the life of the differential equation. (Events are described in detail in the manual on the CD.) In our case, the simple pendulum model includes no events.

All together this page describes, using the editor for ODEs, what will happen as time passes: the differential equations will be solved numerically for an increment of `dt` of the independent variable. The controls you see in the left-hand side of Figure 17.8 are used to tell Ejs how often the evolution should take place when the simulation runs. As the figure shows, we have instructed Ejs to run the evolution 20 times, or frames, per second. This, together with the value of 0.05 that we used for `dt`, results in a simulation which runs (approximately) in real time.

Finally, there is a checkbox labeled Autoplay. This instructs Ejs to run the evolution as soon as the program runs. We left it unchecked because we want to offer the user the possibility of changing the position of the pendulum and then to click on a button that will start the animation.

Constraints Among Variables

The fourth subpanel of the model is called Constraints. This panel is used to write Java code that establishes fixed relationships among variables. Consider again the pendulum example as shown in Figure 17.9.

The evolution of our model solves the ODEs in terms of the angular magnitudes `theta` and `omega`. However, we are interested in displaying on the screen the actual position of the pendulum and its velocity vector, and for this we need the corresponding Cartesian coordinates of both position and velocity. These can be easily derived from the angular magnitudes using the formulas

$$x = l \sin(\theta) \tag{17.4}$$
$$y = -l \cos(\theta) \tag{17.5}$$
$$v_x = \omega l \cos(\theta) \tag{17.6}$$
$$v_y = \omega l \sin(\theta). \tag{17.7}$$

This is what we call "fixed relationships among variables." This means that once we know the values of l, θ, and ω, the other variables can be easily obtained using the expressions above. Our constraints consist of translating these expressions into Java so that Ejs can use them whenever it is needed.

Notice that the page contains only the Java code for our expressions. Ejs will take care of wrapping this code into a Java method and automatically calling this method whenever it is necessary. This simplifies our programming task.

Now comes a subtle point. One of the questions most frequently asked by new users of Ejs is "Why don't we write these equations into the evolution instead of in the constraints?"

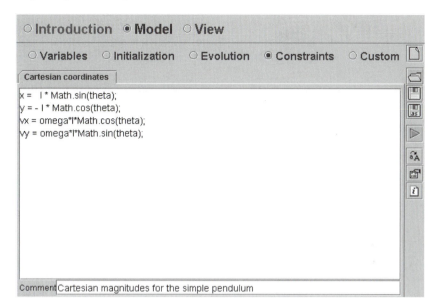

Figure 17.9   Constraints pages that compute the Cartesian magnitudes of a simple pendulum.

The reason is that this relationship among variables must *always* hold, even if the evolution is not running. It could very well happen that the simulation is paused and the user interacts with the simulation to change the angle `theta` (or `omega`, or `l`) variable. If we write the code to compute the Cartesian magnitudes in the evolution, the values for the variables `x`, `y`, `vx`, and `vy` will not be properly updated because the evolution is only evaluated when the simulation is playing.

Constraint pages, on the other hand, are always automatically executed after the initialization (at the beginning of the simulation), after every step of the evolution (when the simulation is playing), and each time the user interacts with the simulation. Therefore, any relationship among variables that we code in here will always be verified.

> Note: You could argue here that since constraints are also evaluated at start-up, there was no reason to initialize the variables `x`, `y`, `vx`, and `vy` in the table of variables, since the evaluation of the constraints will assign them the correct values at start-up. You would be almost right. There is still a reason for doing what we did, though.
>
> Ejs doesn't evaluate constraints itself (the generated simulation does), but it evaluates expressions in the Value column of the table of variables. Hence, we wrote the initial expressions for these variables so that the pendulum's bob appears at the right place in the mock-up of the view displayed. But it is true that, in general, constraints can also be used to initialize variables.

Custom Pages of Code

The final subpanel of the model is called Custom. This panel can be used by the author of the simulation to define his or her own Java methods. Different from the rest of the panels of the model, which play a well-defined role in the structure of the simulation, methods created in this panel must be explicitly used by the author in any of the other parts of the simulation. Again, in our example this panel is empty and is not displayed.

> Note: Although Ejs is designed to make programming as simple as possible and includes the typical tools an author may need, it also opens a way for programmers to use their own Java libraries. The custom panel of the model offers a simple mechanism to add external Java archives of compiled classes, packed in jar or zip form, to your simulation. The procedure is explained in detail in the Ejs manual.

The Model as a Whole

Our description of the model is ready, and we can look at it as one unit to describe the integral behavior of all the subpanels of the model.

To start the simulation, Ejs declares the variables and initializes them, using both the initial values specified in the table of variables and whatever code the user may have written in the initialization subpanel. At this moment Ejs also executes whatever code the user may have written in the constraint pages. This is so that all dependencies between variables are properly evaluated. The system is now correctly initialized.

When the simulation plays, Ejs executes the code provided by the evolution subpanel and, immediately after, the possible constraints (for the same reason as above). Once this is done, the system will be ready for a new step of the evolution, which it will repeat at the prescribed speed (number of frames per second).

As mentioned already, note that methods in the custom subpanel are not automatically included in this process.

This simple mechanism provides a basic, but very effective, structure for novices (and experts!) to build their simulations. The author just fills in the subpanels of the model, usually from left to right, and Ejs handles the pieces automatically taking care of all technical issues required (such as multitasking and synchronization).

The View

We now turn our attention to the view of the simulation. Recall that two new windows appeared when we loaded the simulation (see Figure 17.5). To learn how this view has been constructed, we can inspect the View panel of Ejs. Figure 17.10 displays this panel where two frames each with several icons are shown. The frame on the right-hand side displays the set of graphical elements grouped by functionality that Ejs offers to authors for the creation of a view. The frame on the left-hand side shows the actual elements that have been used for this particular simulation.

The view panel of Ejs can be considered as an advanced drawing tool which specializes in the visualization of scientific phenomena and its data and user interaction. Obviously, to completely master the creation of views, an author needs to become familiar with all the graphical elements offered and what they can do. (A description of all possible view elements is out of the scope of this chapter, but a complete reference can be found on the companion CD.)

Creating a view with *Easy Java Simulations* takes two steps. The first step is to build a tree-like diagram of the objects (elements) that will make up the user interface. Each

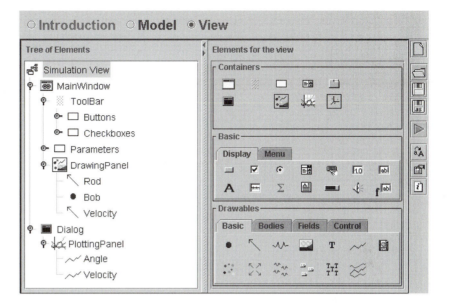

Figure 17.10   Tree of elements for the simulation (left) and set of graphical elements (right) of Ejs.

element is designed for a given graphical or interactivity task, and our job is to select the elements that we need and combine them appropriately to build our view. Selecting and adding new elements to a view is done in a simple "click-and-use" way which we will illustrate in Section 17.5.

Some elements are of a special family called *containers* which can be used to group other elements, thus forming the tree-like structure of the elements shown in the figure.

The second step, less evident but also simple, consists of customizing the selected view elements by editing their so-called *properties*. Properties are internal fields of an element that can be changed to make the element look and behave in a particular way. The key point is that properties can be given constant values (for instance to customize fonts and colors), but they can also be *linked* to variables of the model (typically for positions, sizes, and labels with numerical displays).

Because linking is a two-way mechanism, this second possibility is what really turns the view into a dynamic, interactive visualization of the physical phenomenon. Hence, once an element property is linked to a model variable, any change in the variable (due for instance to the evolution of the model) is automatically reported to the view element which changes its graphical aspect accordingly. But also, if the user interacts with any view element to modify any of its properties (typically doing a gesture with the mouse or keyboard), the change is automatically reported back to the model variable, therefore changing the state of the system.

This basic mechanism is a very simple and effective way to design interactive user interfaces. Let us see an example of how it works in practice. Select the panel for the view and right-click on the element called Bob of the tree of elements of our simulation. This element corresponds to the circle displayed as the bob of the pendulum. The pop-up menu shown in Figure 17.3 will appear.

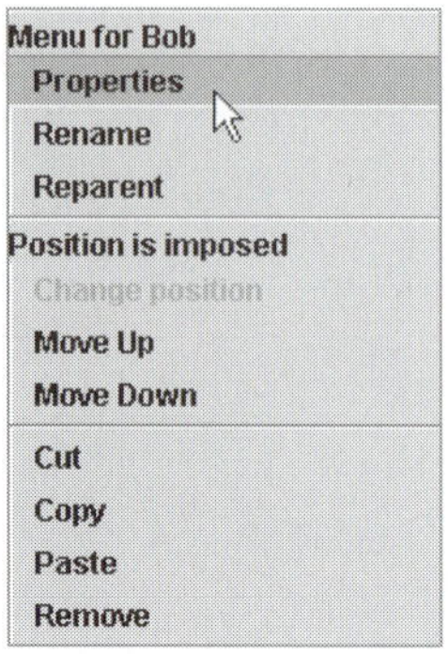

Figure 17.11   Pop-up menu for the element Bob of the view.

Select the Properties option from this menu (the one highlighted in the figure) and a new window will appear with the table of properties for this element.[4] All that is required is to edit this table according to our needs. Figure 17.12 reflects the properties we defined for this particular element.

This table of properties illustrates very well all the possibilities offered by element properties. In the first place, you can see in the figure that some properties are given constant values, for instance, those which specify the color, drawing style, and size of the element (which will be displayed as a cyan-colored ellipse of size (0.2,0.2) units). You can also see that the element is enabled, that is, that it will respond to user interaction.

Secondly, you will observe that other properties of the element are given the value of model variables. In particular, the properties X and Y, which correspond to the center of the circle in its parent drawing panel, have been linked to the model variables x and y by the simple fact of typing their names in the corresponding property field. This connection is the magic. Automatically, the circle will move according to the successive values of the model variables x and y when the simulation runs. Also, if the user drags the pendulum's bob with the mouse, the model variables x and y will be changed accordingly.

Every time the user interacts with the simulation's view, Ejs will automatically execute the constraints of the model. This ensures that any change that the interaction caused to variables linked to properties is correctly propagated to other variables which may depend on them.

Finally, there is a third type of property that needs to be taken care of for this example to work properly. Recall that the model primarily computes the values of the angular magnitudes, and that we wrote constraints to make sure that the Cartesian magnitudes (which include x and y) were readily computed to match them. This means that if we interact with the pendulum's bob to change the values of x and y, the change will be overwritten by the constraints unless the change affects the angular variables. This can be easily taken care of by means of the (somewhat special) *action* properties of the element. These correspond to pieces of Java code that the computer will evaluate whenever the prescribed interaction takes place.

In our case we need to edit the On Drag action property, which is evaluated every time we drag the element on the screen. We have set this property to execute the following sequence

---

[4]Double-clicking on the element's node in the tree is a shortcut for this option.

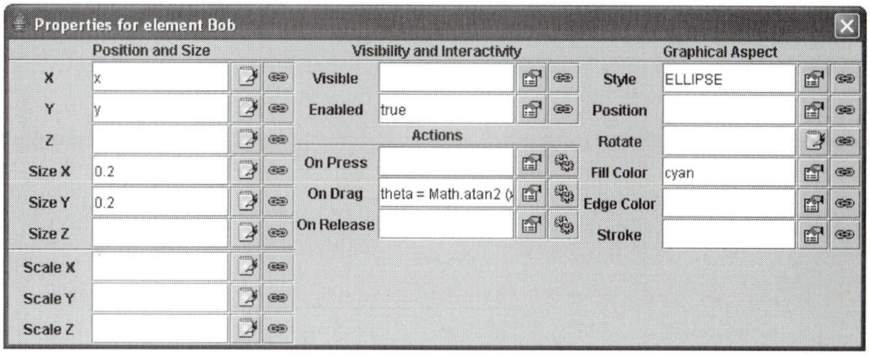

Figure 17.12   Table of properties for the view's Bob element.

of sentences:[5]

```
theta = Math.atan2(x,-y);
l = Math.sqrt(x*x + y*y);
omega = 0.0;
```

The first two statements use the (new) values of the x and y variables to compute the new length of the pendulum and its angular position. We have added a third sentence that sets the angular velocity to zero, because we want the motion to restart from rest. All together, these sentences help set up a new state for the model of the system.

The three types of customization we did to the properties of this view element illustrate how *Easy Java Simulations* combines the creation of the control and the visualization tasks of a simulation in one single process (the view). By using constants, model variables, and Java expressions and statements, we can configure a user interface that is used simultaneously to display data, to provide input, and to execute control actions on the simulation.

You can inspect the properties of other elements of this view to become familiar with the different types of view elements offered by Ejs and their properties. Of particular interest are the elements called Angle and Velocity, which correspond to time plots of these magnitudes. For each element, you can click on , the first button to the right of each property box, to bring up an editor for that property. For example, the Style property editor will allow you to select the shape of the element, and the Fill Color property editor will allow you to select the color from a palette.

17.4 ■ RUNNING A SIMULATION

Once we have inspected the different parts of the simulation, we are ready to run it. Here goes a warning that may arrive too late if you are running Ejs while reading these pages: Don't try to run the simulation by clicking on the buttons of the windows in Figure 17.5! Recall that these were just "Ejs windows." Hence, they are not part of the real simulation but just a mock-up of what the real view will look like. Their purpose is to help the author design the view, but they are not operative.

---

[5]Unlike other properties, action properties can span more than one line. When this happens, the property field changes its background color slightly. To better display this code, we can click on the first button to its right, 🖾 , and an editor window will display it more clearly.

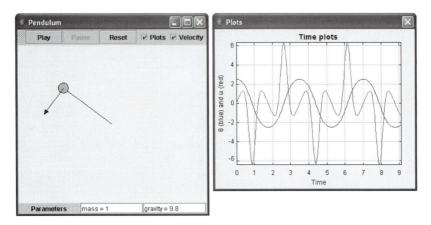

Figure 17.13   The simple pendulum displaying oscillations with a big amplitude. The largest plot corresponds to the angular velocity.

To actually run the simulation, you need to click on the Run icon of the Ejs toolbar, ▷. When you do this, *Easy Java Simulations* collects all the information provided in its panels and subpanels and constructs a complete, independent simulation out of it, taking care of all the required technical subtleties. This includes generating the complete Java source code for the simulation, compiling it, and packing it into a single jar file. Finally, it also runs this file, which will initialize the model and display the simulation's view in the computer screen.

This view is now fully interactive. You can drag the pendulum to any desired position, change any of its parameters, and then click the Play button. The pendulum will oscillate as the model solves the underlying differential equations. The view will display both the pendulum's oscillations and the time plot of the position and (optionally) the velocity (see Figure 17.13).

Simulations created with *Easy Java Simulations* are independent of it once generated. This means that final users don't need to install Ejs to run the simulations. They simply double-click the jar file. Moreover, when you have successfully run a simulation once, you'll find everything you need to run and distribute the simulation in the Simulations directory. We now briefly discuss the different ways to run a simulation created with *Easy Java Simulations* together with some remarks about its distribution.

Running the Simulation as an Application

After running the simulation, you will find in the Simulations directory a jar file with the same name as the simulation file we first loaded but with its first letter in lower-case to conform with Java programming custom. Recall that the name of this file was PendulumBasic.xml. Hence, you will find there a jar file called pendulumBasic.jar.

This is a self-running jar file. Hence, if your operating system is properly configured, you will be able to run the simulation by double-clicking on the jar file icon. (Windows and Mac OS X are usually automatically configured for this if the Java runtime environment is installed.) If double-clicking won't work, then you can still use the launch file called

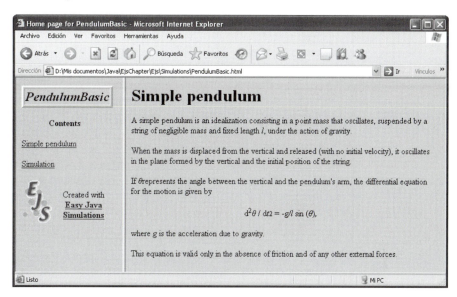

Figure 17.14   The set of html pages created for our simulation.

PendulumBasic.bat. This has been generated by Ejs specifically for your operating system and can be executed like any other batch (Windows) or shell script (Unix-like systems) on your operating system.

This jar and launch file will work correctly assuming that you have the Java runtime environment (JRE)(JRE) installed in your system and that you run them from within the Simulations directory. If you want to move your simulation to a different directory or computer, read the distribution notes below.

Running the Simulation as an Applet

A second possibility to use the simulations created with Ejs is to run them as Java applets from within html pages. This is a very attractive possibility because it opens the world for distributing simulations over the Web (see also the note about Java Web Start technology on page 333).

This possibility is also taken care of by *Easy Java Simulations*. Every time you run a simulation from Ejs, it also generates the html pages required to wrap the simulation in the form of an applet. More precisely, it will create a complete set of html pages, one for each of the introduction pages (those in the introduction panel of Ejs) and one that contains the simulation as an applet. It finally creates a master html file that structures all of the others using a simple set of frames.

All these html files are easily identified because they begin with the same name as your original simulation file. In particular, the name of the master file is the same as that of the simulation file—in our case, PendulumBasic.html. If you load this file in a Java-enabled browser, you will see something like Figure 17.14.

Notice that the frame on the left displays a table of contents that includes the introduction page that we created for our simulation (which is also shown in the right frame) and a second link for the simulation itself. If we click on this link, the frame to the right will change to

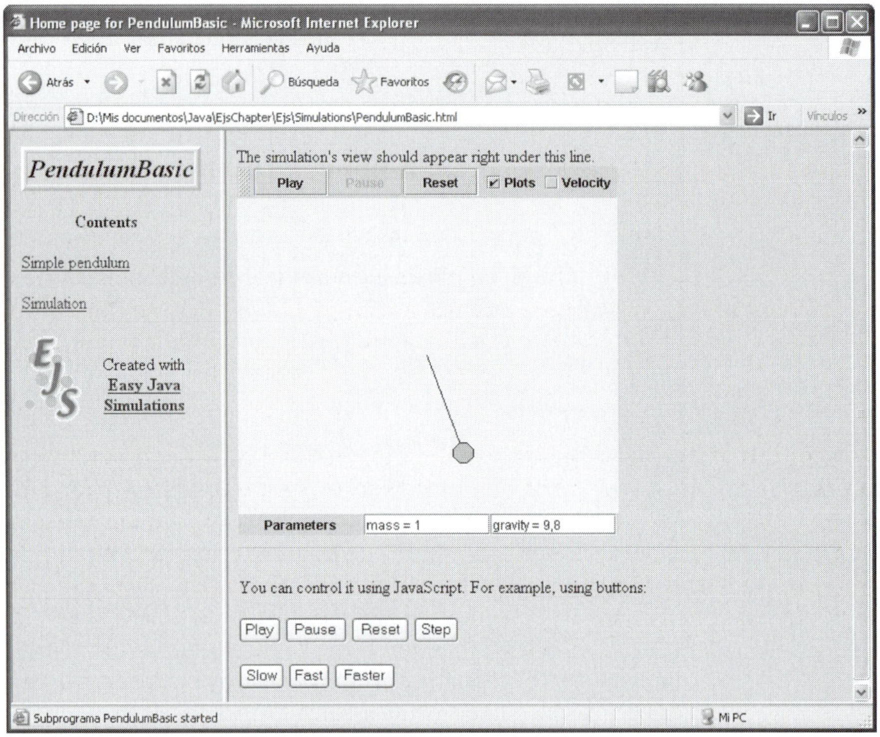

Figure 17.15   The simulation running as an applet.

display an html page with the simulation embedded as an applet (see Figure 17.15). (The dialog window with the time plots is displayed separately. We don't reproduce it here.)

Notice, finally, that the html page that includes the simulation also displays a set of buttons that can be used to control the simulation using JavaScript.[6]

Simulations created with Ejs require Java 2 to run them. Thus, your browser will need to have a recent plug-in installed to display them properly. We recommend the Java plug-in version 1.5.0_04.

Distributing a Simulation

As we said before, simulations created with *Easy Java Simulations* are independent of it once generated. However, the simulation will need to use a set of library files that include all the compiled OSP classes that provide the background functionality for your simulation, from numerical methods to the visualization elements. This library can be freely distributed and is contained in the _library subdirectory of the Simulations directory. Finally, if you designed your simulation to use any additional file, such as a gif image or sound file, you will need to distribute this resource along with the simulation.

With this said, the distribution process is quite simple. You just need to copy the required files and the _library directory to the distribution media or Web server. Simulations created with Ejs can be distributed using any of the following ways:

[6]JavaScript is a scripting language that can be used in html pages for simple programming tasks. The use of JavaScript to access methods of simulations created with Ejs is described in the Ejs manual.

Web server Just copy the jar, html, and auxiliary files for the simulation in a suitable directory of your Web server. Finally, copy the `_library` directory into the same directory as your simulation on the Web server. Recall that your users will need to have a Java 2 plug-in enabled Web browser to properly display the simulation.

CD-ROM This procedure is similar to the previous one. Just copy the jar, html, and auxiliary files, and the `_library` directory into your distribution media. Your users will need to have the JRE installed to run your simulation.

Easy Java Simulations is compatible with `Launcher` (see Chapter 15), and the Ejs installation includes a copy of `LaunchBuilder` that will scan your `Simulations` directory and will automatically generate the xml file that you need to run your simulations using `Launcher`. You can then distribute this xml file along with your simulation files so that your users can use `Launcher` to run them.

Java Web Start This is a technology created by Sun to help deliver Java applications from a Web server. The programs are downloaded from a server at a single mouse click, and they install automatically and run as independent applications. *Easy Java Simulations* is prepared to help you deliver your simulations using this technology, and it can automatically generate a Java Web Start jnlp file for your simulation. Details are provided in the manual. Your users will need to have the JRE installed, which includes Java Web Start. Finally, your server will need to report a MIME type of `application/x-java-jnlp-file` for any file with the extension `jnlp`.[7]

Together with Ejs A final possibility that is worth considering is that of distributing your simulation files together with Ejs, that is, asking your users to run the simulations using Ejs itself. This requires your users to learn how to use Ejs, if only basically to load and run the simulations. But it has, in our opinion, the enormous advantage that your users can not only run the final simulation but also inspect it in detail and learn how you actually simulated a given phenomenon. This possibility also offers a simple way for students to begin programming to simulate physical phenomena, which we find of great pedagogical value.

17.5 ■ MODIFYING A SIMULATION

If you read up to this point, you should already have a general impression of how *Easy Java Simulations* works, how it can be used to create a simulation, and how to run and distribute the simulations you create with it. In this section we will modify the simulation of the simple pendulum to include new features. This will further illustrate the operating procedures required to work with the different panels of Ejs.

In particular:

- We will modify the model to add friction and an external driving force. The resulting second-order differential equation is

$$\frac{d^2\theta}{dt^2} = -\frac{g}{l}\sin(\theta) - \frac{b}{m}\frac{d\theta}{dt} + \frac{1}{ml}f_e(t), \tag{17.8}$$

---

[7]The Multipurpose Internet Mail Extension (MIME) data type specifies the type of content that will be delivered by the Web server.

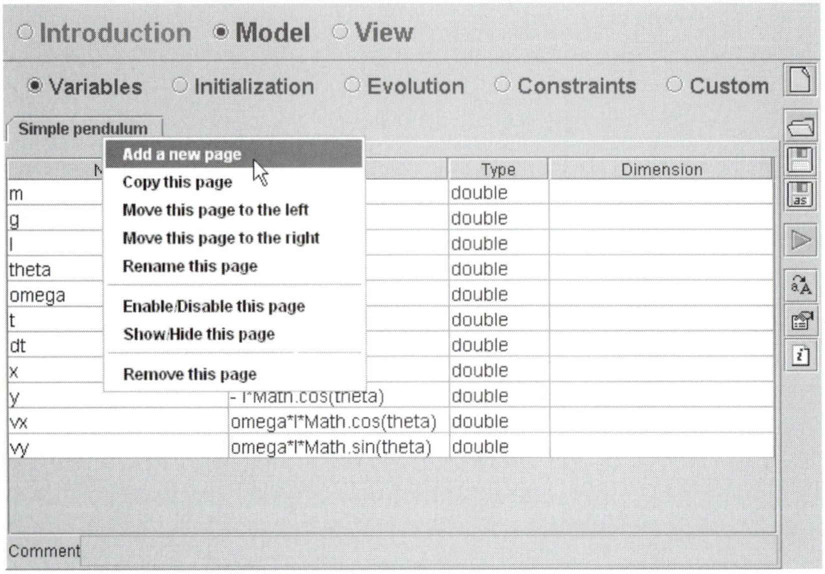

Figure 17.16   The pop-up menu for a page of variables.

where b is the coefficient of dynamic friction, m is the mass of the pendulum, and $f_e(t)$ is a time-dependent external driving force. We will use, in particular, a sinusoidal driving force of the form $f_e(t) = A \sin(Ft)$, where A and F are the amplitude and frequency of this force, respectively.

- We will modify the view so that it displays a phase-space diagram of the system, that is, a plot of angular position versus angular velocity.

- We will modify both the model and the view to compute and plot the potential and kinetic energies of the system and their sum.

- We will show how to modify the introduction pages to update the description of the simulation.

Modifying the Model

We need to revisit the different subpanels for the model and make the neccessary changes to each of them.

Adding New Variables

The introduction of friction and an external driving force requires adding new variables to the model. We do this by creating a second table of variables. Although we could add the new variables to the existing table, it is sometimes preferable, for clarity, to organize the variables into separate tables. For this we select the Variables subpanel of the Model panel of Ejs and right-click on the upper tab of the existing page. A pop-up menu will appear as shown in Figure 17.16.

From this menu, we select the Add a new page option (the one highlighted in the figure), and Ejs will create a page with an empty table of variables. (Before creating it though, Ejs

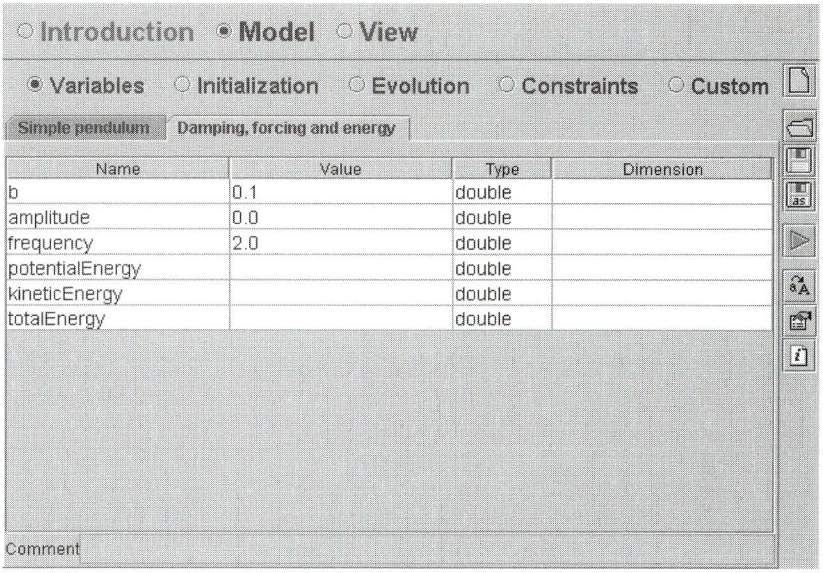

Figure 17.17   The new table of variables for our simulation.

will ask you for the name of this new page. You can choose, for instance, `Damping, forcing,` and `energy`.)

In this new table we can type all of the new variables that help us extend the model in the prescribed way. The mechanism to add a variable is simple. We just need to type a name for the variable in the column `Name`, select one of the possible types in the Type column, and optionally provide an initial value in the Value column. (The column labeled Dimension is used to declare arrays, and we won't use it for this model.)

Double-click a cell in the table in order to edit that cell. Use the Tab key or Arrow keys to move from cell to cell. Ejs will automatically add rows as needed. In this way, create new variables of type `double` called b, `amplitude`, `frequency`, `potentialEnergy`, `kineticEnergy`, and `totalEnergy`. Assign the first three variables the initial values of `0.1`, `0.0`, and `2.0`, respectively. The energy variables will be initialized as a result of the evaluation of constraints. The final new table of variables is displayed in Figure 17.17. Note that Ejs will add an empty row to the end of the table. This can be deleted by right-clicking on the row and selecting "Remove this variable."

Modifying the Evolution
We need to edit the differential equations for the system to add the new forces. For this go to the Evolution subpanel and edit the right-hand side of the second differential equation so that it reads

```
-g/l * Math.sin(theta) - b*omega/m + force(t)/(m*l)
```

The result is shown in Figure 17.18. Notice that we are using in this expression the method `force(t)` that is not yet defined. We will need to create it as a user-defined method when we get to the Custom subpanel.

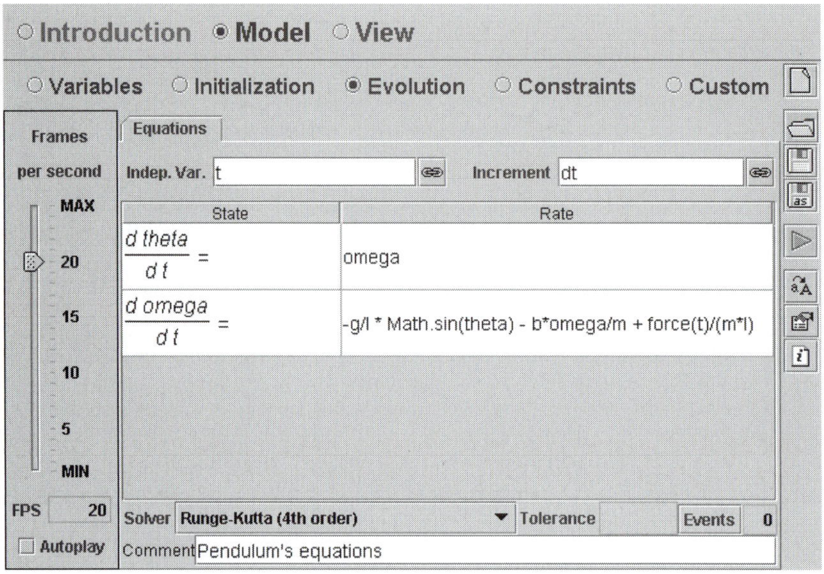

Figure 17.18 The edited differential equations.

Computing the Energy

Select the Constraints subpanel and follow a procedure similar to what we did for the table of variables to create a second page of constraints. Call the new page Energies, and in the blank editor that appears type the following code:

```
potentialEnergy = m*g*(y+1);
kineticEnergy = 0.5*m*(vx*vx+vy*vy);
totalEnergy = potentialEnergy + kineticEnergy;
```

(This sets the potential energy to zero when the pendulum is hanging straight down.) The reasons to compute the energies in a page of constraints (instead of in the evolution) are the same as those for the computation of the Cartesian magnitudes explained in Section 17.3. Any expression that we specify as constraints will be evaluated every time the state of the system changes. Thus, the corresponding relationships (the value of the energy variables) are always kept up to date.

Coding the External Force

To finish our changes to the model, we need to specify the expression for the external force. We do this using a page of Custom code. Move to this subpanel and click in the empty work area to create a new page called External force. The new page that appears looks very much like the other editors of code that we have used previously. But there is an important difference.

Since custom code is not automatically used by Ejs, the code we write here is not wrapped into an internal Java method but must be explicitly defined as a valid Java method (and later invoked in your code). For this reason, type in the editor exactly the following:

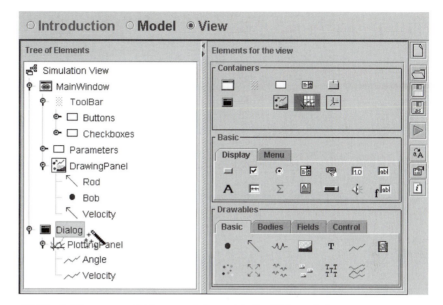

Figure 17.19 Adding a new plotting panel element to the Dialog window.

```
public double force (double time) {
  return amplitude * Math.sin(frequency*time);
}
```

This correctly defines the custom method and concludes our changes to the model.

Modifying the View

Let us start the changes to the view of the simulation by including the phase-space graph of angular velocity versus angular position. For this, go to the View panel and from the right-hand side collection of elements offered by Ejs, click on the icon for an OSP PlottingPanel, .

When you click on it, the icon will be highlighted and the cursor will change to a magic wand, . With this wand go to the left-hand side frame of the view and in the tree of elements, click on the element called Dialog. You are then asking Ejs to create a new plotting panel as child of the Dialog window. This is the simple mechanism used to add new elements to the view of the simulation. Figure 17.19 illustrates this action.

When you do this (and after providing a name for the new element), a new plotting panel will appear in the dialog window, sharing the available space with the previous plotting panel. Because the Dialog window is too small to host both panels, we need to enlarge it. Remove the magic wand by clicking on any blank area of the left frame. Double-click the Dialog element and set its Size property to 372,600. You can also edit the properties of the new plotting panel to customize its title and axis labels by double-clicking on the new plotting panel element that you created.

Now, let's add a Trace element, , to the new plotting panel. A trace is an element that can display a graph consisting of a sequence of points. Follow the creation process just

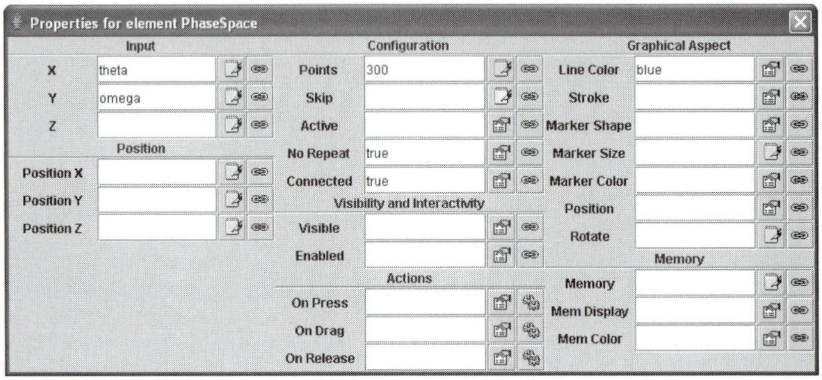

Figure 17.20   Properties for the PhaseSpace trace element.

described (again using the magic wand), name the new element PhaseSpace, and edit the table of properties so that it looks like Figure 17.20. These properties simply instruct the element to add to the graph a new point (θ, ω) after each evolution step, displaying only the last 300 added points and connecting them with a blue line.

The changes we did so far illustrate very well how to add and customize new elements to the view. It is as simple as it looks. Just use the magic wand to add new elements and edit their properties to match your needs. In most cases you will use some of the variables of the model for the properties of the view element. The connection between model and view is then automatically handled by *Easy Java Simulations*.

If we now run the simulation, we would obtain something like Figure 17.21.

We can now proceed similarly to add time plots of the energies of the system. We leave it to you as an exercise to create a third plotting panel as a child of the Dialog window and three traces (with different colors) in this panel that will plot the potential, kinetic, and total energies of the system.

We will end the changes to the view by adding new fields for the user to visualize and edit the values of the variables b, amplitude, and frequency. Click on the icon for view elements of type NumberField, ᵈ, and add three of these to the view element called Parameters. Give the three fields the same name as the corresponding variables, that is, b, amplitude, and frequency. (Names of elements must be unique in the view, but they don't clash with the name of model variables.)

Parameters is a basic Swing[8] panel that has been configured (using its Layout property) to display its children using a grid with one single row. When we add the three new fields, the six children of the panel will look pretty small. To solve this change the Layout property of Parameters to grid:2,3. This will organize the children in two rows of three elements each.

Now we need to edit the table of properties of each of the field elements so that they display the corresponding variables. We show how to do this for the first element. Double-click the element b and edit its properties as shown in Figure 17.22.

The association of the Variable property with the variable b of the model tells the element that the value displayed or edited in this field is that of b.

---

[8]Swing is the standard Java library for graphical components such as panels, buttons, and labels.

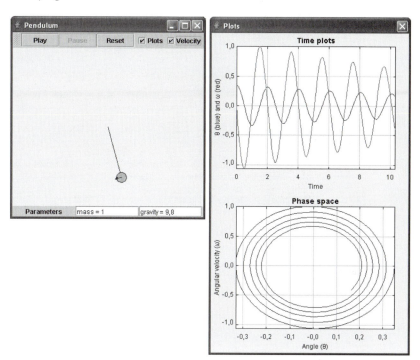

Figure 17.21 The simulation displaying both time and phase-space plots.

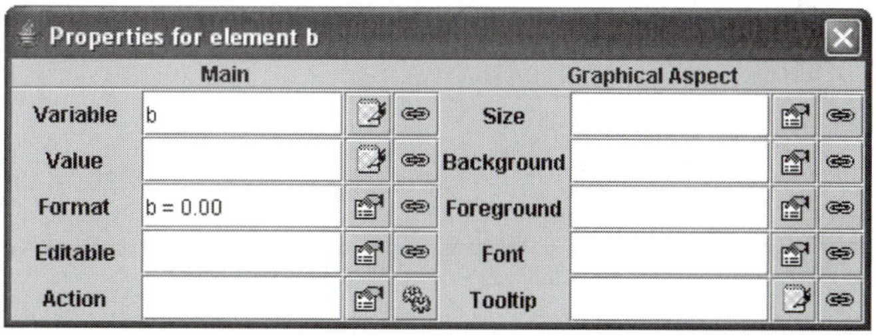

Figure 17.22 Properties for the b field element.

Note: To facilitate the association of properties with variables of the model, you can use the icon ∞ that appears to the right of the text field for the corresponding property. If you click on this icon, a list of all the variables of the model that can be associated to this property will be offered to you. You can then comfortably select the one you want with the mouse.

Also, to help edit some of the technical properties (such as layouts, colors, and fonts, for instance), Ejs offers dedicated editors that can be accessed by clicking on the icon 🖼, when shown.

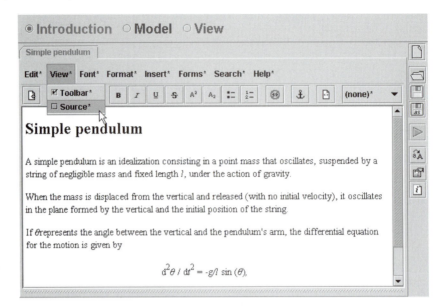

Figure 17.23   Edit mode of the html editor for the introduction page.

The property called Format, to which we assigned the value b = 0.00, has a special meaning. It doesn't indicate that b should take the value of 0, but is interpreted by the element as an instruction to display the value of b with the prefix "b = " and with two decimal digits.

This completes our changes to the view.

Modifying the Introduction

We now want to modify the introduction to reflect the new situation. Select the Introduction panel of Ejs and right-click on the tab of the existing page to bring in its pop-up menu. From this menu, select the "Edit/View this page" option. This will activate the edit mode for the displayed html page (see Figure 17.23).

HTML pages are text pages that include special instructions or tags that allow Web browsers to give a nice format to the text as well as to include several types of multimedia elements. The editor allows you (when in Edit mode) to work in a WYSIWYG (what you see is what you get) mode. However, if you are familiar with html and want to work directly on the code (which sometimes is preferable), you can select the option highlighted in Figure 17.23 to directly access the html source for the page.

You can write as many introduction pages as you want. As we saw earlier, each of the pages will turn into a link in the main html page that Ejs generates for the simulation. To return to View mode, right-click the tab of the existing page and again select the "Edit/View this page" option. In this way you toggle the Edit/View mode.

An Improved Laboratory

Our new simulation is finished. We now need to save it to a disk. Because we don't want to loose our original simulation, click on the Save As icon of the Ejs taskbar, 🖫, and a file dialog will allow you to save the simulation to a new file.

This simulation allows you to explore the behavior of a simple pendulum playing with different possible values of the parameters. Running the simulation can produce situations such as the one displayed in Figure 17.1 at the beginning of this chapter.

You will find the complete improved simulation in the `_ospguide` subdirectory of your `Simulations` directory with the name `PendulumComplete.xml`.

17.6 ■ A GLOBAL VISION

We end this chapter with an overview of what we did. We learned that *Easy Java Simulations* is an authoring and modeling tool that provides a simplified way to design and build complete interactive simulations in Java. For this, it structures a simulation into three main parts and provides, for each of these parts, a specialized editor that helps you implement them using high-level access to many of the OSP classes and utilities.

To investigate in more detail how each of these editors work, we loaded an existing simulation and inspected all of the panels and subpanels of Ejs in turn. This helped us understand how the different parts of the simulation are specified and how all the pieces fit together.

We also learned how to run and distribute the simulation either using physical media or through the Internet. Finally, we modified the simulation in order to learn some of the operating procedures of *Easy Java Simulations*.

The three parts of a simulation are the introduction, the model, and the view. Each of these parts has its function in the simulation and its own panel in the Ejs interface, each with its own look and feel.

The introduction offers an editor for the html pages required to create the multimedia narrative that introduces the simulation. This editor allows us to edit this narrative, either working in WYSIWYG mode or writing the html code directly.

The model is the engine of the simulation and is created using a sequence of subpanels where we specify the different parts of it: definition of variables, initialization, evolution of the system, constraints (or relationships among variables), and custom methods. Each panel provides editing tools that facilitate the job of creation (including sophisticated tasks such as solving differential equations and treatment of events).

Finally, the view contains a set of predefined elements, based on Swing components and OSP graphics classes, that can be used as individual building blocks to construct a structure in the form of a tree for the interface of our simulation. These elements, that can be added to a view through a simple procedure of click and create (our magic wand), have in turn a set of properties that indicate how each element looks and behaves. These properties, when associated to variables from the model (or Java expressions that use them), turn the simulation into a true dynamic and interactive visualization of the phenomenon under study.

And that's it! Though the chapter is long, because we accompanied the description with details and instructions, the process is straightforward. Obviously, learning to manipulate the interface of *Easy Java Simulations* with fluency requires a bit of practice as well as the familiarization with all the possibilities that exist. In particular, with respect to the creation of the view, you'll need to learn the many types of elements offered and what each of them can do for you.

If you want to know more about *Easy Java Simulations* or want to see more examples of simulations created with Ejs, you are cordially invited to read the manual found in the companion CD and to visit the Ejs home page at `http://fem.um.es/Ejs`.

18

Dissemination and Databases

by Anne J. Cox and William Junkin

We describe technologies, such as the BQ-OSP database and ComPADRE, which allow OSP-based materials to be disseminated across the Internet and show how these technologies can be used to adopt, adapt, and run OSP-based curricular materials.

18.1 ■ OVERVIEW

With the resources available from OSP developers, including Ejs (Chapter 17) and Tracker (Chapter 16), and the ease of using `LauncherBuilder` to author curricular material (Chapter 15), there is one final aspect to address: dissemination. As the OSP user community grows and curricular materials are developed, some instructors will want to take existing simulations, change initial conditions, or customize the tutorials and guided activities based on these simulations.[1,2,3,4] This chapter describes how we use Internet technologies to enable others to use the simulations and curricular materials that we have written and how you can add your material to this collection.

18.2 ■ DISSEMINATING PROGRAMS USING THE INTERNET

You have already seen that we can distribute material as a single double-clickable jar file using Launcher packages such as `osp_demo.jar`. This material can then be distributed on a CD or downloaded from a website and then run on the user's computer as an application. But there are other ways to disseminate material over the Internet. Two common technologies are Java Web Start, which installs a program on a local computer, or Java applets, which execute a program from within a Web browser. Both methods impose security restrictions which are necessary precautions to protect the local computer from malicious software.

---

[1]Consider curricular innovations such as Peer Instruction, Just-in-Time Teaching, and Physlets that are widely used in the physics community not only because they are based on sound pedagogical practices but also because they are easily adaptable to a variety of instructional settings. It is this adaptability that is crucial and that the BQ-OSP Database is designed to facilitate.

[2]E. Mazur, *Peer Instruction: A User's Manual*, Prentice Hall (1997).

[3]G. Novak, *et al.*, *Just-in-Time Teaching: Blending Active Learning with Web Technology*, Prentice Hall (1999).

[4]W. Christian and M. Belloni, *Physlets: Teaching Physics with Interactive Curricular Material*, Prentice Hall (2001) or see `http://webphysics.davidson.edu/Applets/Applets.html`.

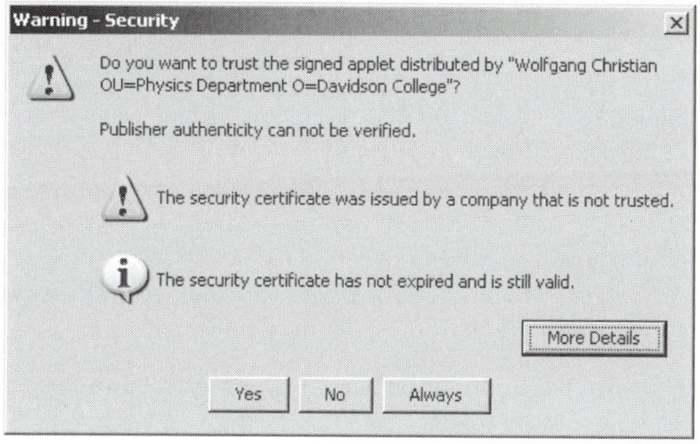

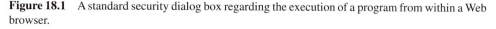

Figure 18.1   A standard security dialog box regarding the execution of a program from within a Web browser.

Java Web Start

Java Web Start allows for the one-click delivery (downloading) of stand-alone applications from the Web. Java Web Start requires that a `jnlp` webpage be posted on a server. This page is similar to an html page except that Java Web Start copies a jar file and other resources into a cache on the local computer. After downloading, a Java Web Start application runs from the desktop just like other Java applications but with restricted access to the operating system. If a Web Start program is signed, the user is prompted by a security dialog box as shown in Figure 18.1. Signed jar files can be given read/write access to a local file system if the user trusts the security certificate and the program author. Several examples can be found and run from the OSP Java Web Start page:

http://www.opensourcephysics.org/webstart

Because Java Web Start automatically checks the server for newer versions of a program, using it ensures that the most current version of the application will always be deployed. Detailed information about Java Web Start is available at

http://java.sun.com/products/javawebstart/

Applets

An applet is a Java program that runs within a Web browser. Because applications and applets have different security requirements and because they use different startup methods, a Java program must be modified to run as an applet. The `ApplicationApplet` is an applet that runs other OSP programs in a Web browser without modification. The applet itself consists of a single button, as shown in Figure 18.2, which when pressed, creates (instantiates) the program. This applet can be used to run the `osp_demo.jar` Launcher package by embedding the following `applet` tag in an html page:

```
<applet codebase="jars"
  code="org.opensourcephysics.davidson.applets.
       ApplicationApplet.class"
    archive="osp_demo.jar" name="demo"
    width="250" height="50">
```

Figure 18.2   A button titled "Launcher" which when pressed uses `ApplicationApplet` to execute a Launcher package.

```
<param name="target"
   value="org.opensourcephysics.tools.Launcher">
</applet>
```

The html page contains a button which runs `Launcher` and reads `launcher_default.xset` packaged in the jar.

`ApplicationApplet` supports the following parameter tags:

htmldata displays an html page that provides the questions or instructions for the animation user (provides the narrative).

xmldata reads an xml file with parameters for the particular application (provides the initial OSP Control parameters) that overrides the program's default parameters.

target is the location of a program, such as

org.opensourcephysics.davidson.nbody.PlanarNBodyApp.

The `xmldata` parameter tag enables us to pass data to a program running within `ApplicationApplet`. The html page again contains a button that runs a program.

```
<applet codebase="jars"
  code="org.opensourcephysics.davidson.applets.
        ApplicationApplet.class"
    archive="osp_demo.jar" name="superposition"
    width="230" height="50">
  <param name="target" value="org.opensourcephysics.
        davidson.qm.QMSuperpositionExpectationXApp">
  <param name="xmldata" value="../sho_3_4x.xml">
</applet>
```

Compare this `applet` tag with the one that runs the entire Launcher package. The `target` parameter signals `ApplicationApplet` to run `QMSuperpositionExpectationXApp` from within `osp_demo.jar`. The optional `xmldata` parameter loads previously saved parameters from the `../sho_3_4x.xml` data file.

Finally, it will be possible with the introduction of Java 1.6 (expected release in late 2006) to script a Java applet in an html page with a JavaScript interpreter *built into* the Java VM.[5] The current version of Java-to-JavaScript communication (LiveConnect) is run by the Web browser, and therefore its functionality is dependent on the operating system, the browser, and the browser version. The OSP API that supports this scripting model will be released in 2006, but it is likely that we will add a `script` parameter tag to `ApplicationApplet` that will load a text file containing JavaScript code.

[5]This interpreter is based on the Mozilla Rhino interpreter.

18.3 ■ THE BQ-OSP DATABASE

Now that we have seen a variety of ways OSP curricular material can be distributed across the Internet, we focus on distribution mechanisms.

The BQ-OSP Database provides our primary mechanism of dissemination for OSP-based curricular materials, which can be used with or without modification.[6] It is located at

```
http://www.bqlearning.org
```

This database simplifies the customization of OSP simulations and curricular material and enables broad OSP collaboration and dissemination. It accomplishes this with the following features:

- content that is searchable by topic and by keyword using the database search or through ComPADRE and other digital libraries,

- narratives that can be customized and associated with one or more simulations to create guided tutorials,

- simulations can be customized by editable xml files that are then saved in the database,

- links to these materials that can easily be embedded in local webpages and presentations for students to access, and

- local (desktop) installation and content that can be exported from and imported to this local database.

For example, the curricular material distributed with the `Launcher` package described in Chapter 15 can be found in the database by entering the word "Launcher" into the search text box. When you select the Launcher package associated with Chapter 15, you will find an html page with an button like that of Figure 18.2. When you click this button, the `Launcher` will open as shown in Figure 18.3. To deliver this to students, you simply give them the url.

However, you may wish to find and use other available curricular materials, edit materials submitted by others (e.g. the "Default Page" in Figure 18.3), or upload your own customized OSP-based curricular materials. The sections that follow will describe how to edit and search for existing materials as well as how to add resources to the database. The goal of the database is to make it easy for others to use or modify resources even if they are not programmers or code developers.

We will begin by describing how to find a single simulation (Sections 18.4 and 18.5), how to customize it (Sections 18.6 and 18.7), and then how to upload your curricular materials including an xml file and a narrative.

18.4 ■ SEARCHING AND BROWSING

Before developing new curricular materials, it is worthwhile to investigate curricular materials that already exist. There are two primary ways of finding materials in the database:

---

[6]The BQ-OSP Database is one part of a collection of instructional resources (BQ Program Suite) to support the use of Peer Instruction, Just-in-Time Teaching and Physlets. Instructors interested in using other aspects of the BQ Program Suite are encouraged to go to `http://www.bqlearning.org` for a demonstration.

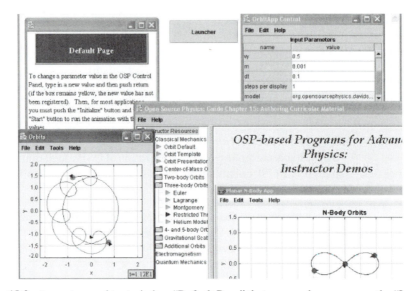

Figure 18.3   `Launcher` and text window "Default Page" that opens when you press the "Launcher" button from an html page delivered by the database.

searching and browsing by topic. Begin at the BQ-OSP homepage (follow the link to the database from `http://www.bqlearning.org`) and click on the "Search" tab (see Figure 18.4).

For example, suppose you want to find three-body orbits such as the one from Section 15.5. You can browse by clicking on the topic: "Classical Mechanics" under "Upper Level Physics." Notice that for each subcategory (see Figure 18.5), you get a list of available resources for each type of curricular materials (Physlets, Open Source (OSP), Ejs, and Tracker).

In this case because you want to find the OSP materials from Section 15.5 on three-body orbits, you would click on the number under the "Open source" heading in the subcategory: "Stability and Multi-body Orbits" in the "Central Forces" category. There you will see a list of resources along with thumbnails of each (see Figure 18.6).

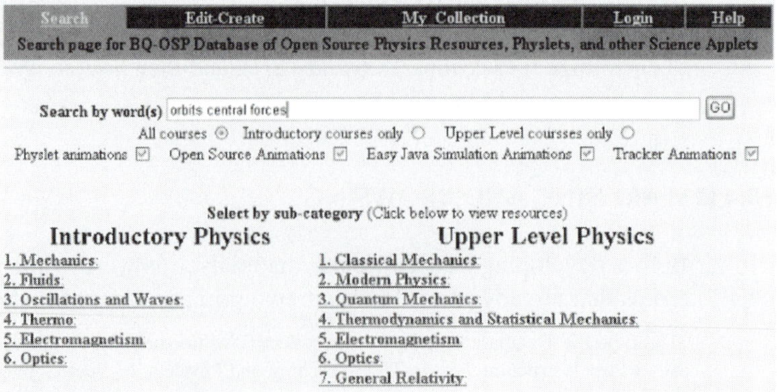

Search	Edit-Create	My Collection	Login	Help

Search page for BQ-OSP Database of Open Source Physics Resources, Physlets, and other Science Applets

Search by word(s) |orbits central forces| [GO]

All courses ⊙   Introductory courses only ○   Upper Level courses only ○
Physlet animations ☑   Open Source Animations ☑   Easy Java Simulation Animations ☑   Tracker Animations ☑

Select by sub-category (Click below to view resources)

Introductory Physics   **Upper Level Physics**

1. Mechanics   1. Classical Mechanics
2. Fluids   2. Modern Physics
3. Oscillations and Waves   3. Quantum Mechanics
4. Thermo   4. Thermodynamics and Statistical Mechanics
5. Electromagnetism   5. Electromagnetism
6. Optics   6. Optics
                                              7. General Relativity

Figure 18.4   Search field available within the database.

	Select by sub-category (Click below to view resources)					
Introductory Physics		physlet	osp	ejs	tracker	**Upper Level Physics**
	A. Mathematics					1. Classical Mechanics:
1. Mechanics:	10. Vector Calculus	(2)	(2)	(0)	(0)	2. Modern Physics:
2. Fluids:	20. Complex Plane	(0)	(1)	(0)	(0)	3. Quantum Mechanics:
3. Oscillations and Waves	B. Newtonian Mechanics					4. Thermodynamics and Statistical Mechanics:
4. Thermo:	10. Conservative Forces	(1)	(2)	(1)	(0)	
5. Electromagnetism:	20. Non-conservative Forces	(2)	(0)	(3)	(2)	5. Electromagnetism:
6. Optics:	C. Oscillations					6. Optics:
	10. Linear Oscillations	(3)	(2)	(1)	(1)	7. General Relativity:
	30. Chaotic Systems	(0)	(4)	(3)	(1)	
	D. Central Forces					
	10. Orbits	(2)	(5)	(1)	(0)	
	20. Stability and Multi-body Orbits	(0)	(6)	(1)	(0)	
	40. Scattering	(2)	(5)	(1)	(0)	

Figure 18.5 List of materials found while browsing the database by topic.

Clicking on the title or the thumbnail opens up a separate window that loads the `ApplicationApplet` allowing you to run the animation and read the text.

Alternatively, you can find these resources from the "Search" page by typing text into the search textbox. Searching supports a search of keywords, description, title, and topic. Searches can also be limited to a particular type of curricular material (OSP, Ejs, Tracker and/or Physlets) and a content level (see Figure 18.4). If you type in the words: "orbits central force" and limit the search to "Upper Level" courses and "Open Source Animations," you will find the same three-body orbit materials.

If you want to use the resource as is, simply click on the Use button and you will be given the url of the resource. If you are logged into the database, you also have the option to add the link to your course webpage stored in the database. In order to customize resources or add your own resources, you are required to log into the database as well.

To log in, simply click on the "Login" page and complete the required fields (if you are the first one from your institution to log in, you will also need to complete the information in that field). Please note that none of the information collected will be shared with anyone except those running this database and ComPADRE (a National Digital Library Project for the physics and astronomy communities; for more on ComPADRE see Section 18.5), which gathers material from this database into its collection.

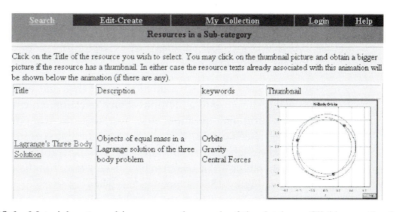

Figure 18.6 Materials returned in an example search of the database. Clicking on the thumbnail image or the title opens a window for running the animation and using or editing it.

18.5 ■ INTEGRATION WITH DIGITAL LIBRARIES

The BQ Database is searchable through ComPADRE (Communities for Physics and Astronomy Digital Resources in Education: `http://www.compadre.org`). The goal of ComPADRE, a joint project of several American physics and astronomy societies, is to organize digital collections of education materials in physics and astronomy. One of ComPADRE's services is a federated, joint search of a number of digital collections. The database is a part of this federated search (available by putting a check mark in the "Physlet/OSP Database" checkbox). If a ComPADRE user chooses to include the database in a federated search, the database provides url links to materials that match the search fields. These are links to a Web Start page or to an html page that contains an applet. The database is designed specifically for organizing and storing Open Source Physics materials, but through the ComPADRE federated search, it provides an easy way to disseminate new animations and make them available to the wider physics community. An instructor can simply submit materials directly through any number of national digital libraries without using the database. However, the advantages of the database are that it is specifically designed to support Open Source Physics resource dissemination, it already has ComPADRE protocols built in, and it is a part of a federated search. An advantage of a federated search with other digital libraries is that other related materials, such as lesson plans, textbooks, and research, can be discovered at the same time as the OSP materials. This puts the simulations in a broader context but still links the instructor to the features and tools of the database program.

18.6 ■ EDITING CURRICULAR MATERIALS

Although you may find resources that fit the instructional needs of your course, at times you or your colleagues may want to customize the materials. As a user of this *Guide*, you will likely be able to customize resources fairly easily, but others may not. Thus, another goal of the database is to lower the technological barrier for other instructors and users of OSP resources. Typical resources in the BQ Database have three parts:

- narrative (or text) that provides the questions students are expected to answer or a tutorial designed to help students use the simulation to understand a given physics concept,

- initial conditions of the simulation (parameter set in OSP Control panel) set by the parameters defined in an xml file, and

- the simulation itself.

You may decide to use an animation as it exists and only change the question asked of the students (turn it into a multiple choice question, give more guidance in solving the problem, ask a different question, etc.). To change the text or narrative associated with the exercise, log in (see Section 18.4) and find the animation of interest (such as the three-body example). Now click on the Edit button next to the text (if you are not logged in, you will not see an Edit button). You can then type in new text, edit the existing text, or upload a new html page (see Figure 18.7). Once loaded, when someone else chooses to use this text, the `ApplicationApplet` will call this new html page from the parameter tag `htmldata` when the applet is loaded.

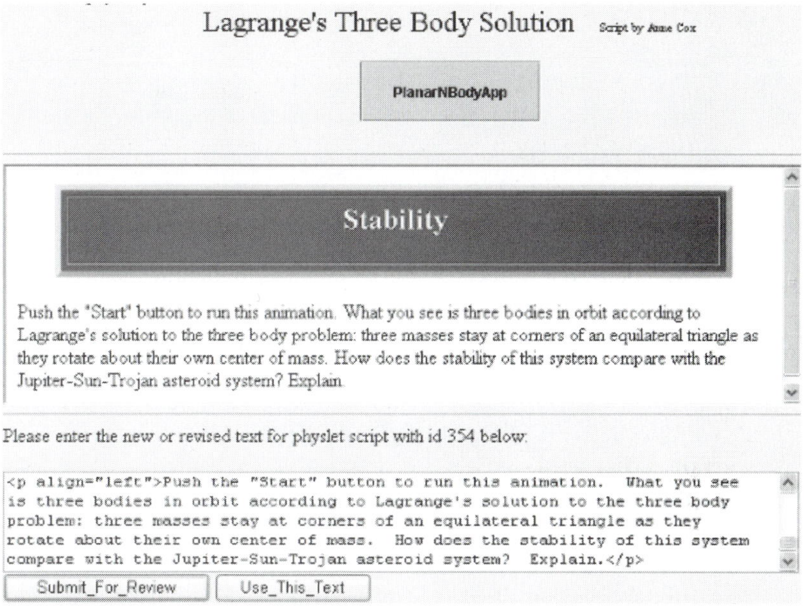

Figure 18.7   Editing the narrative.

Anyone who simply edits the text associated with the animation does not need to know about these parameters. To increase the ease of use, the database program automatically takes care of these details and fills in the proper values for the parameters. After editing a text, you may make these available for others to see (public) or save it to your own private collection so that only you and your students may view it (private).

18.7 ■ EDITING ANIMATION BY CHANGING XML

You may find an animation that you want to use, but you would prefer to deliver it to your students with different initial conditions as set by the OSP Control parameters. In this case you need to change (edit) the xml file. Choosing "Edit this script" opens up a screen that shows the xml file in a frame (see Figure 18.8). Alternatively, you can also upload an xml file directly from your local computer.

After changing xml parameters, a click on the applet's button shows the animation with the changes. When you have made all your changes to the parameter set, you simply save it and it is ready for delivery to your students. You may choose to save it to your private collection or to all users of the database (public). You will also be asked to choose a topic, category, and subcategory for the animation (from a pull-down list). (In order to upload parameter sets from multiple files when using Launcher, see the following section.)

18.8 ■ UPLOADING NEW MATERIALS

Instructors are encouraged to submit new curricular materials to the BQ Database. Some new materials will be based on the original OSP library and the jar files already provided in

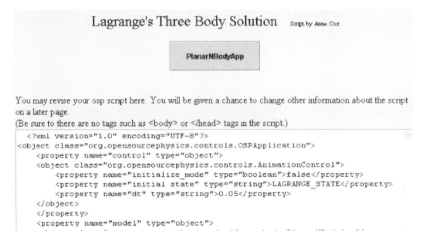

Figure 18.8   XML editing mode.

the database (`osp_demo.jar`, `osp_guide.jar`, etc.). The instructor will provide the target (the application class file in the OSP library), the xml data file (initial parameters for the OSP Control panel), and the html data file (narrative). Consider the `quantum_mechanics.xml` file that you used with the `Launcher` in Section 15.3 shown in Figure 15.8. To upload just that animation (not the entire Launcher package) and its associated xml file (you saved this as `test.xml` in Section 15.3), choose the "Edit/Create" tab, and then the Upload New Materials button to get the page shown in Figure 18.9.

Edit the information `osp.opensourcephysics.davidsonqm.QMSuperpositionApp` for the target in the appropriate jar file, and then either copy the xml file (shown in Figure 18.9) or use the Browse button to upload the associated xml (`test.xml`) file. You can check that the animation runs and then save it to the database. To save it you will need to select a category (from the pull-down list) and give it a title. You can then choose whether it will be part of the public collection (for anyone to see) or your private collection (for you and your students only), pick keywords (from a list maintained by the database administration), and have the option of providing a brief description. Once it has been submitted, you will have the option of providing a thumbnail of a screen shot of the animation and a narrative (an associated html page) for the animation. Providing these data makes it easier for other instructors to find and use the simulations and will support wider dissemination of the resources.

Submitting curricular materials designed using `Launcher` works in much the same way. In that case the target information is `org.opensourcephysics.tools.Launcher` and the xset, xml, and html files must be provided in a single zip file which must be named `launcher_default.zip` as described in Section 15.3.

If you wish to submit an animation based on your own jar file, the database can store signed jar files that are not too big. The current limitation is 3 MB, but this size limitation is likely to change over time. If the jar file is too big to be stored in the database, the jar file must be maintained elsewhere and its url stored in the database. The BQ Database program will use that url to obtain the jar file and deliver it to the user when the animation is run. Of course, this means that these animations can only be run on computers that have Internet access.

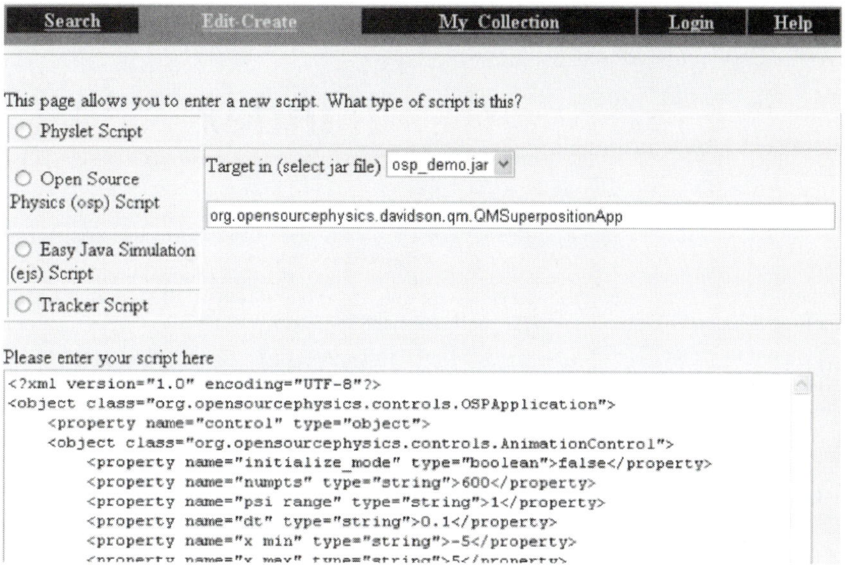

Figure 18.9   Page for uploading new resources into the database.

While the database will store curricular materials and most jar files, it will not store the source files for any jar files. Instead, the database will provide a link to the source files along with a record of the date the source files were last modified and the source file size. Thus, the database will provide a warning to other users if the source files (associated with a jar of interest) have been changed.

18.9 ■ DATABASE INSTALLATION

Although the BQ Database is available online, it can also be run from a local computer, with the local computer serving the webpages. A CD distribution of the database is available, as is a zip file of the CD contents from `http://www.bqlearning.org`. The installation instructions are in the `DB_Instructions.doc` file. On machines running the Windows operating system, installing the database on a local computer requires creating the folder `C:/Util` and then running the required batch file (`InstallBQdbCDrive.bat`). This will install the MySQL database program and the Apache webserver, which allows your local computer to serve webpages using the BQ Program Suite.

As you generate curricular materials based on OSP resources, consider adding them to the online BQ Database as part of the public collection. The goal of the OSP Project is to share resources not only with the community of OSP developers but also to the wider physics instructional community.

Index